T0310016

**Techniques and Methods in Urban
Remote Sensing**

Techniques and Methods in Urban Remote Sensing

Qihao Weng

Indiana State University
Terre Haute, Indiana, USA

IEEE PRESS

WILEY

For general information on our other products and services or for technical support, please contact our Customer Care Department within the United States at (800) 762-2974, outside the United States at (317) 572-3993 or fax (317) 572-4002.

Wiley also publishes its books in a variety of electronic formats. Some content that appears in print may not be available in electronic formats. For more information about Wiley products, visit our web site at www.wiley.com.

Library of Congress Cataloging-in-Publication data is available.

hardback: 9781118217733

Cover Design: Wiley
Cover Image: © kentoh/ Shutterstock

Set in 10/12pt Warnock by SPi Global, Pondicherry, India

Printed in the United States of America.

V10015005_102619

Contents

Preface

Since the 1970s, there has been an increase in using remotely sensed data for various studies of ecosystems, resources, environments, and human settlements. Driven by the societal needs and improvement in sensor technology and image processing techniques, we have witnessed an explosive increase in research and development, technology transfer, and engineering activities worldwide since the turn of the twenty-first century. On one hand, hyperspectral imaging afforded the potential for detailed identification of materials and better estimates of their abundance in the Earth's surface, enabling the use of remote sensing data collection to replace data collection that was formerly limited to laboratory testing or expensive field surveys (Weng 2012). On the other hand, commercial satellites started to acquire imagery at spatial resolutions previously only possible to aerial platforms, but these satellites possess advantages over aerial imageries including their capacity for synoptic coverage, shorter revisit time, and capability to produce stereo image pairs conveniently for high-accuracy 3D mapping thanks to their flexible pointing mechanism (Weng 2012). While radar technology has been reinvented since the 1990s due greatly to the increase of spaceborne radar programs, Light Detection and Ranging (LiDAR) technology is increasingly employed to provide high-accuracy height and other geometric information for urban structures and vegetation (Weng 2012). These technologies are integrated with more established aerial photography and multispectral remote sensing techniques to serve as the catalyst for the development of urban remote sensing research and applications. It should further be noted that the integration of Internet technology with satellite imaging and online mapping have led to the spurt of geo-referenced information over the Web, such as Google Earth and Virtual Globe. These new geo-referenced "worlds," in conjunction with GPS, mobile mapping, and modern telecommunication technologies, have sparked unprecedented interest in the public for remote sensing and imaging science.

A few global initiatives and programs have fostered the development of urban remote sensing. First, Group on Earth Observation (GEO) established in its 2012–2015 Work Plan, a new task called "Global Urban Observation and Information (GUOI)" (Weng et al. 2014). The main objectives of this task were to improve the coordination of urban observations, monitoring, forecasting, and assessment initiatives worldwide, to produce up-to-date information on the status and development of the urban systems at different scales, to fill existing gaps in the integration of global urban land observations, and to develop innovative concepts and techniques for effective and sustainable urban development. These objectives have been morphed into a new initiative of the GEO Work Programme of 2017–2019 (Weng et al. 2014). These GEO objectives support well the United Nations' goals on Sustainable Urban Development (United Nations Development Programme 2015). The sustainable development goals (SDGs) represent the UN's response to numerous societal challenges and efforts to build a sustainable Earth. GEO hopes to make cities and human settlements more inclusive, safe, resilient, and sustainable, through GUOI Initiative, by supplying objective data and information on the footprints of global urbanization and cities, developing indicators for sustainable cities, and developing innovative methods and techniques in support of effective management of urban environment, ecosystems, natural resources, and other assets, and the mitigation of adverse impacts caused by urbanization (Weng 2016). Furthermore, the International Human Dimensions Programme (IHDP) on Global Environmental Change sponsored a project on Urbanization and Global Environmental Change Project (UGEC), between 2006 and 2017, by convening an international network of scholars and practitioners working at the intersection of UGEC. UGEC has improved and expanded knowledge of the key role of these interactions to better understand the dynamic and complex challenges our societies face in the twenty-first century and has been instrumental in shaping some international policy and science agendas, including the IPCC Fifth Assessment Report and Urban Sustainable Development Goal (Sánchez-Rodríguez and Seto 2017). Finally, the World Bank has been supporting several groundbreaking initiatives related to urbanization, including the Urbanization Reviews. Projects have been funded to support green, inclusive, resilient, and competitive cities across the world at various stages of urban development. More importantly, the World Bank has developed its Platform for Urban Mapping and Analysis (PUMA) (puma.worldbank.org), a geospatial tool that allows users with no prior GIS experience to access, analyze, and share urban spatial data, and adapts open-source software to the data needs of various sectors. The "PUMA – Satellite Imagery Processing Consultation" presents up-to-date and retrospective land use information of selected cities obtained by processing of high-resolution satellite imagery. The rapid development of urban remote sensing has greatly benefited from the knowledges, data sets, products, and services that these programs/initiatives have to offer.

Over the past two decades, we have witnessed new opportunities continue to appear for combining ever-increasing computational power, modern telecommunication technologies, more plentiful and capable data, and more advanced algorithms, which allow the technologies of remote sensing and GIS to become mature and to gain wider and better applications in environments, ecosystems, resources, geosciences, and urban studies. To meet the growing interests in applications of remote sensing technology to urban and suburban areas, Dr. Dale A. Quattrochi, NASA Marshall Space Flight Center, and I decided to assemble a team of experts to edit a volume on "Urban Remote Sensing" in 2006. Since its inception, the book had been serving as a reference book for researchers in academia, governmental, and commercial sectors and has also been used as a textbook in many universities. Remote sensing technology has since changed significantly and its applications have emerged worldwide, so it is feasible to write a book on the subject. I decided to write a text book on urban remote sensing, when Mr. Taisuke Soda, Wiley-IEEE Press, Hoboken, New Jersey, sent me an invitation in 2011. It has not been easy to complete it with my increased commitments and numerous interruptions. I found it even more challenged to determine what to include in the book, as researches and applications on urban remote sensing continue to expand rapidly.

Qihao Weng

References

Sánchez-Rodríguez, R. and Seto, K. (2017). *Special Announcement from UGEC Co-Chairs Karen Seto and Roberto Sanchez-Rodriguez.* https://ugec. org/2017/01/18/special-announcement-ugec-co-chairs-karen-seto-roberto-sanchez-rodriguez (accessed 13 January 2018).

United Nations Development Programme (2015). *Sustainable Development Goals (SDGs).* http://www.undp.org/content/undp/en/home/sdgoverview/post-2015-development-agenda.html (accessed 5 March 2016).

Weng, Q. (2012). *An Introduction to Contemporary Remote Sensing.* New York: McGraw-Hill Professional, pp. 320.

Weng, Q. (2016). *Remote Sensing for Sustainability.* Boca Raton, FL: CRC Press/ Taylor and Francis, pp. 366.

Weng, Q., Esch, T., Gamba, P. et al. (2014). Global urban observation and information: GEO's effort to address the impacts of human settlements. In Weng, Q. editor. *Global Urban Monitoring and Assessment Through Earth Observation.* Boca Raton, FL: CRC Press/Taylor and Francis, pp. 15–34.

Synopsis of the Book

This book is settled down with 12 chapters and addresses theories, methods, techniques, and applications in urban remote sensing. Both Chapters 1 and 2 address fundamental theoretical issues in remote sensing of urban areas. Because of the significance of impervious surface as an urban land cover, land use, or material, Chapter 1 examines the general requirements for mapping impervious surfaces, with a particular interest in the impacts of remotely sensed data characteristics, i.e. spectral, temporal, and spatial resolutions. The discussion is followed by a detailed investigation of the mixed pixels issue that often prevails in medium spatial resolution imagery in urban landscapes. This investigation employs linear spectral mixture analysis (LSMA) as a remote sensing technique to estimate and map vegetation – impervious surface – soil components (Ridd 1995) in order to analyze urban pattern and dynamics in Indianapolis, USA. Chapter 2 discusses another basic but pivot issue in urban remote sensing – the scale issue. The requirements for mapping three interrelated substances in the urban space – material, land cover, and land use – and their relationships are first assessed. The categorical scale is closely associated with spectral resolution in urban imaging and mapping, while the observational scale of a remote sensor (i.e. spatial resolution) interacts with the fabric of urban landscapes to create different image scene models (Strahler et al. 1986). Central to the observational scale–landscape relationship is the problem of mixed pixels and various pixel and sub-pixel approaches to urban analysis. Next, the author's two previous studies were discussed, both assessing the patterns of land surface temperature (LST) at different aggregation levels to determine the operational scale/the optimal scale for image analysis. Chapter 2 ends with discussion on the issue of scale dependency of urban phenomena and two case studies, one on LST variability across multiple census levels and the other on multi-scale residential population estimation.

A large portion of the book is dedicated to the methods and techniques in urban remote sensing by showcasing a series of applications of various aspects. Typically, each application area examines with an analysis of the state-of-the-art methodology followed by a detailed presentation of one or two case studies. The application areas include building extraction and impervious surface estimation and mapping (Chapters 3 and 4), LST generation, urban heat island

(UHI) analysis and anthropogenic heat modeling (Chapters 5–7), cities at night (Chapter 8), urban surface runoff and ecology of West Nile Virus (WNV) (Chapters 9 and 10), assessment of urbanization impacts (Chapter 11), and estimation of socioeconomic attributes (Chapter 12).

In Chapter 3, building types are identified by using remote sensing-derived morphological attributes for the City of Indianapolis, Indiana, USA. First, building polygons and remotely sensed data (i.e. high-spatial-resolution ortho-photography and LiDAR point cloud data) in 2012 were collected. Then, morphological attributes of buildings were delineated. Third, a Random Forest (RF) classifier was trained using randomly selected training samples obtained from the City of Indianapolis Geographic Information System (IndyGIS) database, Google Earth Maps, and field work. Finally, the trained classifier was applied to classify buildings into three categories: (i) nonresidential buildings; (ii) apartments/condos; and (iii) single-family houses.

Chapter 4 illustrates a few commonly used methods for estimating and mapping urban impervious surfaces. This chapter begins with a brief review of the various methods of urban impervious surface estimation and mapping, followed by two case studies. The first case study was conducted to demonstrate the capability of LSMA and the multilayer perceptron (MLP) neural network for impervious surface estimation using a single Hyperion imagery. The second case study employed a semi-supervised fuzzy clustering method to extract annual impervious surfaces in the Pearl River Delta, southern China, from 1990 to 2014, aiming to utilizing time series Landsat imagery.

Chapters 5–7 are concerned about urban thermal landscape and surface energy balance. In Chapter 5, the Spatiotemporal Adaptive Data Fusion Algorithm for Temperature mapping (SADFAT) (Weng et al. 2014) is introduced. This algorithm was developed for fusing thermal infrared (TIR) data from two satellites, the high spatial resolution Landsat Thematic Mapper (TM) and high temporal resolution MODIS data, to predict daily LST at 120-m resolution. The second algorithm to be introduced is called "DELTA," which stands for the five modules of Data filtEr, temporaL segmentation, periodic modeling, Trend modeling, and Gaussian, respectively (Fu and Weng 2016). This algorithm is developed to reconstruct historical LSTs at daily interval based solely on irregularly spaced Landsat imagery by taking into account some significant factors, such as cloudy conditions, instrumental errors, and disturbance events, that impact analysis of long-term LST data.

In Chapter 6, two methods for characterizing and modeling UHI using remotely sensed LST data are introduced. A kernel convolution modeling method for two-dimensional LST imagery will be introduced to characterize and model the UHI in Indianapolis, USA, as a Gaussian process model (Weng et al. 2011). The main contribution of this method lies in that UHIs can be examined as a scale-dependent process by changing the smoothing kernel parameter. Furthermore, an object-based image analysis procedure will be introduced to extract hot spots from LST maps in Athens, Greece. The spatial and thermal attributes associated with these objects (hot spots) are then calculated and used for the analyses of the intensity, position, and spatial extent of UHIs.

To gain a better understanding of UHI phenomenon and dynamics, one must examine the temporal and spatial variability of surface heat fluxes in the urban areas, especially of anthropogenic heat flux. In Chapter 7, we develop an analytical protocol, based on the two-source energy balance (TSEB) algorithm, to estimate urban surface heat fluxes by combined use of remotely sensed data and weather station measurements. This method was applied to four Terra's ASTER images of Indianapolis, Indiana, United States, to assess the seasonal, intraurban variations of spatial pattern of surface energy fluxes. In addition, anthropogenic heat discharge and energy use from residential and commercial buildings were estimated. Based on the result, the relationship between remotely sensed anthropogenic heat discharge and building energy consumption was examined across multiple spatial scales.

Nighttime light (NTL) imagery provides a unique source of Earth Observational data to examine human settlements and activities at night. In Chapter 8, a method is proposed for large-scale urban detection and mapping by utilizing spatiotemporally adjusted NTL images across different times. Secondly, this chapter will analyze the spatiotemporal pattern of electricity consumption in the USA and China from 2000 to 2012 by using NTL imagery. This analysis offers a spatially explicit method to characterize the spatial and temporal pattern of energy consumption at regional and global scale.

Chapter 9 relates land-use and land-cover (LULC) change to spatially distributed hydrological modeling in order to study urban surface runoff. Two case studies will be illustrated: one in Guangzhou, China, and the other in Chicago, USA. A model widely used for estimating surface runoff was developed by the United States Soil Conservation Service, that is, the SCS model. The Guangzhou study developed a new procedure to calculate composite Curve Number, a key parameter in SCS model, based on urban compositional vegetation-impervious surface soil (VIS) model, and then simulated surface runoff under different precipitation scenarios. The Chicago study aimed to assess the impact on water quality resulted from LULC changes in an urban watershed over a long time period, by employing the long-term hydrologic impact assessment nonpoint source (L-THIA-NPS) model. This model also used the SCS curve number method to estimate runoff depth and volume.

Chapter 10 focuses on the ecology of WNV in the US urban environments. Two case studies will be presented to illustrate the applications of remote sensing and GIS techniques for analyzing and modeling the spread of WNV. The first study, by a case study of the City of Chicago, aimed to improve the understanding of how landscape, LST, and socioeconomic variables were combined to influence WNV dissemination in the urban setting, and to assess the importance of environmental factors in the spread of WNV. The second study investigated the WNV spread in the epidemiological weeks from May to October in each year of 2007–2009 in the Southern California and modeled and mapped the risk areas.

Urbanization can bring about significant changes to the environment. In Chapter 11, two case studies will be introduced to examine the impact of LULC change on LST and on surface water quality, respectively. The first study utilized 507 Landsat TM/Enhanced Thematic Mapper Plus (ETM+) images of Atlanta, Georgia, United States, between 1984 and 2011, to investigate the impact of urban LULC changes on temporal thermal characteristics by breaking down the time-series LST observations into temporally homogenous segments. The second study simulated future land use/planning scenarios for the Des Plaines River watershed in the Chicago metropolitan area and to evaluate the response of total suspended solids to the combined impacts of future land use and climate scenarios.

The final chapter, Chapter 12, explores the feasibility of using remote sensing to estimate urban socioeconomic attributes and to analyze their changes, spatially and temporally. Specifically, methods for estimating population and assessing urban environmental quality (UEQ) will be demonstrated through two case studies, both conducted in Indianapolis, Indiana, USA. The population estimation study intended to combine the statistical-based and dasymetric-based methods and to redistribute census population. The objectives of this research are to compare the effectiveness of the spectral response based and the land-use-based methods for population estimation of US census block groups and to produce a more accurate presentation of population distribution by combining the dasymetric mapping with land-use-based methods. The UEQ study intended to evaluate the UEQ changes from 1990 to 2000 in Indianapolis by using the integrated techniques of remote sensing and GIS. The physical variables were extracted from Landsat images, while socioeconomic variables derived from US census data to derive a synthetic UEQ indicator.

References

Fu, P. and Q. Weng. 2016. Consistent land surface temperature data generation from irregularly spaced Landsat imagery, *Remote Sensing of Environment*, 184(10), 175–187.

Ridd, M. K. (1995). Exploring a V-I-S (vegetation–impervious surface–soil) model for urban ecosystem analysis through remote sensing: comparative anatomy for cities. *International Journal of Remote Sensing*, 16, 2165–2185.

Strahler, A.H., Woodcock, C.E. and Smith, J.A., 1986, On the nature of models in remote sensing. *Remote Sensing of Environment*, 70, pp. 121–139.

Weng, Q., Fu, P. and F. Gao. 2014. Generating daily land surface temperature at Landsat resolution by fusing Landsat and MODIS data. *Remote Sensing of Environment*, 145, 55–67.

Weng, Q., Rajasekar, U. and X. Hu. 2011. Modeling urban heat islands with multi-temporal ASTER images. *IEEE Transactions on Geosciences and Remote Sensing*, 49(10), 4080–4089.

Acknowledgments

I wish to extend my most sincere appreciation to several former and current students of Indiana State University who have contributed to this book substantially. Listed in alphabetical order of their family name, Dr. Peng Fu, Dr. Xuefei Hu, Dr. Guiying Li, Dr. Bingqing Liang, Dr. Hua Liu, Dr. Dengsheng Lu, Dr. Umamaheshwaran Rajasekar, Dr. Cyril Wilson, and Dr. Yanhua Xie. My collaborators, Dr. Lei Zhang, Dr. Iphigenia Keramitsoglou, and Dr. Fenglei Fan, have contributed to the writing of Chapters 4, 6, and 9, respectively. This book is truly a collective effort of all these great scholars. Finally, I am indebted to my family for their enduring love and support, to whom this book is dedicated. It is my hope that the publication of this book will provide stimulations to students and researchers to conduct more in-depth work and analysis of urban remote sensing and to contribute to global urban observation and sustainable urban development goals. In the course of increased worldwide urbanization and global environment changes and with increased interest in remotely sensed Big Data, urban remote sensing has become a very dynamic field of study.

Qihao Weng
Hawthorn Woods, Indiana

About the Author

Qihao Weng is the Director of the Center for Urban and Environmental Change and a Professor at Indiana State University and worked as a Senior Fellow at the NASA Marshall Space Flight Center from 2008 to 2009. He is currently the Lead of GEO Global Urban Observation and Information Initiative and serves as an Editor-in-Chief of *ISPRS Journal of Photogrammetry and Remote Sensing* and the Series Editor of *Taylor & Francis Series in Remote Sensing Applications* and *Taylor & Francis Series in Imaging Science.* He has been the Organizer and Program Committee Chair of the biennial IEEE-/ISPRS-/GEO-sponsored International Workshop on Earth Observation and Remote Sensing Applications conference series since 2008, a National Director of American Society for Photogrammetry and Remote Sensing from 2007 to 2010, and a panelist of US DOE's Cool Roofs Roadmap and Strategy in 2010. In 2008, Weng received a prestigious NASA senior fellowship and was also the recipient of the Outstanding Contributions Award in Remote Sensing in 2011 and the Willard and Ruby S. Miller Award in 2015 for his outstanding contributions to geography, both from the American Association of Geographers, and a recipient of Taylor & Francis Lifetime Achievement Award in 2019 and a fellowship by 2019 JSPS Invitational Fellowships for Research in Japan (Short-term S). At Indiana State, he was selected as a Lilly Foundation Faculty Fellow in 2005 and in the following year, he received the Theodore Dreiser Distinguished Research Award. In addition, he was the recipient of 2010 Erdas Award for Best Scientific Paper in Remote

Sensing (first place) and 1999 Robert E. Altenhofen Memorial Scholarship Award, both awarded by American Society for Photogrammetry and Remote Sensing. In 1998, he received the Best Student-Authored Paper Award from International Geographic Information Foundation. Weng has been invited to give more than 110 talks by organizations and conferences worldwide and is honored with distinguished/chair/honorary/guest professorship by a dozen of universities. In 2018, he was elected a Fellow of Institute of Electrical and Electronics Engineers and a member of EU Academy of Sciences. Weng's research focuses on remote sensing applications to urban environmental and ecological systems, land-use and land-cover changes, urbanization impacts, environmental modeling, and human–environment interactions, with funded support from NSF, NASA, USGS, USAID, NOAA, National Geographic Society, among others. Weng is the author of more than 230 articles and 14 books. According to Google Scholar, as of June 2019, his SCI citation has reached over 16 000 (H-index of 59), and 39 of his publications had more than 100 citations each.

1

Urban Mapping Requirements

1.1 Introduction

Urban landscapes are typically a complex combination of buildings, roads, parking lots, sidewalks, garden, cemetery, soil, water, and so on. Each of the urban component surfaces possesses unique biophysical properties and relates to their surrounding environment to create the spatial complexity of urban ecological systems and landscape patterns. To understand the dynamics of patterns and processes and their interactions in heterogeneous landscapes such as urban areas, one must be able to quantify accurately the spatial pattern of the landscape and its temporal changes (Wu et al. 2000). In order to do so, it is necessary (i) to have a standardized method to define theses component surfaces and (ii) to detect and map them in repetitive and consistent ways, so that a global model of urban morphology may be developed, and monitoring and modeling their changes over time be possible (Ridd 1995).

Impervious surfaces are anthropogenic features through which water cannot infiltrate into the soil, such as roads, driveways, sidewalks, parking lots, rooftops, and so on. In the past two decades, impervious surface has emerged not only as an indicator of the degree of urbanization, but also a major indicator of environmental quality (Arnold and Gibbons 1996). Impervious surface is a unifying theme for all participants at all watershed scales, including planners, engineers, landscape architects, scientists, social scientists, local officials, and others (Schueler 1994). The magnitude, location, geometry, and spatial pattern of impervious surfaces, and the pervious–impervious ratio in a watershed have hydrological impacts. Although land-use zoning emphasizes roof-related impervious surfaces, transport-related impervious surfaces could have a greater impact. The increase of impervious cover would lead to the increase in the volume, duration, and intensity of urban runoff (Weng 2001), and an overall decrease of groundwater recharge and baseflow but an increase

Techniques and Methods in Urban Remote Sensing, First Edition. Qihao Weng.
© 2020 by The Institute of Electrical and Electronics Engineers, Inc.
Published 2020 by John Wiley & Sons, Inc.

of stormflow and flood frequency (Brun and Band 2000). Furthermore, imperviousness is related to the water quality of a drainage basin and it's receiving streams, lakes, and ponds (Hurd and Civco 2004). In addition, the areal extent and spatial occurrence of impervious surfaces may significantly influence urban climate by altering sensible and latent heat fluxes within the urban canopy and boundary layers (Yang et al. 2003). Therefore, estimating and mapping impervious surface is significant to a range of issues and themes in environmental science central to global environmental change and human–environment interactions and has been regarded as a key variable in urban remote sensing studies. The data sets of impervious surfaces are valuable not only for environmental management, e.g. water quality assessment and storm water taxation, but also for urban planning, e.g. building infrastructure and sustainable urban development.

Remote sensing technology has been widely applied in urban land-use and land-cover (LULC) classification and change detection. However, it is rare that the classification accuracy of greater than 80% can be achieved by using per-pixel classification (so-called "hard classification") algorithms (Mather 1999, p. 10). Therefore, the "soft"/fuzzy approach of LULC classifications has been applied, in which each pixel is assigned a class membership of each LULC type rather than a single label (Wang 1990). Nevertheless, as Mather (1999) suggested, either "hard" or "soft" classifications was not an appropriate tool for the analysis of heterogeneous landscapes. Mather (1999) maintained that identification/description/quantification, rather than classification, should be applied in order to provide a better understanding of the compositions and processes of heterogeneous landscapes such as urban areas. Ridd (1995) proposed a major conceptual model for remote sensing analysis of urban landscapes, i.e. the vegetation–impervious surface–soil (V-I-S) model. It assumes that land cover in urban environments is a linear combination of three components, namely, vegetation, impervious surface, and soil. Ridd believed that this model can be applied to spatial-temporal analyses of urban morphology, biophysical, and human systems. Having realized that the V-I-S model may be used as a method to define standardized urban landscape components, this chapter employs linear spectral mixture analysis (LSMA) as a remote sensing technique to estimate and map V-I-S components in order to analyze urban pattern and dynamics. The case study will be conducted in Indianapolis, United States, from 1991 to 2000, by using multi-temporal satellite images, i.e. Landsat Thematic Mapper (TM)/Enhanced Thematic Mapper Plus (ETM+) imagery of 1991, 1995, and 2000. Because of the significance of impervious surface as an urban land cover, land use, or material, this chapter will start with examining data requirements for remote sensing of impervious surfaces, with a particular interest in the impacts of remotely sensed data characteristics (i.e. spectral, temporal, and spatial resolutions).

1.2   Spectral Resolution Requirement

Remote sensing of impervious surfaces should consider the requirements for mapping three interrelated entities or substances on the Earth surface (i.e. material, land cover, and land use) and their relationships. Mapping of each entity/substance must consider the spectral resolution of a remote sensor. The spectral features include the number, locations, and bandwidths of spectral bands. The number of spectral bands can range from a limited number of multispectral bands (e.g. 4 bands in SPOT data and 7 for Landsat TM) to a medium number of multispectral bands (e.g. Advanced Spaceborne Thermal Emission and Reflection Radiometer [ASTER] with 14 bands and Moderate Resolution Imaging Spectroradiometer [MODIS] with 36 bands) and to hyperspectral data (e.g. AVIRIS and EO-1 Hyperion images with 224 bands). A large number of spectral bands provide the potential to derive detailed information on the nature and properties of different surface materials on the ground, but it also means a difficulty in image processing and a large data redundancy due to high correlation among the adjacent bands. Increase of spectral bands may improve classification accuracy, only when those bands are useful in discriminating the classes (Thenkabail et al. 2004a).

Urban areas are composed of a variety of materials, including different types of artificial materials (i.e. impervious surfaces), soils, rocks and minerals, and green and non-photosynthetic vegetation. These materials comprise land cover and are used in different manners for various purposes by human beings. Land cover can be defined as the biophysical state of the earth's surface and immediate subsurface, including biota, soil, topography, surface and ground water, and human structures (Turner et al. 1995). Land use can be defined as the human use of the land and involves both the manner in which the biophysical attributes of the land are manipulated and the purpose for which the land is used (Turner et al. 1995). Remote sensing technology has often been applied to map land use or land cover, instead of materials. Each type of land cover may possess unique surface properties (material), however, mapping land covers and materials have different requirements (Figure 1.1). Land-cover mapping needs to consider characteristics in addition to those coming from the material (Herold et al. 2006). The surface structure (roughness) may influence the spectral response as much as the intra-class variability (Gong and Howarth 1990; Myint 2001; Shaban and Dikshit 2001; Herold et al. 2006). Two different land covers, for example, asphalt roads and composite shingle/tar roofs, may have very similar materials (hydrocarbons) and thus are difficult to discern, although from a material perspective, these surfaces can be mapped accurately with hyperspectral remote sensing techniques (Herold et al. 2006). Therefore, land-cover mapping requires taking into account of the intra-class variability and spectral separability. On the other hand, analysis of land-use classes would nearly be impossible with spectral information alone.

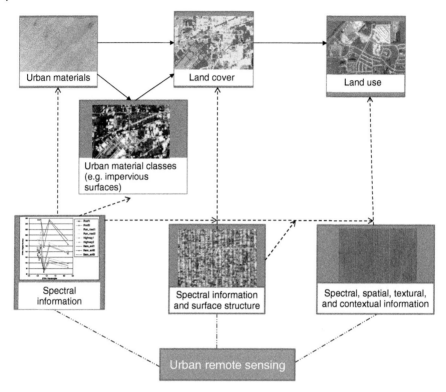

Figure 1.1 Illustration of the relationship among remote sensing of urban materials, land cover, and land use. *Source:* after Weng and Lu (2009). Reproduced with permission of Taylor & Francis.

Additional information, such as spatial, textural, and contextual information, is usually required in order to have a successful land-use classification in urban areas (Gong and Howarth 1992; Stuckens et al. 2000; Herold et al. 2003).

LSMA has been widely used in impervious surface estimation, implying that impervious surface is a type of surface material. This view has much to do with the spectral resolution of a remote sensor. LSMA is a physically deterministic modeling method that decomposes the signal measured at a given pixel into its component parts called end-members (Adams et al. 1986; Boardman 1993; Boardman et al. 1995). End-members are regarded as recognizable surface materials that have homogenous spectral properties all over the image. Impervious surfaces can be extracted and mapped as a single end-member or the combination of two or more end-members (Rashed et al. 2003; Wu and Murray 2003; Lu and Weng 2006a; Weng et al. 2008, 2009). Previous research has largely applied LSMA to medium spatial resolution, multispectral images, such as Landsat TM/ETM+ and Terra's ASTER images, for extraction of impervious surfaces (Weng 2007). However, both spatial and spectral resolution is regarded as too coarse for use in urban environments because of the heterogeneity and complexity of urban impervious surface materials. In the LSMA model, the maximum number of end-members is directly proportional to the number of spectral bands used. Hyperspectral imagery may be more effective in extracting end-members than multispectral imagery. The vastly increased dimensionality of a hyperspectral sensor may remove the sensor-related limit on the number of end-members available. More significantly, the fact that the number of hyperspectral image channels far exceeds the likely number of end-members for most applications readily permits the exclusion from the analysis of any bands with low signal-to-noise ratios or with significant atmospheric absorption effects (Lillesand et al. 2004). In previous research, hyperspectral data have been successfully used for land-use/cover classification (Benediktsson et al. 1995; Hoffbeck and Landgrebe 1996; Platt and Goetz 2004; Thenkabail et al. 2004a, 2004b), vegetation mapping (McGwire et al. 2000; Schmidt et al. 2004; Pu et al. 2008), and water mapping (Bagheri and Yu 2008; Moses et al. 2009). As spaceborne hyperspectral data such as EO-1 Hyperion become available, research and applications with hyperspectral data will increase. Weng et al. (2008) found a Hyperion image was more powerful in discerning low-albedo surface materials, which has been a major obstacle for impervious surface estimation with medium-resolution multispectral images. A sensitivity analysis suggested that the improvement of mapping accuracy in general and the better ability in discriminating low-albedo surfaces resulted largely from additional bands in the mid-infrared region (Weng et al. 2008).

The spectral characteristics of land surfaces are the fundamental principles for land imaging. Previous studies have examined the spectral properties of urban materials (Hepner et al. 1998; Ben-Dor et al. 2001; Herold et al. 2003; Heiden et al. 2007) and spectral resolution requirements for separating them (Jensen and Cowen 1999). When Jensen and Cowen (1999) explained minimum spectral resolution requirements for urban mapping, the discussion focused mainly on multispectral imagery data. They suggested that spatial resolution was more important than spectral resolution in urban mapping. The spectrum from visible to near infrared (NIR), mid-infrared (MIR), and microwave are suitable for LULC classification at coarser categorical resolutions (e.g. Levels I and II of the Anderson classification); however, at the finer categorical resolutions (e.g. Levels III and IV of the Anderson classification) and for extraction of buildings and roads, panchromatic band is needed (Jensen and Cowen 1999). Other urban studies had employed hyperspectral sensing for discriminating among urban surface features based on their diagnostic absorption and reflection characteristics and for detailed identification of urban materials. Hepner et al. (1998) suggested that interferometric synthetic aperture radar (IFSAR) imagery, when combined with AVIRIS data, can provide information on 3D geometry, topography, and impervious surfaces, and other urban surface characteristics. Ben-Dor et al. (2001) examined the feasibility of using detailed spectral information in the spectral region of 0.4–1.1 μm for identifying different features in the urban environment using Compact Airborne Spectral Imager (CASI) data. Herold et al. (2003) found that several bands in the visible, NIR, and SWIR regions were best suited for distinguishing different urban features and emphasized that both reflection and absorption features due to material composition in the SWIR region was significant in urban land-cover classification, especially in separating different types of impervious surfaces. Heiden et al. (2007) presented a hierarchical classification method for the derivation of diagnostic urban spectral features that can be used for an automated identification of spectrally homogeneous end-members from hyperspectral image data.

1.3 Temporal Resolution Requirement

Temporal resolution refers to the amount of time it takes for a sensor to return to a previously imaged location, commonly known as the repeat cycle or the time interval between acquisitions of two successive images. For airborne remote sensing, temporal resolution is less pertinent, since users can schedule flights for themselves. Jensen and Cowen (1999) suggested that studies of urban development, buildings and property infrastructure, and road center lines only need to have an image every one to five years. They further suggested that the temporal resolution requirement should be a bit higher for delineating precise road width, namely, one to two years. However, Herold (2007) believed that road aging and deterioration may have various temporal resolution depending upon the quality of pavement, traffic, and distresses (e.g. cracks, raveling, etc.).

Temporal differences between remotely sensed imageries are not only caused by the changes in spectral properties of the Earth's surface features/objects, but they can also result from atmospheric differences and changes in sun position during the course of a day and during the year. Temporal resolution is a very important consideration in remote sensing of vegetation, because vegetation grows according to daily, seasonal, and annual phenological cycles. Weng et al. (2009) found that a summer ASTER image was better for estimation of impervious surfaces than a spring (April) and a fall (October) one; based on a case study in Indianapolis, USA, it is suggested that mapping of impervious surfaces tended to be more accurate with contrasting spectral response from green vegetation (GV) when LSMA technique was employed. Plant phenology caused changes in the variance partitioning and impacted the mixing space characterization, leading to a less accurate estimation of impervious surfaces in the spring and fall (Weng et al. 2009). When other methods are used, detection of urban buildings and roads may well be suited in the leaf-off season in the temperate regions. In addition, the set overpass times of satellites may coincide with clouds or poor weather conditions. This is especially true in the tropical and coastal areas, where persistent clouds and rains in the wet season offer limit clear views of the Earth's surface and thus prevent from getting good quality images (Lu et al. 2008). In consideration of the impact of street trees on impervious surfaces, Linden and Hostert (2009) suggested that the quantification of this impact was generally possible by performing multi-temporal analyses at leaf-on/leaf-off seasons, but solar illumination geometry and shadowing can presumably preclude leaf-off acquisitions for many temperate cities.

The off-nadir imaging capability of satellite sensors, such as SPOT-5, IKONOS, and Quickbird, reduces the usual revisit time depending on the latitude of the imaged areas. This feature is designed for taking stereoscopic images and for producing digital elevation models, but it obviously also allows for more frequent coverage of selected regions for short periods and provides another means for monitoring and assessing impervious surfaces.

1.4 Spatial Resolution Requirement

Spatial resolution is a function of sensor altitude, detector size, focal size, and system configuration (Jensen 2005). It defines the level of spatial detail depicted in an image, and it is often related to the size of the smallest possible feature that can be detected from an image. This definition implies that only objects larger than the spatial resolution of a sensor can be picked out from an image. However, a smaller feature may sometimes be detectable if its reflectance dominates within a particular resolution cell or it has a unique shape (e.g. linear features). Another meaning of spatial resolution is that a ground feature should be distinguishable as a separate entity in the image. But the separation from neighbors or background is not always sufficient to identify the object. Therefore, the concept of spatial resolution includes both detectability and separability. For any feature to be resolvable in an image, it involves consideration of spatial resolution, spectral contrast, and feature shape. Jensen and Cowen (1999) suggested that the minimum spatial resolution requirement should be one-half the diameter of the smallest object of interest. For two major types of impervious surface, buildings (perimeter, area, height, and property line) and roads (width) are generally detectable with the minimum spatial resolution of 0.25–0.5 m, while road centerline can be detected at a lower resolution of 1–30 m (Jensen and Cowen 1999). Before 1999, lack of high spatial resolution (less than 10 m) images is a main reason for scarce research in remote sensing of impervious surfaces before 2000. Cracknell (1999) asserted that NOAA's AVHRR had been the major instrument for remote sensing studies between 1980 and 1999, and that Landsat or SPOT was not mentioned among the 12 most cited papers published in the International Journal of Remote Sensing during this period. The medium (10–100 m) spatial resolution images, such as

Landsat and SPOT, were not readily available and were expensive. Many researchers employed per-pixel classifiers and applied successful experience in vegetation mapping to remote sensing of impervious surfaces (Gillies et al. 2003; Carlson 2004; Bauer et al. 2007). This approach can avoid a major problem that existed in the medium-resolution imagery, i.e. mixed pixels.

Mixed pixels dominate in coarse resolution images such as AVHRR and MODIS. However, for a remote sensing project, image spatial resolution is not the only factor needed to consider. The relationship between the geographical scale of a study area and the spatial resolution of remote sensing image has to be studied (Quattrochi and Goodchild 1997). For mapping at the continental or global scale, coarse spatial resolution data are usually employed. Gamba and Herold (2009) assessed eight major research efforts in global urban extent mapping and found that most maps were produced at the spatial resolution of 1–2 km. When using coarse resolution images, a threshold has to be defined with respect to what constitute a built-up/ impervious pixel (Lu et al. 2008; Schneider et al. 2010). Reliable impervious surface data that derive from medium-resolution imagery are helpful for validating and predicting urban/built-up extent at the coarse resolution level (Lu et al. 2008).

With the advent of very high-resolution satellite imagery, such as IKONOS (launched 1999), QuickBird (2001), and OrbView (2003) images, great efforts have been made in the applications of these remote sensing images in urban and environmental studies. High-resolution satellite imageries have been applied in impervious surface mapping (Cablk and Minor 2003; Goetz et al. 2003; Lu and Weng 2009; Wu 2009; Hu and Weng 2011). These fine spatial resolution images contain rich spatial information, providing a greater potential to extract much more detailed thematic information (e.g. land use and land cover), cartographic features (buildings and roads), and metric information with stereo-images (e.g. height and area). These information and cartographic characteristics are highly beneficial to estimating and mapping of impervious surfaces. The proportion of mixed pixels is significantly reduced in an image scene. However, some new problems come with these image data, notably shadows caused by topography, tall buildings, or trees (Dare 2005), and the high spectral variation within the same land-cover class (Hsieh et al. 2001). Shadows obscure impervious surfaces underneath and thus increase the difficulty to extract both thematic and cartographic information. These disadvantages may lower image classification accuracy if classifiers used cannot effectively handle them (Irons et al. 1985; Cushnie 1987). In order to make full use of the rich spatial information inherent in fine spatial resolution data, it is necessary to minimize the negative impact of high intra-spectral variation. Algorithms that use the combined spectral and spatial information may be especially effective for impervious surface extraction in the urban areas (Lu and Weng 2007).

1.5 Linear Spectral Mixture Analysis of Urban Landscape

The prevalence of mixed pixels in urban areas implies that the instantaneous field of the view of the medium-resolution sensors does not match with the operational scale of the landscapes. Such landscapes are better viewed as a continuum formed by continuously varying proportions of idealized materials, just as soils may be described in terms of the proportions of sand, silt, and clay (Mather 1999). Agricultural land in the Midwest United States, residential areas, and semiarid areas are typical examples of continuum-type landscapes.

LSMA is a method that has been widely employed to handle the mixed pixel problem, besides the fuzzy classification. Instead of using statistical methods, LSMA is based on physically deterministic modeling to unmix the signal measured at a given pixel into its component parts called end-members (Adams et al. 1986; Boardman 1993, Boardman et al. 1995). End-members are recognizable surface materials that have homogenous spectral properties all over the image. LSMA assumes that the spectrum measured by a sensor is a linear combination of the spectra of all components within the pixel (Boardman 1993). Because of its effectiveness in handling spectral mixture problem and ability to provide continuum-based biophysical variables, LSMA has been widely used in (i) estimation of vegetation cover (Asner and Lobell 2000; McGwire et al. 2000; Small 2001; Weng et al. 2004; Lee and Lathrop 2005), (ii) impervious surface estimation and/or urban morphology analysis (Phinn et al. 2002; Wu and Murray 2003; Rashed et al. 2003; Lu and Weng 2006a, 2006b; Wu et al. 2005), (iii) vegetation or land-cover classification (Adams et al. 1995; Cochrane and Souza 1998; Aguiar et al. 1999; Lu and Weng 2004), and (iv) change detection (Rashed et al. 2005; Powell et al. 2007). However, with a few exceptions, these studies have focused on technical specifics and on the examination of the effectiveness of LSMA. Only a few studies have explicitly adopted the V-I-S model as the conceptual model to explain urban land-cover patterns (Phinn et al. 2002; Wu and Murray 2003; Wu et al. 2005; Lu and Weng 2006a, 2006b; Powell et al. 2007), while others implicitly (Rashed et al. 2003, 2005). Rashed et al. (2005) and Powell et al. (2007) are the only research attempts to examine urban land-cover "change" with the V-I-S model. Rashed et al. (2005) assessed changes between landscape components aggregated to census tracts in Cairo, Egypt, but determination of the thresholds of change may be problematic since they may vary from image to image. Powell et al. (2007) identified the stages of urban development by selecting four neighborhoods from an image of Manaus, Brazil, which did not involve change detection from multi-temporal satellite images. This chapter applies the V-I-S concept for a spatial-temporal analysis of the urban morphology in Indianapolis with LSMA, and by doing so, the potentials and limitations of this model for characterizing and quantifying urban landscape components can further be examined.

The mathematical model of LSMA can be expressed as

$$R_i = \sum_{k=1}^{n} f_k R_{ik} + E_i \tag{1.1}$$

where $i = 1, ..., m$ (number of spectral bands); $k = 1, ..., n$ (number of end-members); R_i is the spectral reflectance of band i of a pixel which contains one or more end-members; f_k is the proportion of end-member k within the pixel; R_{ik} is the known spectral reflectance of end-member k within the pixel in band i; and E_i is the error for band i. To solve f_k, the following conditions must be satisfied: (i) selected end-members should be independent of each other, (ii) the number of end-members should not be larger than the spectral bands used, and (iii) selected spectral bands should not be highly correlated. A constrained least-square solution assumes that the following two conditions are satisfied simultaneously:

$$\sum_{k=1}^{n} f_k = 1 \text{ and } 0 \le f_k \le 1 \tag{1.2}$$

$$RMSE = \sqrt{\frac{\left(\sum_{i=1}^{m} ER_i^2\right)}{m}} \tag{1.3}$$

Estimation of end-member fraction images involves four steps, i.e. image processing, end-member selection, unmixing solution, and evaluation of fraction images. Of these steps, selecting suitable end-members is the most critical one in the development of high-quality fraction images. Two types of end-members may be applied: image end-members and reference end-members. The former are derived directly from the image itself, while the latter are derived from field measurements or the laboratory spectra of known materials (Roberts et al. 1998). Many remote sensing applications have employed image end-members, since they can be easily obtained and are capable of representing the spectra measured at the same scale as the image data itself (Roberts et al. 1998). Image end-members may be derived from the extremes of the image feature space, based on the assumption that they represent the purest pixels in the image (Boardman 1993).

1.5.1 Image Preprocessing

The study area is located at 39°46′N and 86°09′W, Marion County (the city proper of Indianapolis), Indiana, USA. According to the US Census Bureau, the county has a total area of 1044 km^2, including 1026 km^2 of land and 18 km^2 of water. The average annual temperature is 11.7 °C with the highest temperature of 24.6 °C in July and the lowest temperature of −1.9 °C in January. Annual precipitation is evenly distributed throughout the year and average annual rainfall is 1021 mm, with the least amount of precipitation occurring in February. It is core of Indianapolis metropolitan area located in the State of Indiana. The county seat in Indianapolis was called "plain city" because of its flat topography (elevation ranges from 218 to 276 m above sea level), which provides the possibility of expansion in all directions. In recent decades, the city has been experiencing areal expansion through encroachment on agricultural land and other nonurban land due to population increases and economic growth. According to the 2010 census results, there were 904 668 people. Indianapolis has the highest concentration of major employers and manufacturing, professional, technical, and educational services in the state. With its moderate climate, rich history, excellent education, social services, arts, leisure, and recreation, Indianapolis was named as one of America's Best Places to Live and Work (Employment Review's August 1996).

Landsat TM images of 6 June 1991 (acquisition time: approximately 10 : 45 a.m.) and 3 July 1995 (approximately 10 : 28 a.m.) and a Landsat ETM+ image of 22 June 2000 (approximately 11 : 14 a.m.) were used in this study. Although the images purchased were geometrically corrected, its geometrical accuracy was determined not high enough for combining them with other high-resolution data sets. The images were therefore further rectified to a common Universal Transverse Mercator coordinate system based on 1 : 24 000 scale topographic maps and were resampled to a pixel size of 30 m for all bands using the nearest neighbor algorithm. A root-mean-square error (RMSE) of less than 0.2 pixels was obtained for all the rectifications. These Landsat images were acquired under clear sky conditions, and an improved image-based dark object subtraction model was applied to implement atmospheric corrections (Chavez 1996; Lu et al. 2002).

1.5.2 Image End-Member Development

In order to identify effectively image end-members and to achieve high-quality end-members, different image transform approaches, such as principal component analysis (PCA) and minimum noise fraction (MNF), may be applied to transform the multispectral images into a new data set (Green et al. 1988; Boardman and Kruse 1994). In this research, image end-members were selected from the feature spaces formed by the MNF components (Garcia-Haro et al. 1996; Cochrane and Souza 1998; Van der Meer and de Jong 2000; Small 2001, 2002, 2004). The MNF transform contains two steps: (i) de-correlation and rescaling of the noise in the data based on an estimated noise covariance matrix, producing transformed data in which the noise has unit variance and no band-to-band correlations, and (ii) implementation of a standard PCA of the noise-whitened data. The result of MNF is a two-part data set, one part associated with large eigenvalues and coherent eigen-images and a complementary part associated with near-unity eigenvalues and noise-dominated images (ENVI 2000). By performing an MNF transform, noise can be separated from the data by saving only the coherent portions, thus improving spectral processing results. In this research, the MNF procedure was applied to transform the Landsat ETM+ (the 2000 image) six reflective bands into a new coordinate set. The first three MNF components accounted for the majority of the information (99%) and were used for the selection of end-members. The scatterplots between the MNF components were illustrated in Figure 1.2, showing the potential end-members. Four end-members, namely, GV, high albedo, low albedo, and soil, were finally selected. Next, a constrained

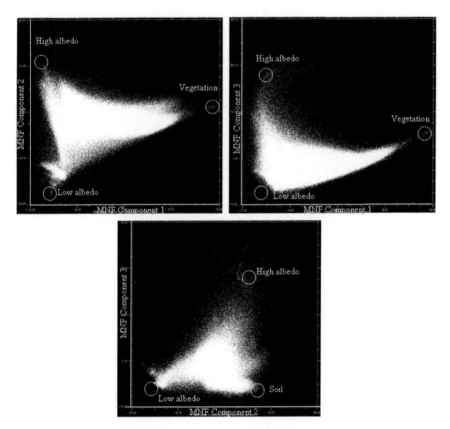

Figure 1.2 Feature spaces between the minimum noise fraction (MNF) components, illustrating potential end-members derived from a Landsat Enhanced Thematic Mapper Plus (ETM+) image. *Source:* adapted from Weng and Lu (2009). Reproduced with permission of Taylor & Francis.

least-square solution was applied to unmix the six ETM+ reflective bands into four fraction images. The same procedures were employed for derivation of fraction images from the Landsat TM 1991 and 1995 images. The first three MNF components computed from the 1991 and 1995 images also accounted for more than 99% of the scene variance, and the topologies of the triangular mixing space were consistent with that shown in Figure 1.2. Figure 1.3 shows four fraction images for the three years.

1.5.3 Extraction of Impervious Surfaces

Previous research indicated that impervious surface can be computed by adding the high- and low-albedo fractions (Wu and Murray 2003), but this method did not consider the impact of pervious surfaces on the low-albedo and high-albedo fraction images, which often resulted in overestimation of impervious surface. Our experiment with Landsat ETM+ imagery indicates that although the high-albedo fraction image related mainly to impervious surface information such as buildings and roads, it also related to other covers such as dry soils. On the other hand, the low-albedo fraction image was found to associate with water and shadows, such as water body, shadows from forest canopy and tall buildings, and moistures in crops or pastures. However, some impervious surfaces, especially dark impervious surfaces, were also linked to the low-albedo fraction image. Therefore, it is important to develop a suitable analytical procedure for removal of non-impervious information from the fraction images. In this study, we developed a procedure by using land surface temperature data to isolate non-impervious from impervious surfaces and by using soil fraction images as the thresholds to purify the high-albedo fraction images.

For the high-albedo fraction images, impervious surface was predominantly confused with dry soils. Therefore, the soil fraction images may be used to remove soils from the high-albedo fraction images. For the low-albedo fraction images, dark impervious surface was confused with water and shadows. Therefore, the critical step was to separate impervious surface from pervious pixels, including water, vegetation (forest, pasture, grass, and crops), and soils. In this study, we developed some expert rules in order to remove pervious pixels. The impervious surface image was then developed by the addition of adjusted low-albedo and high-albedo fraction images. Figure 1.4 provided a comparison of the impervious surface images before and after the adjustment. Our accuracy assessment of Landsat ETM+ image indicated that the overall RMSE of 9.22% and system error of 5.68% were obtained (Lu and Weng 2006a).

1.5.4 Image Classification

Fraction images were used for thematic land classification via a hybrid procedure that combined maximum likelihood and decision tree classifiers (Lu and

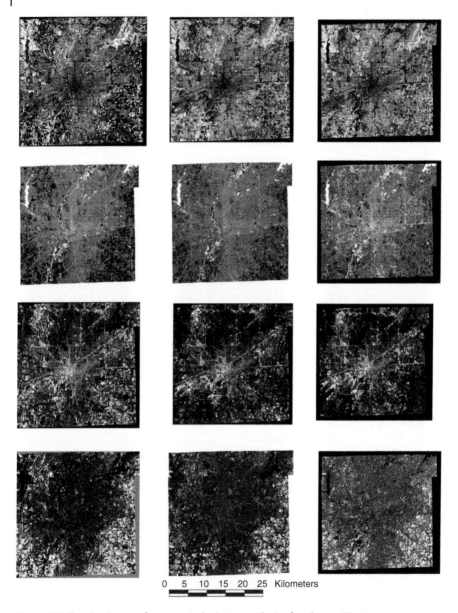

0 5 10 15 20 25 Kilometers

Figure 1.3 Fraction images from spectral mixture analysis of each year (First row: green vegetation; second row: low albedo; third row: high albedo; and fourth row: soil).
Source: adapted from Weng and Lu 2006. Reproduced with permission of Taylor & Francis.

(a)

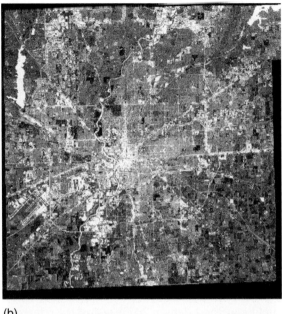

(b)

5   0   5   10 Kilometers

Figure 1.4 Comparison of impervious surface images developed from different methods. The image values range from 0 to 1, with lowest values in black and highest values in white. (a) Based on the direct addition of high-albedo and low-albedo fraction images. (b) Based on the addition of modified high-albedo and low-albedo fraction images, in which other covers were removed in the impervious surface image through the combined use of land surface temperature and fraction images. *Source:* adapted from Lu and Weng (2006a). Reproduced with permission of Taylor & Francis.

Weng 2004). Sample plots were identified from high-resolution aerial photographs, covering initially 10 LULC types: commercial and industrial, high-density residential, low-density residential, bare soil, crop, grass, pasture, forest, wetland, and water. On average, 10–16 sample plots for each class were selected. A window size of three by three was applied to extract the fraction value for each plot. The mean and standard deviation values were calculated for each LULC class. The characteristics of fractional composition for selected LULC types were then examined. Next, the maximum likelihood classification algorithm was applied to classify the fraction images into 10 classes, generating a classified image and a distance image. A distance threshold was selected for each class to screen out the pixels that probably do not belong to that class, which was determined by examining interactively the histogram of each class in the distance image. Pixels with a distance value greater than the threshold were assigned a class value of zero. A decision tree classifier was then applied to reclassify these pixels. The parameters required by the decision tree classifier were identified based on the mean and standard deviation of the sample plots for each class. Finally, the accuracy of the classified image was checked with a stratified random sampling method (Jensen 2005) against the reference data of 150 samples collected from large-scale aerial photographs. To simplify urban landscape analysis, 10 classes were merged into six LULC types, including (i) commercial and industrial urban land, (ii) residential land, (iii) agricultural and pasture land, (iv) grassland, (v) forest, and (vi) water (Lu and Weng 2004). Figure 1.5 shows the classified LULC maps in the three years.

The overall accuracy, producer's accuracy, and user's accuracy were calculated based on the error matrix for each classified map, as well as the KHAT statistic, kappa variance, and Z statistic. The overall accuracy of LULC map for 1991, 1995, and 2000 were determined to be 90, 88, and 89%, respectively. Apparently, LULC data derived from the LSMA procedure have a reasonably high accuracy and are sufficient for urban landscape analysis.

1.5.5 Urban Morphologic Analysis Based on the V-I-S Model

Three images in the first row of Figure 1.3 shows the geographic patterns of GV fractions. These images display a large dark area (low fraction values) at the center of the study area corresponding to the central business district of Indianapolis City. Bright areas of high GV values were found in the surrounding areas. Various types of crops were still at the early stage of growth or were before emergence, as indicated by medium gray to dark tone of the GV fraction images in the southeastern and southwestern parts of the city. Table 1.1 indicates that forest had the highest GV fraction values, followed by grassland. In contrast, commercial and industrial land displayed the lowest GV values. Little vegetative amount was found in water bodies, as indicated by the GV fraction values. Both residential land and pasture-agricultural land yielded a mediate

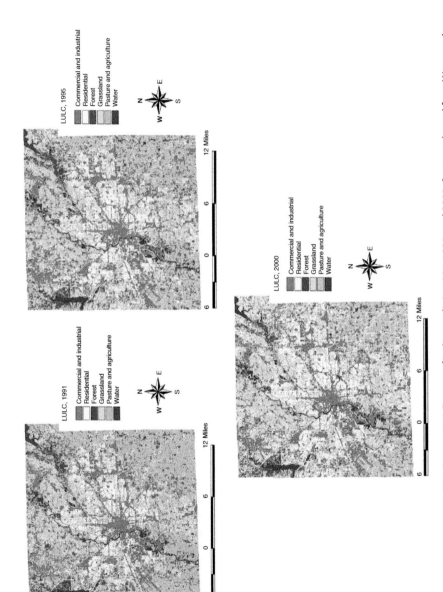

Figure 1.5 Land-use and land-cover (LULC) maps of Indianapolis in 1991, 1995, and 2000. *Source:* adapted from Weng and Lu (2006). Reproduced with permission of Taylor & Francis.

Table 1.1 V-I-S compositions of land-use and land-cover (LULC) types in Indianapolis in 1991, 1995, and 2000.

Land-cover type	1991 TM image			1995 TM image			2000 Enhanced Thematic Mapper Plus (ETM+) image		
	Mean vegetation (standard deviation)	Mean impervious surface (standard deviation)	Mean soil (standard deviation)	Mean vegetation (standard deviation)	Mean impervious surface (standard deviation)	Mean soil (standard deviation)	Mean vegetation (standard deviation)	Mean impervious surface (standard deviation)	Mean soil (standard deviation)
Commercial and industrial	0.167 (0.128)	0.709 (0.190)	0.251 (0.193)	0.127 (0.097)	0.679 (0.178)	0.273 (0.177)	0.125 (0.092)	0.681 (0.205)	0.276 (0.191)
Residential	0.314 (0.132)	0.558 (0.138)	0.198 (0.152)	0.371 (0.115)	0.508 (0.108)	0.149 (0.092)	0.298 (0.095)	0.467 (0.124)	0.247 (0.137)
Grassland	0.433 (0.176)	0.451 (0.135)	0.268 (0.208)	0.553 (0.145)	0.366 (0.096)	0.155 (0.131)	0.442 (0.099)	0.276 (0.083)	0.305 (0.119)
Agriculture and pasture	0.304 (0.213)	0.374 (0.112)	0.602 (0.285)	0.388 (0.191)	0.291 (0.091)	0.378 (0.236)	0.371 (0.168)	0.275 (0.072)	0.407 (0.222)
Forest	0.654 (0.162)	0.436 (0.128)	0.182 (0.166)	0.716 (0.085)	0.388 (0.065)	0.046 (0.052)	0.584 (0.075)	0.327 (0.074)	0.175 (0.055)
Water	0.226 (0.186)	0.730 (0.197)	0.188 (0.178)	0.176 (0.210)	0.805 (0.167)	0.094 (0.068)	0.111 (0.120)	0.891 (0.136)	0.078 (0.071)

Source: Weng and Lu (2009). Reproduced with permission of Taylor & Francis.

GV fraction value, subject to the impact of the dates of image acquired. In all the years observed, pasture-agricultural land exhibited a large standard deviation value, suggesting that pasture and agricultural land may hold various amount of vegetation.

The percentage of land covered by impervious surface may vary significantly with LULC categories and subcategories (Soil Conservation Service 1975). This study shows a substantially different estimate for each LULC type, as this study applied a spectral unmixing model to the remote sensing images, and the modeling had introduced some errors as expected. For example, a high impervious fraction value was found in water, since water related to the low-albedo fraction, and the latter were included in the computation of impervious surface. Generally speaking, an LULC type with a higher GV fraction appeared to have a lower impervious fraction. Commercial and industrial land detected very high impervious fraction values around 0.7 in all years. Residential land came after with fraction values around 0.5. Grassland, agricultural-pasture land, and forestland detected lower values of impervious surface, owing largely to their exposure to bare soil, confusion with commercial/industrial and residential land, and modeling errors.

Soil fraction values were generally low in the majority of the urban area, but high in the surroun'ing areas. Especially, in agricultural fields located in the southeastern and southwestern parts of the city, soil fraction images appeared very bright since various types of crops were still at the early stage of growth. Table 1.1 shows that agricultural-pasture land observed a fraction value close to 0.4 at all times. Grassland possessed medium fraction values averaging 0.25, substantially higher than the fraction values of forestland and residential land. Commercial and industrial land displayed similar fraction values as grassland, which had much to do with its confusion with dry soils in the high-albedo images. Water generally possessed a minimal impervious fraction value. Like GV fraction, soil fraction displayed the highest standard deviation values in agricultural-pasture land due to various amount of emerging vegetation.

The V-I-S composition may be examined by taking samples along transects. Figure 1.6a shows ternary plots of four transects across the geometry center of the city, sampled from the 2000 Landsat ETM+ image. Sample 1 runs from west to east, Sample 2 from north to south, Sample 3 from southwest to northeast, and Sample 4 from southeast to northwest. Errors from the spectral unmixing modeling are not included in these diagrams due to their low values clustering to near zero. Along the east–west transect, nearly all pixels sampled showed a GV fraction of less than 0.6, while soil fraction values ranged from 0.1 to 0.7. A clustering pattern was apparent, if impervious fraction values were observed in the range from 0.2 to 0.7, and GV fraction values in the range of 0.5–0.8. A more clustered pattern can be observed in the ternary diagrams based on the north–south and the southwest–northeast transects. However, the southeast–northwest transect exhibited clearly a more dispersed pattern of

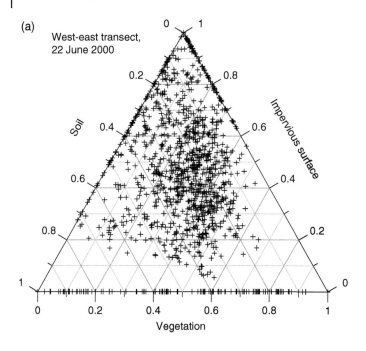

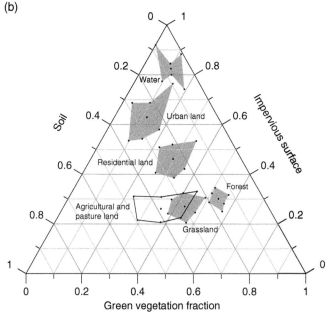

Figure 1.6 (a) Ternary plot of V-I-S composition along a sampled transect derived from Landsat ETM+ image in Indianapolis; (b) Quantitative relationships among the LULC types in respect to the V-I-S model. *Source:* adapted from Weng (2009). Reproduced with permission of the McGraw-Hill Companies, Inc.

pixel distribution, suggesting a variety of V-I-S composition types. GV along this transect yielded a fraction value from 0.3 to 1, whereas impervious fraction might have any value between 0 and 1. Soil fraction values continued to increase up to 0.8. When mean signature values of the fractions for each LULC type is plotted, quantitative relationships among the thematic LULC types in terms of the V-I-S composition can be examined. Figure 1.6b shows the V-I-S composition by LULC in 2000, with an area delineating one standard deviation from the mean fraction value.

1.5.6 Landscape Change and the V-I-S Dynamics

Table 1.2 shows the composition of LULC by year and changes occurred between two intervals. In 1991, residential use and pasture-agriculture accounted equally for 27% of the total land, while grassland shared another 20%. The combination of commercial and industrial land used 13% of the total area, and forestland had a close match, yielding another 10%. Water bodies occupied the remaining 3%, and this percentage kept unchanged from 1991 to 2000. However, LULC dynamics occurred in all other categories, as seen in the last three columns of Table 1.2. The most notable increment was observed in residential use, which grew from 27% in 1991 to 33% in 1995 reaching 38% in 2000. Associated with this change, grassland was increased from 20 to 23%. Highly developed land, mainly for commercial, industrial uses, transportation, and utilities, continued to expand. In 2000, it accounted for over 15 000 ha, or 15%, generating a 2% of increase over the nine years. These results suggest that urban land dispersal in Indianapolis was related both to population increase and to economic growth. In contrast, a pronounced decrease in pasture and agricultural land was discovered from 1991 (27%) to 1995 (20%). This decrease was also evident between 1995 and 2000, when pasture and agricultural land was further shrunk by 6581.30 ha (31.56%). Forestland in a city like Indianapolis was understandably limited in size. Our remote sensing geographic information system (GIS) analysis indicates, however, that forestland continued to disappear with a stable, marked rate. Between 1991 and 2000, forestland was reduced by 2864.81 ha (i.e. 28.75%), leveling down to approximately 7100 ha. The cross-tabulation of the 1991 and 2000 LULC maps reveals that most of the losses in pasture, agricultural, and forestland were converted to residential and other urban uses, owing to the continued process of urbanization and suburbanization. GIS overlay of the two maps further shows the spatial occurrence of urban expansion to be mostly in the edges of the city. These changes in LULC have led to changes in the composition of image fractions.

Figure 1.7 shows the pattern of changes in the V-I-S components of the LULC classes. Impervious surface, as an important urban land-cover feature, not only indicates the degree of urbanization, but also a major contributor to the environmental impacts of urbanization (Arnold and Gibbons 1996).

Table 1.2 Changes in LULC in Indianapolis, 1991–2000.

LULC type	Area, 1991 (ha)	Area, 1995 (ha)	Area, 2000 (ha)	Change, 1991–1995	Change, 1995–2000	Change, 1991–2000
Commercial and industrial	13 322.10	16 706.50	15 489.00	3 384.40 (25.40%)	−1 217.50 (−7.29%)	2 166.90 (16.27%)
Residential	28 708.90	34 123.70	40 771.70	5 414.80 (18.86%)	6 648 (19.48%)	12 062.80 (42.02%)
Grassland	21 132.50	21 356.60	23 976.40	224.10 (1.06%)	2 619.80 (12.27%)	2 843.90 (13.46%)
Pasture and agriculture	28 466.00	20 853.80	14 272.50	−7 612.20 (−26.74%)	−6 581.30 (−31.56%)	−14 193.50 (−49.86%)
Forest	9 965.58	8 547.71	7 100.77	−1 417.87 (−14.23%)	−1 446.94 (−16.93%)	−2 864.81 (−28.75%)
Water	2 894.21	2 903.96	2 903.06	9.75 (0.34%)	−0.90 (−0.03%)	8.85 (0.31%)

Source: Weng (2009). Reproduced with permission of Taylor & Francis.

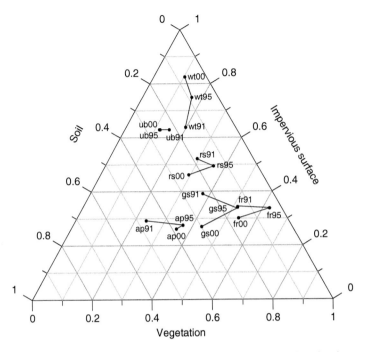

Figure 1.7 V-I-S dynamics. *Source:* Weng and Lu (2009). Reproduced with permission of Taylor & Francis.

To examine how impervious surfaces in Indianapolis had changed from 1991 to 2000, Figure 1.8 was created to show the distribution of impervious surfaces for the observed years at four categories. Result shows that the amount of pixels with values of greater than zero increased from 42 501 in 1991 to 45 804 in 1995 and further increased to 46 560 in 2000. A comparison between 1991 and 2000 indicates that this increase in the amount of pixels took place over the entire range of impervious fraction values (except for the category of 0.1–0.2). This analysis further substantiates the above findings, i.e. Indianapolis underwent an extensive urbanization process, during which impervious or impenetrable surfaces, such as rooftops, roads, parking lots, driveways, and sidewalks, were widely generated. In other words, significant amount of non-urban pixels became urbanized during the study period. Furthermore, in 1991, only 7.03% of the urbanized pixels (pixels that contain some impervious surface areas) had a value of impervious fraction greater than 0.6. In 1995 and 2000, they were 7.41 and 9.24%, respectively. This increase of pixel counts in the higher percentage categories of impervious fraction suggests that even more construction had taken place in previously urbanized pixels, i.e. there existed infill type of urban development (Wilson et al. 2003).

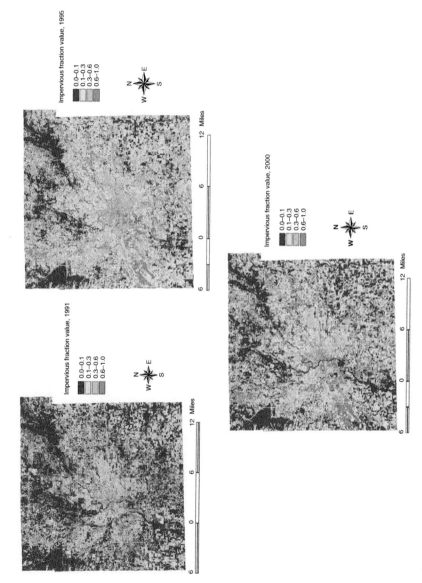

Figure 1.8 Distribution of impervious coverage by year. *Source:* adapted from Weng and Lu (2009). Reproduced with permission of Taylor & Francis.

1.6 Summary

Remote sensing of urban material, land use, and land cover has different requirements. This study intended to approach their relationship by using a continuum field model. This model was developed based on the reconciliation of the V-I-S model (Ridd 1995) and LSMA of Landsat imagery. The case study demonstrated successfully that the continuum model was effective for characterizing and quantifying the spatial and temporal changes of the landscape compositions in Indianapolis between 1991 and 2000.

The linkage between the Ridd's model and LSMA lies in the urban material. Ridd (1995) suggested that urban landscape can be decomposed into three classes of urban materials (vegetation, impervious surfaces, and soil), but did not implement the model using digital image processing algorithms. LSMA has been widely used to analyze spectrally heterogeneous urban reflectance based on "endmembers," which are recognizable surface materials that have homogenous spectral properties over the entire image. The number of end-members that can be extracted from a satellite image is determined by the data dimensionality reflected in the image mixing space, which, in turn, is subject to the number of spectral bands available. Theoretically speaking, Landsat TM/ETM+ images, with six spectral bands (excluding the thermal infrared band), can derive as many as seven end-members. However, our study indicated that the three lower order principal components accounted for more than 99% of the image variance, implying that the mixing space was perfectly three-dimensional. The RMSE was small (less than 0.1 of mean pixel reflectance) for all three unmixings. Therefore, the end-members selected for this study (high albedo, low albedo, vegetation, and soil) were regarded proper to account for the observed radiance. When high-albedo and low-albedo end-members were used to model impervious surface, the RMSE was also found to be reasonably small for the LSMA models (less than 0.02). As such, LSMA with the four-end-member model offered a simple, robust, physically based solution to quantify the urban reflectance. Our study further indicated that although the spectral end-members were different over time because of changes in urban materials and land cover/use, the topology of the triangular mixing space was consistent. It became apparent that LSMA can provide a repetitive way to derive consistent image end-members from Landsat images. Therefore, by representing Ridd's V-I-S components as image fractions, the continuum model develop in this study provided an effective approach for quantifying urban landscape patterns as standardized component surfaces.

Existing remote sensing literature has regarded impervious surface as a type of surface material, land cover, or land use. The discrepancy in conceptual view has stimulated research into three major directions. Various sub-pixel algorithms applied largely to medium resolution but less frequently to high-resolution imagery to estimate and map impervious surfaces as a type of surface material. Per-pixel algorithms were employed for all sorts of images at various spatial resolutions to classify impervious surfaces as a type of land cover or land use. Feature extraction methods were applied mainly to high-resolution satellite imagery, aerial photographs, and Light Detection and Ranging (LiDAR) data to extract roads and buildings, implicitly suggesting impervious surface as a special type of land use/cover. These research directions are sometime intermingled in a study, but clearly they represent different research traditions and have been approached from different perspectives. Many research endeavors have been oriented towards the spatial heterogeneity of urban landscapes, ideal spatial resolution for urban mapping, and the strengths and limitations of existing remote sensors. In contrast, less research efforts have been devoted to the spectral diversity of impervious surfaces and the spectral requirements for remote sensing of impervious surfaces. Hyperspectral imaging has been applied most extensively in the studies of vegetation and water but little to impervious surfaces. Similarly, the geometric properties (especially the 3D nature) of urban environments have been understudied. The least attention was paid to temporal resolution, change, and evolution of impervious surface over time, and temporal requirement for urban mapping. Therefore, there is a great need to address the temporal resolution requirement in urban remote sensing and how it relates to spectral resolution, spatial resolution, and the geometric characteristics of urban features and objects.

References

Adams, J.B., Sabol, D.E., Kapos, V. et al. (1995). Classification of multispectral images based on fractions of endmembers: application to land cover change in the Brazilian Amazon. *Remote Sensing of Environment* 52: 137–154.

Adams, J.B., Smith, M.O., and Johnson, P.E. (1986). Spectral mixture modeling: a new analysis of rock and soil types at the Viking Lander site. *Journal of Geophysical Research* 91: 8098–8112.

Aguiar, A.P.D., Shimabukuro, Y.E., and Mascarenhas, N.D.A. (1999). Use of synthetic bands derived from mixing models in the multispectral classification of remote sensing images. *International Journal of Remote Sensing* 20: 647–657.

Arnold, C.L. Jr. and Gibbons, C.J. (1996). Impervious surface coverage: the emergence of a key environmental indicator. *Journal of the American Planning Association* 62: 243–258.

Asner, G.P. and Lobell, D.B. (2000). A biogeophysical approach for automated SWIR unmixing of soils and vegetation. *Remote Sensing of Environment* 74: 99–112.

Bagheri, S. and Yu, T. (2008). Hyperspectral sensing for assessing nearshore water quality conditions of Hudson/Raritan estuary. *Journal of Environmental Informatics* 11 (2): 123–130.

Bauer, M.E., Loffelholz, B.C., and Wilson, B. (2007). Estimating and mapping impervious surface area by regression analysis of Landsat imagery. In: *Remote Sensing of Impervious Surfaces* (ed. Q. Weng), 3–19. Boca Raton, FL: CRC Press.

Ben-Dor, E., Levin, N., and Saaroni, H. (2001). A spectral based recognition of the urban environment using the visible and near-infrared spectral region (0.4–1.1 m) – a case study over Tel-Aviv. *International Journal of Remote Sensing* 22 (11): 2193–2218.

Benediktsson, J.A., Sveinsson, J.R., and Arnason, K. (1995). Classification and feature extraction of AVIRIS data. *IEEE Transactions on Geoscience and Remote Sensing* 33: 1194–1205.

Boardman, J.M., Kruse, F.A., and Green, R.O. (1995). Mapping target signature via partial unmixing of AVIRIS data. *Summaries of the Fifth JPL Airborne Earth Science Workshop*, JPL Publication 95-1, NASA Jet Propulsion Laboratory, Pasadena, California, pp. 23–26.

Boardman, J.W. (1993). Automated spectral unmixing of AVIRIS data using convex geometry concepts. *Summaries of the Fourth JPL Airborne Geoscience Workshop*, JPL Publication 93-26, NASA Jet Propulsion Laboratory, Pasadena, California, pp. 11–14.

Boardman, J.W. and Kruse, F.A. (1994). Automated spectral analysis: a geological example using AVIRIS data, north Grapevine Mountains, Nevada. *Proceedings, ERIM Tenth Thematic Conference on Geologic Remote Sensing*, Ann Arbor, MI, pp. 407–418.

Brun, S.E. and Band, L.E. (2000). Simulating runoff behavior in an urbanizing watershed. *Computers, Environment and Urban Systems* 24: 5–22.

Cablk, M.E. and Minor, T.B. (2003). Detecting and discriminating impervious cover with high resolution IKONOS data using principal component analysis and morphological operators. *International Journal of Remote Sensing* 24: 4627–4645.

Carlson, T.N. (2004). Analysis and prediction of surface runoff in an urbanizing watershed using satellite imagery. *Journal of the American Water Resources Association* 40 (4): 1087–1098.

Chavez, P.S. Jr. (1996). Image-based atmospheric corrections – revisited and improved. *Photogrammetric Engineering & Remote Sensing* 62: 1025–1036.

Cochrane, M.A. and Souza, C.M. Jr. (1998). Linear mixture model classification of burned forests in the eastern Amazon. *International Journal of Remote Sensing* 19: 3433–3440.

Cracknell, A.P. (1999). Twenty years of publication of the *International Journal of Remote Sensing*. *International Journal of Remote Sensing* 20: 3469–3484.

Cushnie, J.L. (1987). The interactive effect of spatial resolution and degree of internal variability within land-cover types on classification accuracies. *International Journal of Remote Sensing* 8: 15–29.

Dare, P.M. (2005). Shadow analysis in high-resolution satellite imagery of urban areas. *Photogrammetric Engineering & Remote Sensing* 71: 169–177.

ENVI (2000). *ENVI User's Guide*. Boulder, CO: Research Systems Inc.

Gamba, p. and Herold, M. (2009). *Global Mapping of Human Settlements: Experiences, Datasets, and Prospects*. Boca Raton, FL: CRC Press.

Garcia-Haro, F.J., Gilabert, M.A., and Melia, J. (1996). Linear spectral mixture modeling to estimate vegetation amount from optical spectral data. *International Journal of Remote Sensing* 17: 3373–3400.

Gillies, R.R., Box, J.B., Symanzik, J., and Rodemaker, E.J. (2003). Effects of urbanization on the aquatic fauna of the Line Creek watershed, Atlanta – a satellite perspective. *Remote Sensing of Environment* 86: 411–422.

Goetz, S.J., Wright, R.K., Smith, A.J. et al. (2003). IKONOS imagery for resource management: tree cover, impervious surfaces, and riparian buffer analyses in the mid-Atlantic region. *Remote Sensing of Environment* 88: 195–208.

Gong, p. and Howarth, P.J. (1990). The use of structure information for improving land-cover classification accuracies at the rural-urban fringe. *Photogrammetric Engineering & Remote Sensing* 56 (1): 67–73.

Gong, p. and Howarth, P.J. (1992). Frequency-based contextual classification and gray-level vector reduction for land-use identification. *Photogrammetric Engineering & Remote Sensing* 58 (4): 423–437.

Green, A.A., Berman, M., Switzer, P., and Craig, M.D. (1988). A transformation for ordering multispectral data in terms of image quality with implications for noise removal. *IEEE Transactions on Geoscience and Remote Sensing* 26: 65–74.

Heiden, U., Segl, K., Roessner, S., and Kaufmann, H. (2007). Determination of robust spectral features for identification of urban surface materials in hyperspectral remote sensing data. *Remote Sensing of Environment* 111: 537–552.

Hepner, G.F., Houshmand, B., Kulikov, I., and Bryant, N. (1998). Investigation of the integration of AVIRIS and IFSAR for urban analysis. *Photogrammetric Engineering & Remote Sensing* 64 (8): 813–820.

Herold, M. (2007). Spectral characteristics of asphalt road surfaces. In: *Remote Sensing of Impervious Surfaces* (ed. Q. Weng), 237–247. Boca Raton, FL: CRC Press.

Herold, M., Liu, X., and Clark, K.C. (2003). Spatial metrics and image texture for mapping urban land use. *Photogrammetric Engineering & Remote Sensing* 69 (9): 991–1001.

Herold, M., Schiefer, S., Hostert, P., and Roberts, D.A. (2006). Applying imaging spectrometry in urban areas. In: *Urban Remote Sensing* (eds. Q. Weng and D. Quattrochi), 137–161. Boca Raton, FL: CRC/Taylor & Francis.

Hoffbeck, J.P. and Landgrebe, D.A. (1996). Classification of remote sensing having high spectral resolution images. *Remote Sensing of Environment* 57: 119–126.

Hsieh, P.-F., Lee, L.C., and Chen, N.-Y. (2001). Effect of spatial resolution on classification errors of pure and mixed pixels in remote sensing. *IEEE Transactions on Geoscience and Remote Sensing* 39: 2657–2663.

Hu, X. and Weng, Q. (2011). Impervious surface area extraction from IKONOS imagery using an object-based fuzzy method. *Geocarto International* 26: 3–20.

Hurd, J.D. and Civco, D.L. (2004). Temporal characterization of impervious surfaces for the State of Connecticut. *ASPRS Annual Conference Proceedings, Denver, Colorado, May 2004* (Unpaginated CD ROM).

Irons, J.R., Markham, B.L., Nelson, R.F. et al. (1985). The effects of spatial resolution on the classification of Thematic Mapper data. *International Journal of Remote Sensing* 6: 1385–1403.

Jensen, J.R. (2005). *Introductory Digital Image Processing: A Remote Sensing Perspective*, 3e. Upper Saddle River, NJ: Prentice Hall.

Jensen, J.R. and Cowen, D.C. (1999). Remote sensing of urban/suburban infrastructure and socioeconomic attributes. *Photogrammetric Engineering & Remote Sensing* 65: 611–622.

Lee, S. and Lathrop, R.G. Jr. (2005). Sub-pixel estimation of urban land cover components with linear mixture model analysis and Landsat Thematic Mapper imagery. *International Journal of Remote Sensing* 26 (22): 4885–4905.

Lillesand, T.M., Kiefer, R.W., and Chipman, J.W. (2004). *Remote Sensing and Image Interpretation*, 614. New York: Wiley.

van der Linden, S. and Hostert, p. (2009). The influence of urban structures on impervious surface maps from airborne hyperspectral data. *Remote Sensing of Environment* 113: 2298–2305.

Lu, D., Mausel, P., Brondizio, E., and Moran, E. (2002). Assessment of atmospheric correction methods for Landsat TM data applicable to Amazon basin LBA research. *International Journal of Remote Sensing* 23: 2651–2671.

Lu, D., Tian, H., Zhou, G., and Ge, H. (2008). Regional mapping of human settlements in southeastern China with multisensor remotely sensed data. *Remote Sensing of Environment* 112 (9): 3668–3679.

Lu, D. and Weng, Q. (2004). Spectral mixture analysis of the urban landscape in Indianapolis with Landsat ETM+ imagery. *Photogrammetric Engineering & Remote Sensing* 70: 1053–1062.

Lu, D. and Weng, Q. (2006a). Use of impervious surface in urban land use classification. *Remote Sensing of Environment* 102 (1–2): 146–160.

Lu, D. and Weng, Q. (2006b). Spectral mixture analysis of ASTER imagery for examining the relationship between thermal features and biophysical descriptors in Indianapolis, Indiana. *Remote Sensing of Environment* 104 (2): 157–167.

Lu, D. and Weng, Q. (2007). A survey of image classification methods and techniques for improving classification performance. *International Journal of Remote Sensing* 28 (5): 823–870.

Lu, D. and Weng, Q. (2009). Extraction of urban impervious surfaces from IKONOS imagery. *International Journal of Remote Sensing* 30 (5): 1297–1311.

Mather, P.M. (1999). Land cover classification revisited. In: *Advances in Remote Sensing and GIS* (eds. P.M. Atkinson and N.J. Tate), 7–16. New York: Wiley.

McGwire, K., Minor, T., and Fenstermaker, L. (2000). Hyperspectral mixture modeling for quantifying sparse vegetation cover in arid environments. *Remote Sensing of Environment* 72: 360–374.

Moses, W.J., Gitelson, A.A., Berdnikov, S., and Povazhnyy, V. (2009). Satellite estimation of chlorophyll-a concentration using the red and NIR bands of MERIS – The Azov Sea case study. *IEEE Geoscience and Remote Sensing Letters* 6: 845–849.

Myint, S.W. (2001). A robust texture analysis and classification approach for urban land-use and land-cover feature discrimination. *Geocarto International* 16: 27–38.

Phinn, S., Stanford, M., Scarth, p. et al. (2002). Monitoring the composition of urban environments based on the vegetation-impervious surface-soil (VIS) model by subpixel analysis techniques. *International Journal of Remote Sensing* 23: 4131–4153.

Platt, R.V. and Goetz, A.F.H. (2004). A comparison of AVIRIS and Landsat for land use classification at the urban fringe. *Photogrammetric Engineering & Remote Sensing* 70: 813–819.

Powell, R.L., Roberts, D.A., Dennison, P.E., and Hess, L.L. (2007). Sub-pixel mapping of urban land cover using multiple endmember spectral mixture analysis: Manaus, Brazil. *Remote Sensing of Environment* 106 (2): 253–267.

Pu, R., Kelly, M., Anderson, G.L., and Gong, p. (2008). Using CASI hyperspectral imagery to detect mortality and vegetation stress associated with a new hardwood forest disease. *Photogrammetric Engineering & Remote Sensing* 74 (1): 65–75.

Quattrochi, D.A. and Goodchild, M.F. (1997). *Scale in Remote Sensing and GIS*. New York City, NY: Lewis Publishers.

Rashed, T., Weeks, J.R., Roberts, D.A. et al. (2003). Measuring the physical composition of urban morphology using multiple endmember spectral mixture models. *Photogrammetric Engineering & Remote Sensing* 69: 1011–1020.

Rashed, T., Weeks, J.R., Stow, D., and Fugate, D. (2005). Measuring temporal compositions of urban morphology through spectral mixture analysis: toward a soft approach to change analysis in crowded cities. *International Journal of Remote Sensing* 26 (4): 699–718.

Ridd, M.K. (1995). Exploring a V-I-S (vegetation-impervious surface-soil) model for urban ecosystem analysis through remote sensing: comparative anatomy for cities. *International Journal of Remote Sensing* 16 (12): 2165–2185.

Roberts, D.A., Batista, G.T., Pereira, J.L.G. et al. (1998). Change identification using multitemporal spectral mixture analysis: applications in eastern Amazônia. In: *Remote Sensing Change Detection: Environmental Monitoring Methods and Applications* (eds. R.S. Lunetta and C.D. Elvidge), 137–161. Ann Arbor, MI: Ann Arbor Press.

Schmidt, K.S., Skidmore, A.K., Kloosterman, E.H. et al. (2004). Mapping coastal vegetation using an expert system and hyperspectral imagery. *Photogrammetric Engineering & Remote Sensing* 70: 703–715.

Schneider, A., Friedl, M.A., and Potere, D. (2010). Mapping global urban areas using MODIS 500-m data: New methods and datasets based on "urban ecoregions". *Remote Sensing of Environment* 114: 1733–1746.

Schueler, T.R. (1994). The importance of imperviousness. *Watershed Protection Techniques* 1: 100–111.

Shaban, M.A. and Dikshit, O. (2001). Improvement of classification in urban areas by the use of textural features: the case study of Lucknow city, Uttar Pradesh. *International Journal of Remote Sensing* 22: 565–593.

Small, C. (2001). Estimation of urban vegetation abundance by spectral mixture analysis. *International Journal of Remote Sensing* 22: 1305–1334.

Small, C. (2002). Multitemporal analysis of urban reflectance. *Remote Sensing of Environment* 81: 427–442.

Small, C. (2004). The Landsat ETM+ spectral mixing space. *Remote Sensing of Environment* 93: 1–17.

Soil Conservation Service (1975). *Urban Hydrology for Small Watersheds, USDA Soil Conservation Service Technical Release No. 55.* Washington, DC: U.S. Department of Agriculture.

Stuckens, J., Coppin, P.R., and Bauer, M.E. (2000). Integrating contextual information with per-pixel classification for improved land cover classification. *Remote Sensing of Environment* 71: 282–296.

Thenkabail, P.S., Enclona, E.A., Ashton, M.S. et al. (2004a). Hyperion, IKONOS, ALI, and ETM+ sensors in the study of African rainforests. *Remote Sensing of Environment* 90: 23–43.

Thenkabail, P.S., Enclona, E.A., Ashton, M.S., and van der Meer, B. (2004b). Accuracy assessments of hyperspectral waveband performance for vegetation analysis applications. *Remote Sensing of Environment* 91: 354–376.

Turner II, B.L., Skole, D., Sanderson, S. et al. (1995). Land-use and land-cover change: science and research plan. *International Geosphere-Biosphere Program and the Human Dimensions of Global Environmental Change Programme* (IGBP Report No. 35 and HDP Report No. 7), Stockholm and Geneva.

Van der Meer, F. and de Jong, S.M. (2000). Improving the results of spectral unmixing of Landsat Thematic Mapper imagery by enhancing the orthogonality of end-members. *International Journal of Remote Sensing* 21: 2781–2797.

Wang, F. (1990). Fuzzy supervised classification of remote sensing images. *IEEE Transactions on Geoscience and Remote Sensing* 28 (2): 194–201.

Weng, Q. (2001). Modeling urban growth effect on surface runoff with the integration of remote sensing and GIS. *Environmental Management* 28: 737–748.

Weng, Q. (2007). Remote sensing of impervious surfaces: an overview. In: *Remote Sensing of Impervious Surfaces*, xv–xxvii. Boca Raton, FL: CRC Press.

Weng, Q. (2009). Building extraction from LiDAR data. In: *Remote Sensing and GIS Integration: Theories, Methods, and Applications*, 183–208. New York: McGraw-Hill.

Weng, Q., Hu, X., and Liu, H. (2009). Estimating impervious surfaces using linear spectral mixture analysis with multi-temporal ASTER images. *International Journal of Remote Sensing* 30 (18): 4807–4830.

Weng, Q., Hu, X., and Lu, D. (2008). Extracting impervious surface from medium spatial resolution multispectral and hyperspectral imagery: a comparison. *International Journal of Remote Sensing* 29 (11): 3209–3232.

Weng, Q. and Lu, D. (2006). Sub-pixel analysis of urban landscapes. In: *Urban Remote Sensing* (eds. Q. Weng and D. Quattrochi), 71–90. Boca Raton, FL: CRC/Taylor & Francis.

Weng, Q. and Lu, D. (2009). Landscape as a continuum: an examination of the urban landscape structures and dynamics of Indianapolis city, 1991–2000. *International Journal of Remote Sensing* 30 (10): 2547–2577.

Weng, Q., Lu, D., and Schubring, J. (2004). Estimation of land surface temperature-vegetation abundance relationship for urban heat island studies. *Remote Sensing of Environment* 89: 467–483.

Wilson, E.H., Hurd, J.D., Civco, D.L. et al. (2003). Development of a geospatial model to quantify, describe and map urban growth. *Remote Sensing of Environment* 86: 275–285.

Wu, C. (2009). Quantifying high-resolution impervious surfaces using spectral mixture analysis. *International Journal of Remote Sensing* 30 (11): 2915–2932.

Wu, C. and Murray, A.T. (2003). Estimating impervious surface distribution by spectral mixture analysis. *Remote Sensing of Environment* 84: 493–505.

Wu, J.G., Jelinski, D.E., Luck, M., and Tueller, P.T. (2000). Multiscale analysis of landscape heterogeneity: scale variance and pattern metrics. *Geographic Information Sciences* 6 (1): 6–19.

Wu, J.W., Xu, J.H., and Yue, W.Z. (2005). V-I-S model for cities that are experiencing rapid urbanization and development. *Geoscience and Remote Sensing Symposium, IGARSS'05. Proceedings* 3: 1503–1506.

Yang, L., Huang, C., Homer, C.G. et al. (2003). An approach for mapping large-scale impervious surfaces: synergistic use of Landsat-7 ETM+ and high spatial resolution imagery. *Canadian Journal of Remote Sensing* 29: 230–240.

2

The Scale Issue

2.1 Introduction

In a review article on the scale and resolution effects in remote sensing and geographic information system (GIS), Cao and Lam (1997) raised two questions: (i) "how large an area should be covered to appropriately examine a geographic phenomenon, or at what scale and resolution should the study be conducted," and (ii) "whether or not the results of the study at one scale can be extrapolated to other scales." These two questions are still valid today for most studies in urban remote sensing.

A key to scale- and resolution-related issues is to develop proper methods for "determining the most appropriate scale and resolution of study and assessing the effects of scale and resolution" (Cao and Lam 1997). The majority of research efforts in urban remote sensing over the past decade have been made for mapping urban landscapes at various scales and on the spatial resolution requirements of such mapping (Weng 2012). Previous models, methods, and image analysis algorithms in urban remote sensing have been largely developed for the imagery of medium resolution (10–100 m). In contrast, there is less interest in spectral and geometric properties of urban features. The advent of high spatial resolution satellite images and increased interests in spaceborne hyperspectral images, Light Detection and Ranging (LiDAR) sensing, and their synergy with existing technologies are stimulating new research idea in urban remote sensing and are driving the current research trends with new models and algorithms. The urban remote sensing community will need to address how these new frontiers in Earth Observation technology since 1999 and advances in remote sensing imaging science – such as object-oriented image analysis, data fusion, and artificial neural networks – have impacted our understanding of the scale and resolution issues. Because little has been done in the past, more researches are also needed to better understand temporal resolution, change and evolution of urban features over time, and temporal

Techniques and Methods in Urban Remote Sensing, First Edition. Qihao Weng.
© 2020 by The Institute of Electrical and Electronics Engineers, Inc.
Published 2020 by John Wiley & Sons, Inc.

requirements for urban mapping (Weng 2012; Zhang et al. 2017). There is not a simple answer any of the questions discussed above. This review starts with the spectral requirement for urban mapping, a common consideration in the imaging and mapping, and its relationship with categorical scale. Then, the relationship between spatial resolution – which is termed the observational scale of a remote sensing in this paper – and the fabric of urban landscape is examined. Central to this relationship is the problem of mixed pixels in the urban areas. The pixel and sub-pixel approaches to urban analyses are thus discussed. Next, two studies were discussed, both assessing the patterns of land surface temperature (LST) at different aggregation levels in order to find out the operational scale/the optimal scale for the studies. Section 2.5 is developed to review the issue of scale dependency of urban phenomena and to discuss two case studies, one on LST variability across multiple census levels (block, block group, and tract) and the other on multi-scale residential population estimation modeling. Section 2.5 is followed by the conclusion section, providing a summary of the discussions and reflecting on the future developments.

2.2 Urban Land Mapping and Categorical Scale

Urban remote sensing should consider the requirements for mapping three interrelated entities or substances on the Earth surface (i.e. material, land cover, and land use) and their relationships (Weng and Lu 2009; Weng 2012). Urban areas are composed of a variety of materials, including different types of artificial materials (i.e. impervious surfaces), soils, rocks and minerals, and green and non-photosynthetic vegetation. These materials comprise land cover and are used in different manners for various purposes by human beings. Land cover can be defined as the biophysical state of the earth's surface and immediate subsurface, including biota, soil, topography, surface and ground water, and human structures (Turner et al. 1995). Land use can be defined as the human use of the land and involves both the manner in which the biophysical attributes of the land are manipulated and the purpose for which the land is used (Turner et al. 1995). Remote sensing technology has been applied to map urban land use, land cover, and materials. Their relationships are illustrated in Figure 1.1. Each type of land cover may possess unique surface properties (material), however, mapping land covers and materials have different requirements. Land-cover mapping needs to consider characteristics in addition to those coming from the material (Herold et al. 2006). The surface structure (roughness) may influence the spectral response as much as the intra-class variability (Gong and Howarth 1990; Myint 2001; Shaban and Dikshit 2001; Herold et al. 2006). Two different land covers, for example, asphalt roads and composite shingle/tar roofs, may have very similar materials (hydrocarbons) and thus are difficult to discern, although from a material perspective, these surfaces can be mapped accurately with hyperspectral remote sensing techniques (Herold et al. 2006). Therefore, land-cover mapping requires taking into account of the intra-class variability and spectral separability. On the other hand, analysis of land-use classes would nearly be impossible with spectral information alone. Additional information, such as spatial, textural, and contextual information, is usually required in order to have a successful land-use classification in urban areas (Gong and Howarth 1992; Stuckens et al. 2000; Herold et al. 2003).

The spectral characteristics of land surfaces are the fundamental principles for land imaging. Previous studies have examined the spectral properties of urban materials (Hepner et al. 1998; Ben-Dor et al. 2001; Herold et al. 2003; Heiden et al. 2007) and spectral resolution requirements for separating them (Jensen and Cowen 1999). Jensen and Cowen (1999) focused mainly their discussion on multispectral imagery data and suggested that spatial resolution was more important than spectral resolution in urban mapping. The spectrum from visible to near infrared (NIR), mid-infrared (MIR), and microwave are suitable for land-use and land-cover (LULC) classification at coarser categorical scales (e.g. Levels I and II of the Anderson classification); however, at the finer categorical scales (e.g. Levels III and IV of the Anderson classification) and for extraction of buildings and roads, panchromatic band is needed (Jensen and Cowen 1999). Hyperspectral imagery data have been successfully used for urban land-use/cover classification (Benediktsson et al. 1995; Hoffbeck and Landgrebe 1996; Platt and Goetz 2004; Thenkabail et al. 2004a, 2004b), extraction of impervious surfaces (Weng et al. 2008), vegetation mapping (McGwire et al. 2000; Schmidt et al. 2004; Pu et al. 2008), and water mapping (Bagheri and Yu 2008; Moses et al. 2009). A large number of spectral bands provide the potential to derive detailed information on the nature and properties of different surface materials on the ground, but it also means a difficulty in image processing and a large data redundancy due to high correlation among the adjacent bands. Increase of spectral bands may improve classification accuracy, only when those bands are useful in discriminating the classes (Thenkabail et al. 2004b).

Traditional classification methods of LULC based on detailed fieldwork suffered two major common drawbacks: confusion between LULC and the lack of uniformity or comparability in classification schemes, leaving behind a sheer difficulty for comparing land-use patterns over time or between areas (Mather 1986). The use of aerial photographs and satellite images after the late 1960s does not solve these problems, since these techniques are based on the formal expression of land use rather than on the actual activity itself (Mather 1986). In fact, many land-use types cannot be identified from the air. As a result, mapping of the Earth's surface tends to present a mixture of LULC data with an emphasis on the latter (Lo 1986). This problem is reflected in the title of the classification developed in the United States for the mapping of the country at a scale of 1 : 100 000 or 1 : 250 000, commencing in 1974 (Anderson et al. 1976). Moreover, this US Geological Survey (USGS) LULC Classification System (so-called "Anderson scheme") has been designed as a resource-oriented one. Therefore, eight out of nine in the first-level categories relate to nonurban areas. The success of most land-use or land-cover mapping efforts has typically been measured by the ability to match remote sensing spectral signatures to the Anderson scheme, which, in the urban areas, is mainly land use (Ridd 1995). The confusion between land use and land cover contributes to the low classification accuracy (Foody 2002). In addition, the spatial scale and categorical scale is not explicitly linked in the classification scheme. The former refers to the manner in which image information content is determined by spatial resolution and the way the spatial resolution is handled in the image processing, while the latter refers to the level of detail in classification categories (Ju et al. 2005). This disconnection leads to the problem of simply lumping classes into more general classes in multi-scale LU/LC classifications, which may cause a great loss of categorical information. Since most classifications are conducted at a single spatial and categorical scale, there remains an important issue of matching an appropriate categorical scale of the Anderson scheme with the spatial resolution of the satellite image used (Welch 1982; Jensen and Cowen 1999). However, the nature of some applications requires LULC classification to be conducted at multiple spatial and/or categorical scales, because a single scale cannot delineate all classes due to contrasting sizes, shapes, and internal variations of different landscape patches (Wu and David 2002; Raptis et al. 2003). This is especially true for complex, heterogeneous landscapes, such as urban ecosystems. When statistical clusters are grouped into LULC classes, in which smaller areas (e.g. pixels) are combined into larger ones (e.g. patches), and both spatial resolution and statistical information are lost (Clapham 2003).

2.3 Observational Scale and Image Scene Models

Spatial resolution is a function of sensor altitude, detector size, focal size, and system configuration (Jensen 2005). Spatial resolution is closely related to the term of spatial scale (Ju et al. 2005). As a matter of fact, spatial resolution defines the "measurement scale" (Lam and Quattrochi 1992) or the "observational scale" of a sensor.

Spatial resolution defines the level of spatial detail depicted in an image, and it is often related to the size of the smallest possible feature that can be detected from an image. This definition implies that only objects larger than the spatial resolution of a sensor can be picked out from an image. However, a smaller feature may sometimes be detectable if its reflectance dominates within a particular resolution cell or it has a unique shape (e.g. linear features). Another meaning of spatial resolution is that a ground feature should be distinguishable as a separate entity in the image. But the separation from neighbors or background is not always sufficient to identify the object. Therefore, the concept of spatial resolution includes both detectability and separability. For any feature to be resolvable in an image, it involves consideration of spatial resolution, spectral contrast, and feature shape. Jensen and Cowen (1999) suggested that the minimum spatial resolution requirement should be one-half the diameter of the smallest object of interest. For two major types of impervious surface, buildings (perimeter, area, height, and property line), and roads (width) are generally detectable with the minimum spatial resolution of 0.25–0.5 m, while road centerline can be detected at a lower resolution of 1–30 m (Jensen and Cowen 1999). Before 1999, lack of high spatial resolution (less than 10 m) images is a main reason for scarce research on urban remote sensing before 2000 (Weng 2012). The medium (10–100 m) spatial resolution images, such as Landsat and SPOT, were not readily available and were expensive to most researchers from developing countries. For a remote sensing project, image spatial resolution should not be the only factor needed to be considered. The relationship between the geographical scale/extent of a study area and the spatial resolution of remote sensing image has to be studied (Quattrochi and Goodchild 1997). For mapping at the continental or the global scale, coarse spatial resolution data are usually employed. Gamba and Herold (2009) assessed eight major research efforts in global urban extent mapping and found that most maps were produced at the spatial resolution of 1–2 km. When using coarse resolution images, a threshold has to be defined with respect to what constitute an urban/built-up pixel (Lu et al. 2008; Schneider et al. 2010).

With the advent of very high-resolution satellite imagery, such as IKONOS (launched 1999), QuickBird (2001), and OrbView (2003) images, great efforts have been made in the applications of these remote sensing images in urban studies. High-resolution satellite imagery have been applied in mapping of impervious surfaces in urban areas (Cablk and Minor 2003; Goetz et al. 2003; Lu and Weng 2009; Wu 2009; Hu and Weng 2011). These fine spatial resolution images contain rich spatial information, providing a greater potential to extract much more detailed thematic information (e.g. LULC), cartographic features (buildings and roads), and metric information with stereo-images (e.g. height and area). These information and cartographic characteristics are highly beneficial to estimating and mapping of impervious surfaces in the urban areas. However, some new problems come with these image data, notably shadows caused by topography, tall buildings, or trees (Dare 2005), and the high spectral variation within the same land-cover class (Hsieh et al. 2001). Shadows obscure impervious surfaces underneath and thus increase the difficulty to extract both thematic and cartographic information. These disadvantages may lower image classification accuracy if classifiers used cannot effectively handle them (Irons et al. 1985; Cushnie 1987). In order to make full use of the rich spatial information inherent in fine spatial resolution data, it is necessary to minimize the negative impact of high intra-spectral variation. Algorithms that use the combined spectral and spatial information may be especially effective for impervious surface extraction in the urban areas (Lu and Weng 2007).

Per-pixel classifications prevail in the previous remote sensing literature, in which each pixel is assigned to one category and land-cover (or other themes) classes are mutually exclusive. Per-pixel classification algorithms are sometimes referred to as "hard" classifiers. Due to the heterogeneity of landscapes (particularly in urban landscapes) and the limitation in spatial resolution of remote sensing imagery, mixed pixels are common in medium and coarse spatial resolution data. However, the proportion of mixed pixels is significantly reduced in a high-resolution satellite image scene. The presence of mixed pixels has been recognized as a major problem, affecting the effective use of per-pixel classifiers (Fisher 1997; Cracknell 1998). The mixed pixel problem results from the fact that the observational scale (i.e. spatial resolution) fails to correspond to the spatial characteristics of the target (Mather 1999). Strahler et al. (1986) defined H- and L-resolution scene models based on the relationship between the size of the scene elements and the resolution cell of the sensor. The scene elements in the L-resolution model are smaller than the resolution cells and are thus not detectable. When the objects in the scene become increasingly smaller than the resolution cell size, they may no longer be regarded as individual objects. Hence, the reflectance measured by the sensor may be treated as the sum of interactions among various types of scene elements as weighted by their relative proportions (Strahler et al. 1986). This is what happens with medium resolution imagery, such as those of Landsat thematic mapper (TM) or Enhanced Thematic Mapper Plus (ETM+), Advanced Spaceborne Thermal Emission and Reflection Radiometer (ASTER), SPOT, and Indian satellites, applied for urban mapping. As the spatial resolution interacts with the fabric of urban landscapes, the problem of mixed pixels is created. Such a mixture becomes especially prevalent in residential areas where buildings, roads, trees, lawns, and water can all lump together into a single pixel (Epstein et al. 2002). The low accuracy of image classification in urban areas reflects, to a certain degree, the inability of traditional per-pixel classifiers to handle composite signatures. Therefore, the "soft" approach of image classifications has been developed, in which each pixel is assigned a class membership of each land-cover type rather than a single label (Wang 1990). Different approaches have been used to derive a soft classifier, including fuzzy set theory, Dempster–Shafer theory, certainty factor (Bloch 1996), and neural network (Foody 1999; Mannan and Ray 2003). Nevertheless, as Mather (1999) suggested, either "hard" or "soft" classification was not an appropriate tool for the analysis of heterogeneous landscapes. To provide a better understanding of the compositions and processes of urban landscapes, Ridd (1995) proposed an interesting conceptual model for remote sensing analysis of urban landscapes, i.e. the vegetation–impervious surface–soil (V-I-S) model (Figure 2.1). It assumes that land cover in urban environments is a linear combination of three components, namely, vegetation, impervious surface, and soil. Ridd suggested that this model can be applied to spatial–temporal analyses of urban morphology, biophysical systems, and human systems. While urban

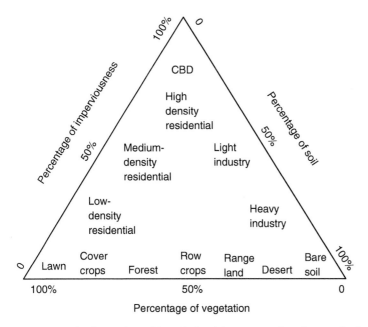

Figure 2.1 The illustration of the relationship among urban features in the Vegetation–Impervious surface–Soil (V-I-S) model. *Source:* adapted from Ridd (1995). Reproduced with permission of Taylor & Francis.

land-use information may be more useful in socioeconomic and planning applications, biophysical information that can be directly derived from satellite data is more suitable for describing and quantifying urban structures and processes (Ridd 1995). The V-I-S model was developed for Salt Lake City, Utah, but has been tested in other cities (Ward et al. 2000; Madhavan et al. 2001; Setiawan et al. 2006). All of these studies employed the V-I-S model as the conceptual framework to relate urban morphology to medium-resolution satellite imagery, but "hard classification" algorithms were applied. Therefore, the problem of mixed pixels cannot be addressed, and the analysis of urban landscapes was still based on "pixels" or "pixel groups." Weng and Lu (2009) suggested that linear spectral mixture analysis (LSMA) provided a suitable technique to detect and map urban materials and V-I-S component surfaces in repetitive and consistent ways and to solve the spectral mixing of medium spatial resolution imagery. The reconciliation between the V-I-S model and LSMA provided a continuum field model, which offered an alternative, effective approach for characterizing and quantifying the spatial and temporal changes of the urban landscape compositions. However, Weng and Lu (2009) warned that the applicability of this continuum model must be further examined in terms of its spectral, spatial, and temporal variability.

2.4 Operational Scale

Urban landscape processes appear to be hierarchical in pattern and structure. A study of the relationship between the patterns at different levels in the hierarchy may help in obtaining a better understanding of the scale and resolution problem (Cao and Lam 1997; Weng et al. 2004) and in finding the optimal scale for examining the relationship, i.e. the operational scale (Frohn 1998; Liu and Weng 2009). Lo et al. (1997) suggested that the urban surface characteristics required a minimum thermal mapping resolution of 5–10 m based on a study in Huntsville, Alabama. Nichol (1996) confirmed that satellite-derived LST image data at the scale of 10^2 m was adequate for depicting most of the intra-urban LST variations related to urban morphology based on a study in Singapore. The length of spatial scale that characterizes the overall distribution of scales of all objects in a collection may be compared among different cities. Small (2009) suggested that the modal scale length was 10–20 m based on a comparative analysis of 14 cities in the world.

Weng et al. (2004) suggested that the operational scale for the urban thermal landscape analysis yielded around 120 m in Indianapolis, Indiana. Remote sensing of urban heat islands has traditionally used the Normalized Difference Vegetation Index (NDVI) as the indicator of vegetation abundance to estimate the LST – vegetation relationship. Weng et al. (2004) investigated the applicability of vegetation fraction derived from LSMA as an alternative indicator of vegetation abundance. An experiment was conducted with a Landsat ETM+ image of Indianapolis City, Indiana, USA, acquired on 22 June 2002. They found that LST possessed a slightly stronger negative correlation with the unmixed vegetation fraction than with NDVI for all land-cover types across the spatial resolution from 30 to 960 m. Correlations reached their strongest at the 120 m resolution. Fractal analysis of image texture further showed that the complexity of these images increased initially with pixel aggregation and peaked around 120 m, but decreased with further aggregation. It was, therefore, suggested that the operational scale for examining the relationship between LST and NDVI or vegetation fraction was around 120 m.

In another Indianapolis study, Liu and Weng (2009) examined the scaling effect between LST and LULC in Indianapolis and found that the optimal spatial resolution for assessing this unique relationship was 90 m. Four Terra's ASTER images were used to derive LULC map and LST patterns in four seasons. Each LULC and LST image was resampled to eight aggregation levels: 15, 30, 60, 90, 120, 250, 500, and 1000 m. The scaling-up effect on the spatial and ecological characteristics of landscape patterns and LSTs were examined by the use of landscape metrics. Optimal spatial scale was determined on the basis of the minimum distance in the landscape metric spaces. Their results showed that the patch percentages of LULC and LST patches were not strongly affected by the scaling-up process. The patch densities and landscape shape indices and LST patches kept decreasing across the scales without distinct seasonal differences. Ninety meters was found to be the optimal spatial resolution for assessing the landscape-level relationship between LULC and LST patterns.

2.5 Scale Dependency of Urban Phenomena

The majority of urban phenomena are scale dependent, meaning that the urban patterns change with scale of observation. In reality, very few geographical phenomena are scale independent, in which the patterns do not change across scales (Cao and Lam 1997). Geographic studies are frequently conducted on the basis of areal units such as states, counties, census tracts, block groups, and blocks. Across-scale analyses with such areal units may produce scale-related problems, most notably, the "modifiable areal unit problem" (MAUP). Fotheringham and Wong (1991) defined MAUP as "the sensitivity of analytical results to the definition of units for which data are collected." Specifically, the results may vary with the aggregation level (the "scale effect") and with the aggregation schemes (the "zoning effect"). When a researcher attempts extrapolate to the result of a study at one scale to other scales, one may meet three kinds of erroneous inferences – individualistic fallacy, cross-level fallacy, and ecological fallacy (Alker 1969). There is no ideal solution to solve MAUP; it is, however, mainly examined by conducting statistical correlation and regression analysis (Fotheringham and Wong 1991). Empirical studies may be the only possible way to explore the nature of MAUP (Fotheringham and Wong 1991). Recently, there has been an increase in using spatial autocorrelation indexes to indicate geographical patterns or geospatial data distributions. The underlying fundamental of spatial autocorrelation is best explained by Tobler's First Law of Geography – "everything is related to everything else, but near things are more related than distance things" (Lo and Yeung 2002). The index is devised to measure the spatial ordering as well as the spatial covariance structure of geospatial data and provides the insights into the degree of clustering, randomness, or fragmentation of a pattern (Read and Lam 2002). However, spatial autocorrelation is subject to the influence of scale, meaning that at one scale its index may suggest a concentrated pattern while at another it may suggest a scattered pattern (Cao and Lam 1997).

2.5.1 Spatial Variations of LST at Multiple Census Scales

Liang and Weng (2008) conducted a multi-scale analysis of census-based LST variations and determinants in Indianapolis, Indiana, USA. Urban temperatures have a close relationship with many environmental, economic, and social issues in the urban areas. This study utilized LST data, derived from a Landsat ETM+ image of Indianapolis city, United States, to examine census-based variations and to model their relationships with the parameters of urban morphology. NDVI, buildings, roads, and water bodies were selected as the variables of the urban morphology. Correlation analysis and stepwise regression modeling at each census level, i.e. block, block group, and tract, were performed. The sensitivity of the relationship to aggregation and thus the scale effect of MAUP were examined. Their results showed that LST had a strongest positive correlation with buildings, but was negatively correlated with water at all scales. The correlation between LST and the four variables tended to become stronger as the scale increased. Table 2.1 indicates that adjusted R-squared value increased with the scale, suggesting that more variations in LST could be explained by the regression models. The regression model for the tract level possessed the closest goodness of fit in the population of LST with the least estimation error. More independent variables are needed to predict LST at finer scales. Meanwhile, when the analytical scale altered, the contribution of each independent variable to the models changed. For example, an increase of 10 in P_{bildg} generated an increase in LST of 2.95 K at the block level. The same increase at the block group and tract levels generated an increase of 5.21 and 5.95 K in the predicted LST values, respectively. However, the strength of the correlation between LST and four biophysical factors were not always consistent with their contributions to LST regression modeling. For example, roads contributed the least to the LST estimation, although it correlated stronger with LST than water. This is because in multivariate modeling, regression estimates attached to any single independent variable could become inflated or deflated, since changes in the variable may be confused by variations in other independent variables.

In order to validate the relationship between input data and LST models, a residual map was produced for each model (Figure 2.2). The range of residuals decreased from 16.81 to 12.24 to 5.61 with the increase in geographical scale. For any LST model, LST was consistently underestimated over the low-temperature areas; while in the high-temperature areas, the distribution of residuals varied with scale. LST was predominantly overestimated at the block scale. As the scale increased, LST became generally underestimated. At the tract scale, LST was equally over- and underestimated. In the downtown area where high temperatures existed, predicted LST values may be either over- or underestimated at any scale, implying the heterogeneity in the thermal landscape. The spatial patterns of residuals appear analogous at the block group and tract levels.

2.5.2 Population Estimation

Many remotely sensed images collected from different sensors have been utilized to estimate population. With various spatial resolutions, they are especially applicable at a certain scale for the study. For instance, high spatial resolution aerial photography is useful for population estimation at the

Table 2.1 Regression coefficients for the three land surface temperature (LST) models and the four variables of urban morphology.

Models			LST1	LST2	LST3
Adjusted			0.752	0.921	0.948
Sig. (significance value) of the F statistic			0.000	0.000	0.000
Factors	M_{NDVI}	Standardized beta coefficients	−0.654	−0.518	−0.429
		t	−138.549	−35.450	−19.756
		Sig.	0.000	0.000	0.000
	P_{bildg}	Standardized beta coefficients	0.295	0.421	0.394
		t	63.218	27.590	12.741
		Sig.	0.000	0.000	0.000
	P_{road}	Standardized beta coefficients	0.059	0.139	0.228
		t	13.742	11.108	8.970
		Sig.	0.000	0.000	0.000
	P_{water}	Standardized beta coefficients	−0.276	−0.234	−0.241
		t	−63.517	−19.624	−14.236
		Sig.	0.000	0.000	0.000

Source: Liang and Weng (2014). Reproduced with permission of John Wiley & Sons.

microscale (Lo and Welch 1977; Lo 1986; Cowen et al. 1995), while low spatial resolution data, such as those from Defense Meteorological Satellite Program Operational Linescan System (DMSP-OLS), are suitable for modeling at the global or regional scale. However, if a medium scale such as at a city level is concerned, images with medium spatial resolution, such as those obtained from Landsat TM/ETM+ and Terra's ASTER sensors, should be considered. Research has proved that such data are efficient and effective in predicting population in city or county levels (Harvey 2002a, 2002b; Qiu et al. 2003; Li and Weng 2005; Lu et al. 2006). Lo (1986) summarized several approaches commonly used in population estimation with remotely sensed data: counting the dwelling units, using per pixel spectral reflectance, measuring urban areas, and using land-use information. The application of these methods may be, in fact, considered to respond to different analytical scales, with the first two applicable for small area (1 km^2 or less) and the last two for larger (or regional) and medium scales, respectively (Harvey 2002a).

Because remotely sensed images are scale dependent, population models derived from such data are also subject to the impact of scale. Lo (2001) used DMSP-OLS nighttime lights data to model the Chinese population and population densities at three different spatial scales: province, county, and city. Either

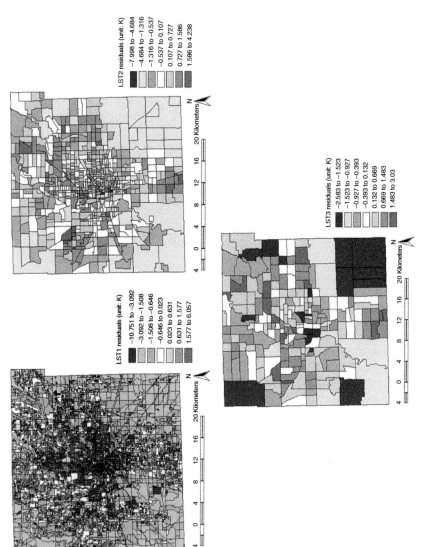

Figure 2.2 Distribution of land surface temperature (LST) residuals at the black, black group, and census tract level. *Source:* Weng (2014). Reproduced with permission of John Wiley & Sons.

allometric growth or linear regression models were found to be promising in estimating population at all three levels, but the best models were obtained at the city level. Qiu et al. (2003) carried out a bi-scale study of the decennium urban population growth from 1990 to 2000 in the north Dallas-Fort Worth Metropolis using models developed with remote sensing and GIS techniques. Both models yielded comparable results with that obtained from a more complex commercial demographics model at the city as well as the census tract levels, yet the GIS model remained robust to the scale change because of its insensitiveness to the spatial scale. The remote sensing model was attenuated when moved to the census tract from the city level. Liang et al. (2007) estimated residential population of Indianapolis, USA, at multiple scales of census units (block, block group, and tract) using remote sensing derived impervious surfaces. Impervious surface has been emerging as a key indicator of urbanization in recent years. The map of impervious surfaces may be useful in revealing socioeconomic characteristics of a city. They found the impervious surface was an effective variable in modeling residential population at all three census scales. Table 2.2 indicates that by any measure (R^2, mean value of relative error [MRE], and median value of relative error [MedRE]), three best population models were all found at the tract level. The difference between MRE and MedRE shows the

Table 2.2 Population estimation models for Indianapolis at each census level.

Census levels	Corr.	R^2	Regression model	MRE	MedRE	ET (%)
Block	0.428	0.183	PD = $-1219.335 + 6987.572 \times$ MRImp	396.15	37.12	-12.69
	0.480	0.230	SPD = $2.096 + 85.737 \times$ MRImp	51.48	19.45	-3.16
	0.466	0.218	LPD = $5.410 + 0.466 \times$ MRImp	30.71	25.52	$+1.42$
BG	0.595	0.354	PD = $-1995.631 + 7378.838 \times$ MRImp	275.80	32.77	-3.91
	0.592	0.351	SPD = $-12.403 + 104.295 \times$ MRImp	44.75	18.71	$+0.62$
	0.555	0.308	LPD = $4.354 + 5.785 \times$ MRImp	10.41	6.16	$+1.14$
Tract	0.710	0.504	PD = $-3021.746 + 9550.366 \times$ MRImp	55.96	27.15	$+4.01$
	0.706	0.499	SPD = $-29.567 + 141.658 \times$ MRImp	22.74	14.32	$+3.53$
	0.672	0.452	LPD = $3.134 + 8.492 \times$ MRImp	6.36	4.28	$+1.33$

Source: Liang et al. (2007). Reproduced with permission of Taylor & Francis.
Notes: Corr., the Pearson correlation coefficient between PD (SPD, LPD) and MRImp; MRE, mean value of relative error; MedRE, median value of relative error; ET (%), error of total, which is the total population estimation error based on the overall data set in the study area (an addition [+] symbol means overestimated while a minus [–] symbol means underestimated); PD, population density; MRImp, mean value of impervious surfaces in residential areas; SPD, square root of population density; LPD, natural log of population density; BG, block group.

performance of population density models was strongly affected by extreme values (extreme low and high population density). A common problem found in all three models was that they significantly underestimated the high-density areas but overestimated the low-density areas. This is because population modeling developed in this study assumed a general uniform population density per census unit. Hence, none of the models predicted well for the extreme values.

2.6 Summary

This review starts with discussion on the requirements for mapping urban materials, land cover, and land use, and their relationships. This examination allows us to understand what spectral property can and cannot be considered in the imaging and mapping. The level of detail in classification categories is essentially modulated by spectral resolution. But, another important consideration in urban remote sensing is spatial resolution. Its interaction with the fabric of urban landscape the problem of mixed pixels are the content of the debate between the pixel and sub-pixel approaches. Research has yet to clearly reveal the linkage between categorical scale and spatial resolution.

Scale influences the examination of landscape patterns (Liu and Weng 2009). The change of scale is relevant to the issues of data aggregation, information transfer, and the identification of appropriate scales for analysis (KrÖnert et al. 2001; Wu and Hobbs 2002). Sections 2.4 and 2.5 relate to this theme of study. I assessed the patterns of LST at different aggregation levels in order to find out the operational scale/the optimal scale through two case studies. The issues of data aggregation and information transfer are addressed by reviewing the concept of scale dependency and by discussing LST variability and residential population estimation modeling across multiple census levels.

Remote sensing technology has been evolving rapidly in the twenty-first century. New frontiers such as very high-resolution sensing, hyperspectral sensing, and LiDAR and their synergy with existing technologies and advances in image processing techniques (such as object-oriented image analysis, data fusion, and artificial neural network) are changing image information content we obtain and the way we handle the image processing. Both aspects will change our understanding of this basic but pivotal issue – scale, and therefore, future research should be warranted in these aspects. For example, an increase in spatial resolution may not lead to a better observation since objects may be oversampled and their features may vary and be confusing (Hsieh et al. 2001; Aplin and Atkinson 2004). Moreover, imagery with too fine resolution for specific purpose can be degraded in the process of image resampling (Ju et al. 2005). In the object-based image analysis, the extraction, representation, modeling, and analyses of image objects at multiple scales have become the common concern of many researchers (Hay et al. 2002a, 2002b; Tzotsos et al. 2011).

References

Alker, H.R. (1969). A typology of ecological fallacies. In: *Quantitative Ecological Analysis in the Social Sciences* (eds. M. Dogan and S. Rokkan), 69–86. Cambridge, MA: MIT Press.

Anderson, J.R., Hardy, E.E., Roach, J.T., and Witmer, R.E. (1976). A land use and land cover classification systems for use with remote sensing data. USGS Professional Paper 964.

Aplin, P. and Atkinson, P.M. (2004). Predicting missing field boundaries to increase per-field classification accuracy. *Photogrammetric Engineering & Remote Sensing* 70 (1): 141–149.

Bagheri, S. and Yu, T. (2008). Hyperspectral sensing for assessing nearshore water quality conditions of Hudson/Raritan estuary. *Journal of Environmental Informatics* 11 (2): 123–130.

Ben-Dor, E., Levin, N., and Saaroni, H. (2001). A spectral based recognition of the urban environment using the visible and near-infrared spectral region (0.4–1.1 m) – a case study over Tel-Aviv. *International Journal of Remote Sensing* 22 (11): 2193–2218.

Benediktsson, J.A., Sveinsson, J.R., and Arnason, K. (1995). Classification and feature extraction of AVIRIS data. *IEEE Transactions on Geoscience and Remote Sensing* 33: 1194–1205.

Bloch, I. (1996). Information combination operators for data fusion: a comparative review with classification. *IEEE Transactions on Systems, Man, and Cybernetics* 26: 52–67.

Cablk, M.E. and Minor, T.B. (2003). Detecting and discriminating impervious cover with high resolution IKONOS data using principal component analysis and morphological operators. *International Journal of Remote Sensing* 24: 4627–4645.

Cao, C. and Lam, N.S.-N. (1997). Understanding the scale and resolution effects in remote sensing. In: *Scale in Remote Sensing and GIS* (eds. D.A. Quattrochi and M.F. Goodchild), 57–72. Boca Raton, FL: CRC Press.

Clapham, W.B. Jr. (2003). Continuum-based classification of remotely sensed imagery to describe urban sprawl on a watershed scale. *Remote Sensing of Environment* 86: 322–340.

Cowen, D.J., Jensen, J.R., Bresnahan, G. et al. (1995). The design and implementation of an integrated GIS for environmental applications. *Photogrammetric Engineering & Remote Sensing* 61 (11): 1393–1404.

Cracknell, A.P. (1998). Synergy in remote sensing – what's in a pixel? *International Journal of Remote Sensing* 19: 2025–2047.

Cushnie, J.L. (1987). The interactive effect of spatial resolution and degree of internal variability within land-cover types on classification accuracies. *International Journal of Remote Sensing* 8: 15–29.

Dare, P.M. (2005). Shadow analysis in high-resolution satellite imagery of urban areas. *Photogrammetric Engineering & Remote Sensing* 71: 169–177.

Epstein, J., Payne, K., and Kramer, E. (2002). Techniques for mapping suburban sprawl. *Photogrammetric Engineering & Remote Sensing* 63: 913–918.

Fisher, P. (1997). The pixel: a snare and a delusion. *International Journal of Remote Sensing* 18: 679–685.

Foody, G.M. (1999). Image classification with a neural network: from completely-crisp to fully-fuzzy situation. In: *Advances in Remote Sensing and GIS Analysis* (eds. P.M. Atkinson and N.J. Tate), 17–37. New York: Wiley.

Foody, G.M. (2002). Status of land cover classification accuracy assessment. *Remote Sensing of Environment* 80: 185–201.

Fotheringham, A.S. and Wong, D.W.S. (1991). The modifiable areal unit problem in multivariate statistical analysis. *Environment and Planning A* 23: 1025–1044.

Frohn, R.C. (1998). *Remote Sensing for Landscape Ecology: New Metric Indicators for Monitoring, Modelling, and Assessment of Ecosystems* (112 p). Boca Raton, FL: Lewis Publishers.

Gamba, P. and Herold, M. (2009). *Global Mapping of Human Settlements: Experiences, Datasets, and Prospects*. Boca Raton, FL: CRC Press.

Goetz, S.J., Wright, R.K., Smith, A.J. et al. (2003). IKONOS imagery for resource management: tree cover, impervious surfaces, and riparian buffer analyses in the mid-Atlantic region. *Remote Sensing of Environment* 88: 195–208.

Gong, P. and Howarth, P.J. (1990). The use of structure information for improving land-cover classification accuracies at the rural-urban fringe. *Photogrammetric Engineering & Remote Sensing* 56 (1): 67–73.

Gong, P. and Howarth, P.J. (1992). Frequency-based contextual classification and gray-level vector reduction for land-use identification. *Photogrammetric Engineering & Remote Sensing* 58 (4): 423–437.

Harvey, J.T. (2002a). Estimating census district populations from satellite imagery: some approaches and limitations. *International Journal of Remote Sensing* 23 (10): 2071–2095.

Harvey, J.T. (2002b). Population estimation models based on individual TM pixels. *Photogrammetric Engineering & Remote Sensing* 68 (11): 1181–1192.

Hay, G.J., Dube, P., Bouchard, A., and Marceau, D.J. (2002a). A scale-space primer for exploring and quantifying complex landscapes. *Ecological Modelling* 153 (1–2): 2–49.

Hay, G.J., Marceau, D.J., and Bouchard, A. (2002b). Modelling multiscale landscape structure within a hierarchical scale-space framework. *International Archives of Photogrammetry and Remote Sensing* 34 (Part 4): 532–536.

Heiden, U., Segl, K., Roessner, S., and Kaufmann, H. (2007). Determination of robust spectral features for identification of urban surface materials in hyper-spectral remote sensing data. *Remote Sensing of Environment* 111: 537–552.

Hepner, G.F., Houshmand, B., Kulikov, I., and Bryant, N. (1998). Investigation of the integration of AVIRIS and IFSAR for urban analysis. *Photogrammetric Engineering & Remote Sensing* 64 (8): 813–820.

Herold, M., Liu, X., and Clark, K.C. (2003). Spatial metrics and image texture for mapping urban land use. *Photogrammetric Engineering & Remote Sensing* 69 (9): 991–1001.

Herold, M., Schiefer, S., Hostert, P., and Roberts, D.A. (2006). Applying imaging spectrometry in urban areas. In: *Urban Remote Sensing* (eds. Q. Weng and D.A. Quattrochi), 137–161. Boca Raton, FL: CRC Press.

Hoffbeck, J.P. and Landgrebe, D.A. (1996). Classification of remote sensing having high spectral resolution images. *Remote Sensing of Environment* 57: 119–126.

Hsieh, P.-F., Lee, L.C., and Chen, N.-Y. (2001). Effect of spatial resolution on classification errors of pure and mixed pixels in remote sensing. *IEEE Transactions on Geoscience and Remote Sensing* 39: 2657–2663.

Hu, X. and Weng, Q. (2011). Impervious surface area extraction from IKONOS imagery using an object-based fuzzy method. *Geocarto International* 26 (1): 3–20.

Irons, J.R., Markham, B.L., Nelson, R.F. et al. (1985). The effects of spatial resolution on the classification of Thematic Mapper data. *International Journal of Remote Sensing* 6: 1385–1403.

Jensen, J.R. (2005). *Introductory Digital Image Processing: A Remote Sensing Perspective*, 3e. Upper Saddle River, NJ: Prentice Hall.

Jensen, J.R. and Cowen, D.C. (1999). Remote sensing of urban/suburban infrastructure and socioeconomic attributes. *Photogrammetric Engineering & Remote Sensing* 65: 611–622.

Ju, J., Gopal, S., and Kolaczyk, E.D. (2005). On the choice of spatial and categorical scale in remote sensing land cover classification. *Remote Sensing of Environment* 96: 62–77.

Krönert, R., Steinhardt, U., and Volk, M. (2001). *Landscape Balance and Landscape Assessment* (304 p). New York: Springer Verlag.

Lam, N.S.-N. and Quattrochi, D.A. (1992). On the issues of scale, resolution, and fractal analysis in the mapping sciences. *The Professional Geographer* 44 (1): 88–98.

Li, G. and Weng, Q. (2005). Using Landsat ETM+ imagery to measure population density in Indianapolis, Indiana, USA. *Photogrammetric Engineering & Remote Sensing* 71 (8): 947–958.

Liang, B. and Weng, Q. (2008). A multi-scale analysis of census-based land surface temperature variations and determinants in Indianapolis, United States. *Journal of Urban Planning and Development* 134 (3): 129–139.

Liang, B. and Weng, Q. (2014). Multiscale fractal characteristics of urban landscape in Indianapolis, USA. In: *Scale Issues in Remote Sensing* (ed. Q. Weng), 230–252. Hoboken, NJ: Wiley.

Liang, B., Weng, Q., and Lu, D. (2007). Census-based multiple scale residential population modeling with impervious surface data. In: *Remote Sensing of Impervious Surfaces* (ed. Q. Weng), 409–430. Boca Raton, FL: CRC/Taylor & Francis.

Liu, H. and Weng, Q. (2009). Scaling-up effect on the relationship between landscape pattern and land surface temperature. *Photogrammetric Engineering & Remote Sensing* 75 (3): 291–304.

Lo, C.P. (1986). *Applied Remote Sensing*. New York: Longman Inc.

Lo, C.P. (2001). Modeling the population of China using DMSP operational linescan system nighttime data. *Photogrammetric Engineering & Remote Sensing* 67 (9): 1037–1047.

Lo, C.P., Quattrochi, D.A., and Luvall, J.C. (1997). Application of high-resolution thermal infrared remote sensing and GIS to assess the urban heat island effect. *International Journal of Remote Sensing* 18: 287–304.

Lo, C.P. and Welch, R. (1977). Chinese urban population estimates. *Annals of the Association of American Geographers* 67 (2): 246–253.

Lo, C.P. and Yeung, A.K.W. (2002). *Concepts and Techniques of Geographic Information Systems*. Upper Saddle River, NJ: Prentice Hall.

Lu, D., Tian, H., Zhou, G., and Ge, H. (2008). Regional mapping of human settlements in Southeastern China with multisensor remotely sensed data. *Remote Sensing of Environment* 112 (9): 3668–3679.

Lu, D. and Weng, Q. (2007). A survey of image classification methods and techniques for improving classification performance. *International Journal of Remote Sensing* 28 (5): 823–870.

Lu, D. and Weng, Q. (2009). Extraction of urban impervious surfaces from IKONOS imagery. *International Journal of Remote Sensing* 30 (5): 1297–1311.

Lu, D., Weng, Q., and Li, G. (2006). Residential population estimation using a remote sensing derived impervious surface approach. *International Journal of Remote Sensing* 27 (16): 3553–3570.

Madhavan, B.B., Kubo, S., Kurisaki, N., and Sivakumar, T.V.L.N. (2001). Appraising the anatomy and spatial growth of the Bangkok metropolitan area using a vegetation-impervious-soil model through remote sensing. *International Journal of Remote Sensing* 22: 789–806.

Mannan, B. and Ray, A.K. (2003). Crisp and fuzzy competitive learning networks for supervised classification of multispectral IRS scenes. *International Journal of Remote Sensing* 24: 3491–3502.

Mather, A.S. (1986). *Land Use*. London: Longman.

Mather, P.M. (1999). Land cover classification revisited. In: *Advances in Remote Sensing and GIS* (eds. P.M. Atkinson and N.J. Tate), 7–16. New York: Wiley.

McGwire, K., Minor, T., and Fenstermaker, L. (2000). Hyperspectral mixture modeling for quantifying sparse vegetation cover in arid environments. *Remote Sensing of Environment* 72: 360–374.

Moses, W.J., Gitelson, A.A., Berdnikov, S., and Povazhnyy, V. (2009). Satellite estimation of chlorophyll-a concentration using the red and NIR bands of MERIS – the Azov Sea case study. *IEEE Geoscience and Remote Sensing Letters* 6: 845–849.

Myint, S.W. (2001). A robust texture analysis and classification approach for urban land-use and land-cover feature discrimination. *Geocarto International* 16: 27–38.

Nichol, J.E. (1996). High-resolution surface temperature patterns related to urban morphology in a tropical city: a satellite-based study. *Journal of Applied Meteorology* 35 (1): 135–146.

Platt, R.V. and Goetz, A.F.H. (2004). A comparison of AVIRIS and Landsat for land use classification at the urban fringe. *Photogrammetric Engineering & Remote Sensing* 70: 813–819.

Pu, R., Kelly, M., Anderson, G.L., and Gong, P. (2008). Using CASI hyperspectral imagery to detect mortality and vegetation stress associated with a new hardwood forest disease. *Photogrammetric Engineering & Remote Sensing* 74 (1): 65–75.

Qiu, F., Woller, K.L., and Briggs, R. (2003). Modeling urban population growth from remotely sensed imagery and TIGER GIS road data. *Photogrammetric Engineering & Remote Sensing* 69: 1031–1042.

Quattrochi, D.A. and Goodchild, M.F. (1997). *Scale in Remote Sensing and GIS*. New York City, NY: Lewis Publishers.

Raptis, V.S., Vaughan, R.A., and Wright, G.G. (2003). The effect of scaling on land cover classification from satellite data. *Computers & Geosciences* 29: 705–714.

Read, J.M. and Lam, N.S.-N. (2002). Spatial methods for characterising land cover and detecting land-cover changes for the tropics. *International Journal of Remote Sensing* 12 (2): 2457–2474.

Ridd, M.K. (1995). Exploring a V-I-S (vegetation–impervious surface–soil) model for urban ecosystem analysis through remote sensing: comparative anatomy for cities. *International Journal of Remote Sensing* 16 (12): 2165–2185.

Schmidt, K.S., Skidmore, A.K., Kloosterman, E.H. et al. (2004). Mapping coastal vegetation using an expert system and hyperspectral imagery. *Photogrammetric Engineering & Remote Sensing* 70: 703–715.

Schneider, A., Friedl, M.A., and Potere, D. (2010). Mapping global urban areas using MODIS 500-m data: new methods and datasets based on "urban ecoregions". *Remote Sensing of Environment* 114: 1733–1746.

Setiawan, H., Mathieu, R., and Thompson-Fawcett, M. (2006). Assessing the applicability of the V-I-S model to map urban land use in the developing world: case study of Yogyakarta, Indonesia. *Computers, Environment and Urban Systems* 30 (4): 503–522.

Shaban, M.A. and Dikshit, O. (2001). Improvement of classification in urban areas by the use of textural features: the case study of Lucknow city, Uttar Pradesh. *International Journal of Remote Sensing* 22: 565–593.

Small, C. (2009). The color of cities: an overview of urban spectral diversity. In: *Global Mapping of Human Settlement: Experiences, Datasets, and Prospects* (eds. P. Gamba and M. Herold), 59–105. Boca Raton, FL: CRC Press.

Strahler, A.H., Woodcock, C.E., and Smith, J.A. (1986). On the nature of models in remote sensing. *Remote Sensing of Environment* 70: 121–139.

Stuckens, J., Coppin, P.R., and Bauer, M.E. (2000). Integrating contextual information with per-pixel classification for improved land cover classification. *Remote Sensing of Environment* 71: 282–296.

Thenkabail, P.S., Enclona, E.A., Ashton, M.S. et al. (2004b). Hyperion, IKONOS, ALI, and ETM+ sensors in the study of African rainforests. *Remote Sensing of Environment* 90: 23–43.

Thenkabail, P.S., Enclona, E.A., Ashton, M.S., and van der Meer, B. (2004a). Accuracy assessments of hyperspectral waveband performance for vegetation analysis applications. *Remote Sensing of Environment* 91: 354–376.

Turner, B.L. II, Skole, D., Sanderson, S. et al. (1995). *Land-Use and Land-Cover Change: Science and Research Plan*. Stockholm and Geneva: International Geosphere-Biosphere Program and the Human Dimensions of Global Environmental Change Programme (IGBP Report No. 35 and HDP Report No. 7).

Tzotsos, A., Karantzalos, K., and Argialas, D. (2011). Object-based image analysis through nonlinear scale-space filtering. *ISPRS Journal of Photogrammetry and Remote Sensing* 66: 2–16.

Wang, F. (1990). Fuzzy supervised classification of remote sensing images. *IEEE Transactions on Geoscience and Remote Sensing* 28 (2): 194–201.

Ward, D., Phinn, S.R., and Murray, A.T. (2000). Monitoring growth in rapidly urbanizing areas using remotely sensed data. *The Professional Geographer* 53: 371–386.

Welch, R.A. (1982). Spatial resolution requirements for urban studies. *International Journal of Remote Sensing* 3: 139–146.

Weng, Q. (2012). Remote sensing of impervious surfaces in the urban areas: requirements, methods, and trends. *Remote Sensing of Environment* 117 (2): 34–49.

Weng, Q. (2014). On the issue of scale in urban remote sensing. In: *Scale Issues in Remote Sensing* (ed. Q. Weng), 61–78. Hoboken, NJ: Wiley.

Weng, Q., Hu, X., and Lu, D. (2008). Extracting impervious surface from medium spatial resolution multispectral and hyperspectral imagery: a comparison. *International Journal of Remote Sensing* 29 (11): 3209–3232.

Weng, Q. and Lu, D. (2009). Landscape as a continuum: an examination of the urban landscape structures and dynamics of Indianapolis city, 1991–2000. *International Journal of Remote Sensing* 30 (10): 2547–2577.

Weng, Q., Lu, D., and Schubring, J. (2004). Estimation of land surface temperature-vegetation abundance relationship for urban heat island studies. *Remote Sensing of Environment* 89 (4): 467–483.

Wu, C. (2009). Quantifying high-resolution impervious surfaces using spectral mixture analysis. *International Journal of Remote Sensing* 30 (11): 2915–2932.

Wu, J. and Hobbs, R. (2002). Key issues and research priorities in landscape ecology: an idiosyncratic synthesis. *Landscape Ecology* 17 (4): 355–365.

Wu, J.G. and David, J.L. (2002). A spatially explicit hierarchical approach to modeling complex ecological systems: theory and application. *Ecological Modelling* 153: 7–26.

Zhang, L., Weng, Q., and Shao, Z.F. (2017). An evaluation of monthly impervious surface dynamics by fusing Landsat and MODIS time series in the Pearl River Delta, China, from 2000 to 2015. *Remote Sensing of Environment* 201 (11): 99–114.

3

Building Extraction and Classification

3.1 Introduction

Building information, such as size, height, age, and type, is valuable for many socioeconomic applications, especially in urban management and planning and demographic applications. The construction of new neighborhoods, transportation networks, and drainage systems are a few examples when the information about buildings is essential. The information of building usage type (e.g. residential building, apartment, and commercial building) is one of the essential variables for demographic, socioeconomic, and microclimate models in urban studies. However, the collection of building information is usually costly and cannot be obtained directly from existing data sources, but by using extraction techniques. Conventional field surveying methods are costly. Planning maps are widely used in various urban applications because of their high accuracy, although producing and updating them is time-consuming.

Remotely sensed data provides the potential for being a relatively convenient and cost-effective mean to derive the attributes of buildings, enabling us to obtain building usage information. Fine spatial resolution image (e.g. aerial photography) and Light Detection and Ranging (LiDAR) point cloud data have shown great potential to map buildings and other urban features. Aerial photographs are a popular data source for building extraction. However, they cover selected areas only and lack coverage of multiple times, making it difficult for a building database to create and to update. Satellite imagery of high resolution has also been frequently used in building extraction, but the issues of shadowing and distortion can affect the accuracy of extraction. LiDAR data provides land surface elevation information by emitting a laser pulse. As a unique data collection and structure, it has potential over satellite imagery and aerial photographs in accurate capture of urban features with absolute height information, especially for building extraction.

In this chapter, a methodology is developed to identify building types by using remote sensing-derived morphological attributes, based on a case study

Techniques and Methods in Urban Remote Sensing, First Edition. Qihao Weng.
© 2020 by The Institute of Electrical and Electronics Engineers, Inc.
Published 2020 by John Wiley & Sons, Inc.

of Marion County (i.e. City of Indianapolis), Indiana, USA. First, remotely sensed data (i.e. high-spatial-resolution orthophotography and LiDAR point cloud data) and Geographic Information System (GIS) data layer of building polygons in 2012 were collected. Then, morphological attributes of buildings were delineated. Third, a Random Forest (RF) classifier was trained using randomly selected training samples from GIS data layers created by the City of Indianapolis GIS (IndyGIS), Google Earth Maps, and field work. Finally, the trained classifier was applied to categorize the building types.

3.2 Building Reconstruction

3.2.1 LiDAR Data and Preprocessing

LiDAR data of the first and last returns dated in March and April 2003 were used to derive Normalized Height Model (NHM). The data were acquired by Laser Mapping Specialists, Inc. (Raymond, MS) for the area of Marion County, Indiana, covering the city proper of Indianapolis. The LiDAR sensor is known as the Optech Airborne Laser Terrain Mapper (ALTM) 2033, and for this survey, was operated at 2700 ft above ground level. A subset of the data covering 4 square miles of downtown Indianapolis was used as the main data source in this study. The data had a resolution of one cloud point per square meter on 45° flight line. Indiana State Plane East (NAD83, NAVD88) system was the coordinate system.

The 2005 Statewide Digital Elevation Model (DEM) of Marion County with the resolution of 5 ft was used to eliminate the elevation influence from Digital Surface Model (DSM) derived from LiDAR data, so that an NHM can be computed. The DEM was created from the 2005 Orthophotography of Indiana (6-in. resolution) imagery collected during March and April. The DEM was coordinated to the State Plane Coordinate System 1983.

The Indianapolis planning map of 2000 was used as the reference for accuracy assessment (in a shape file format), which contained information of each buildings and houses in the downtown area. A comparison between this map and the raw LiDAR data indicated that the buildings and houses in the area had little change from 2000 to 2003. Thus, in this research, it was assumed that the planning map can be used as the "ground truth" in assessing the accuracy of building extraction result.

DSM derived from LiDAR data provides the height, shape, and context information, which has great advantages to distinguish building objects from other features on the ground due to their height differences. The building extraction in this research was performed based on the NHM, which was calculated by subtracting statewide DEM from DSM (derived from LiDAR data). In the NHM image, only objects above the terrain were considered. The following steps detailed the derivation of NHM.

The LiDAR system has the ability to capture more than one return for each height point (Alharthy and Bethel 2002). Among all the returns, the first and last return points are the most important. Their individual characteristics also determine their respective applications. The first return points reflect the surface of all ground objects, including solid objects and transparent objects. This property is extremely useful for detecting penetrable objects, especially trees (Popescue et al. 2003; Secord and Zakhor 2007; Voss and Sugumaran 2008). In contrast, the last return points hit through the leakage among tree branches and leaves to non-penetrable objects such as the ground and buildings. Urban trees were not completely invisible in the last return DSM image because thick branches and leaves of trees stopped the laser pulse from hitting the ground. The DSM image derived from the last return points was selected to produce an NHM image due to its minimized height information of urban trees. Before obtaining the DSM image, a process of rasterizing the height information from the raw LiDAR data was performed. A 3-ft^2 (approximately $0.91 \times 0.91\,m^2$) neighborhood function was selected for converting vector points data to a DSM grid image, so that the DSM had the same resolution as the DEM.

NHM was then calculated using the following formula:

$$NHM = DSM - DEM \tag{3.1}$$

The NHM image contained the absolute height information of the objects that were above the ground surface. Some pixel values of NHM were less than 0. According to Figure 3.1, it is clear that most of them were located near water bodies and flyover highways (where the pixels are light yellow). Water pixels usually had negative values because laser points were absorbed and did not reflect back to the scanner, causing the problem of point lost. Moreover, due to a variable scanning angle of 1–75°, points emitting to flyover highways can easily hit the under-bridge areas where reflected range values were usually lower than the surface elevation. Pixel values in the range of −2 to 0 ft were regarded as common errors and thus considered acceptable. A correction process was needed to filter out abnormal points through the following criteria:

If IV of *NHM* < 0 then IV of *NHM* = 0;

Else IV of *NHM* = IV of *NHM*.

where IV is the image value. After the correction, the resultant NHM image was used in the building reconstruction and classification.

3.2.2 Methodology for Building Reconstruction

The procedure of building points extraction consists of three steps: (i) ground point filtering (interpolation based); (ii) vegetation point filtering (Normalized

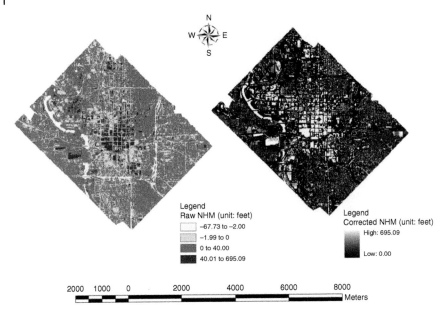

Figure 3.1 The original and corrected Normalized Height Model (NHM) image of Indianapolis. *Source:* Weng (2009). Reproduced with permission of The McGraw-Hill Companies, Inc.

Difference Vegetation Index [NDVI] and geometry based); and (iii) assigning building data points to each specific building and building reconstruction. The first step is to remove ground points according to its relatively low elevation. The multi-resolution hierarchical classification (MHC) algorithm proposed by Chen et al. (2013) was used to filter ground points due to its simplicity (Figure 3.2).

Vegetation (i.e. trees) points were removed through two methods. One was applying a mask of NDVI based on the fact that tree pixels had higher NDVI values than other objects. NDVI was computed from an IKONOS image by using its red and near-infrared bands. Another method was geometry based, which was based on the assumption that the geometric properties of tree points were significantly different from those of other urban objects. Selected geometric properties included regularity (the distance between sample point and the centroid of neighbor points), horizontality (the angle between normal vector of the sample point and the vertical vector [0,0,1]), flatness (a smaller flatness value indicates more possibility for a point to be a roof point or ground point), and normal vector distribution (the distribution of normal vectors). Support Vector Machine (SVM) was applied to train classification rules to separate tree points from others

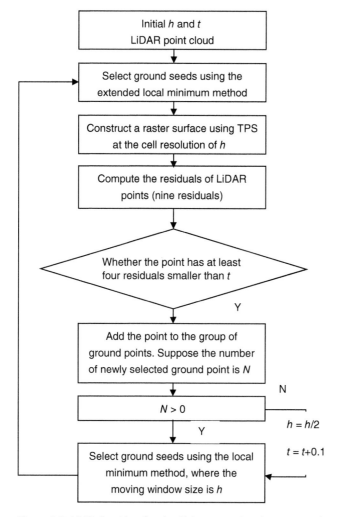

Figure 3.2 MHC algorithm for classifying ground and non-ground points (Chen et al. 2013).

(Figure 3.3). It was found that NDVI-based approach was useful for removal of trees away from buildings and that geometry-based method was effective to filter out points near buildings.

After removing ground and tree points, the remaining LiDAR points were assumed from buildings. This was reasonable for residential area (small objects like cars would be removed in the following step, according to its relatively small area and volume). Building points were then assigned to each specific building using a region growing algorithm. The procedure for building boundary tracing and regularizing is displayed in Figure 3.4.

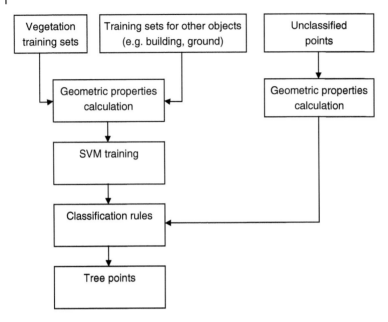

Figure 3.3 The procedure for removing trees from light detection and ranging (LiDAR) data.

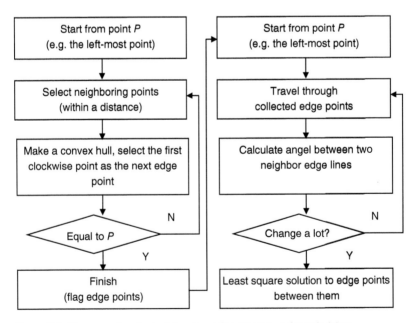

Figure 3.4 The procedure for building boundary tracing and regularizing.

3.2.3 Building Reconstruction Results

Figure 3.5 illustrates the results of ground point filtering, training sets development, and tree points detection. Building reconstruction for the central business district and a residential area in Indianapolis is shown in Figures 3.6 and 3.7, in which buildings were assumed to have flat roofs.

The result of building reconstruction experiment is shown in Table 3.1. It is found that the number of buildings was slightly different from the actual number. This is because some adjacent buildings were merged into a larger

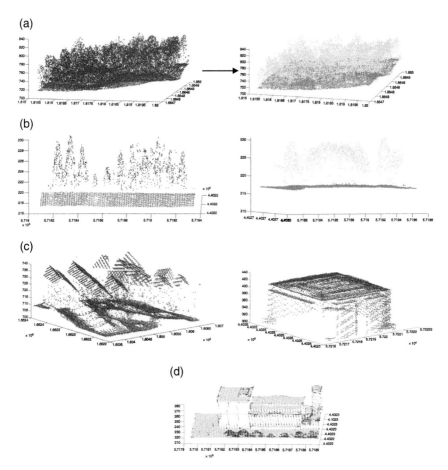

Figure 3.5 Results of ground point filtering, training sets development, and tree points detection. (a) Ground point filtering. (b) Vegetation training set. (c) Building and ground training sets. (d) Detected tree points using the training sets.

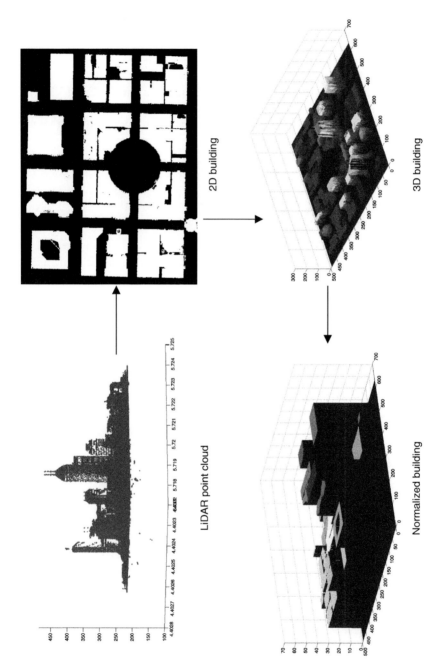

Figure 3.6 Building reconstruction for the central business district of Indianapolis.

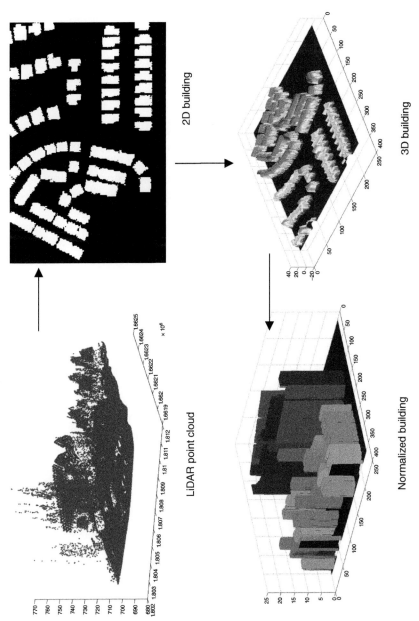

Figure 3.7 Building reconstruction for a residential area in Indianapolis.

Table 3.1 Building statistics for two study scenes.

Scene	Number of buildings	Total area of buildings	Total volume of buildings
Central business district	24	$45\,714\,107\,\text{ft}^2$	$2.510 \times 10^{14}\,\text{ft}^3$
Residential area	56	$130\,890\,\text{ft}^2$	$2\,371\,500\,\text{ft}^3$

one. However, it had little impact on the calculation of building area and volume. In the residential area, buildings were not typically as close to each other as in the central business district. Based on the estimated building area and volume, we can determine the number of housing units. Since buildings were assumed to have flat roofs, which might not seem reasonable in some applications, it would not impact population estimation.

3.3 Building Classification

3.3.1 Classification Method

Buildings were classified into three categories: (i) nonresidential buildings (including commercial/industrial buildings, churches, etc.); (ii) apartments; and (iii) single-family housing (including buildings with less than three units since they had similar morphologic characteristics). The selection of morphological metrics for building type identification followed studies by Lu et al. (2014) and Du et al. (2015). The metrics used for training the classifier were basic statistics of LiDAR-derived surfaces, geometric features of buildings, and landscape background information. Landscapes around buildings were classified into three types, i.e. grass, tree, and impervious surface by using high-spatial-resolution orthophotography image.

The RF classifier was used to classify building types. One of the most appealing aspects of this algorithm is its reasonable accuracy and insensitivity to the configuration of training data sets. The number of classification trees and mtry were set to 500 and 5, respectively. The training samples were collected by using a random sampling strategy (Figure 3.8).

3.3.2 Classification Results

The slope of buildings and certain shape indexes (such as building area, perimeter, volume, compactness) significantly contributed to building type classification (Figure 3.9). Although building background metrics were found useful

Figure 3.8 The distribution of collected samples for classifier training and accuracy evaluation.

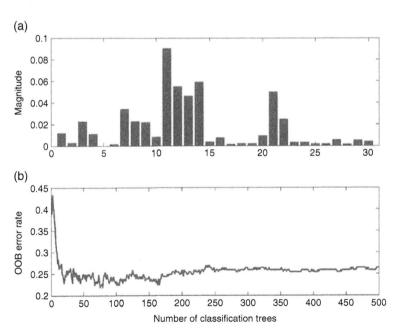

Figure 3.9 Sensitivity of morphological metrics mean decrease in accuracy (a) and out-of-bag (OOB) error rate for building type classification (b).

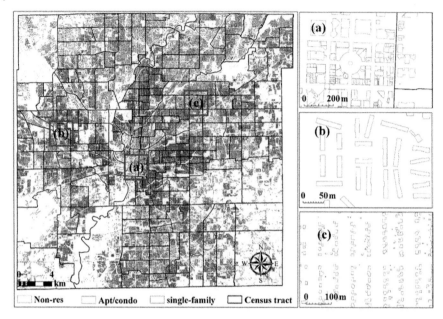

Figure 3.10 Building type classification result. Note that buildings were classified as nonresidential usage (a), apartment/condo (b), and single-family house (c).

to building type identification, most of them contributed less than 1%. Thus, these metrics were used for future building type identification. In total, there were 263, 201, and 259 buildings identified for nonresidential, apartment/condo, and single-family housing, respectively (Figure 3.10).

References

Alharthy, A. and Bethel, J. (2002). Heuristic filtering and 3D feature extraction from LIDAR data. *International Archives of Photogrammetry and Remote Sensing (IAPRS)*, Graz, Austria, Vol. XXXIV(Part 3A), pp. 29–34.

Chen, C., Li, Y., Li, W., and Dai, H. (2013). A multiresolution hierarchical classification algorithm for filtering airborne LiDAR data. *ISPRS Journal of Photogrammetry and Remote Sensing* 82: 1–9.

Du, S., Zhang, F., and Zhang, X. (2015). Semantic classification of urban buildings combining VHR image and GIS data: an improved random forest approach. *ISPRS Journal of Photogrammetry and Remote Sensing* 105: 107–119.

Lu, Z., Im, I., Rhee, J., and Hodgson, M. (2014). Building type classification using spatial and landscape attributes derived from LiDAR remote sensing data. *Landscape and Urban Planning* 130: 134–148.

Popescue, S.C., Wynne, R.H., and Nelson, R.F. (2003). Measuring individual tree crown diameter with LiDAR and assessing its influence on estimating forest volume and biomass. *Canadian Journal of Remote Sensing* 29: 564–577.

Secord, J. and Zakhor, A. (2007). Tree detection in urban regions using aerial lidar and image data. *IEEE Geoscience and Remote Sensing Letters* 4: 196–200.

Voss, M. and Sugumaran, R. (2008). Seasonal effect on tree species classification in an urban environment using hyperspectral data, LiDAR, and an object-oriented approach. *Sensors* 8: 3020–3036.

Weng, Q. (2009). *Remote Sensing and GIS Integration: Theories, Methods, and Applications*, 397. New York: McGraw-Hill Professional.

4

Estimation and Mapping of Impervious Surfaces

4.1 Introduction

Impervious surface is the anthropogenic surface that prevents water from infiltrating into soil (Arnold and Gibbons 1996), and it is widely distributed in urban areas. Impervious surface can be categorized into two primary types: rooftops and transportation systems (e.g. roads, sidewalks, and parking lots) (Schueler 1994). Impervious surfaces are made of materials impenetrable for water such as asphalt, concrete, bricks, or stones. Compacted soils are also highly impervious and considered another type of impervious surface (Arnold and Gibbons 1996).

Impervious surface is associated with various adverse environmental outcomes and is a crucial environmental indicator (Arnold and Gibbons 1996). Impervious surface decreases the recharge of underground water, increases the velocity and volume of the surface runoff, and as a result, increases the risk of flooding. Moreover, increased runoff further erodes construction sites and river banks (Arnold and Gibbons 1996). In addition, impervious surface leads to nonpoint source pollution and threats to the surface water quality (Civico and Hurd 1997; Sleavin et al. 2000). Nonpoint source pollutants include pathogens, nutrients, toxic contaminants, and sediments, which degrade water qualities and are harmful to both animals and humans. Impervious surface also affects the energy balance in the urban areas and is one of the important factors that cause the urban heat island effect. Previous studies showed that impervious surface was positively related to increased surface temperatures in the urban areas (Lu and Weng 2006; Yuan and Bauer 2007). Likewise, the increase of impervious surface inevitably results in the decrease of vegetation cover, which reduces ecological productivity, interrupts atmospheric carbon cycling, and degrades air quality. Hence, impervious surface is crucial for urban environmental management (Arnold and Gibbons 1996; Flanagan and Civco 2001; Wu and Murray 2003).

Techniques and Methods in Urban Remote Sensing, First Edition. Qihao Weng.
© 2020 by The Institute of Electrical and Electronics Engineers, Inc.
Published 2020 by John Wiley & Sons, Inc.

4.2 Methods for Impervious Surface Extraction

To extract impervious surface cover, traditional methods include ground survey, global positioning system (GPS), aerial photo interpretation, and satellite remote sensing interpretation (Stocker 1998). Ground measurement is both time and cost inefficient, while GPS is not feasible for mapping large areas. Extraction of impervious surfaces from aerial photos is expensive. Satellite remote sensing has been widely used in impervious surface estimation studies due to its relatively low cost and capability for mapping large areas (Bauer et al. 2004). A challenge in extracting impervious surfaces from medium-resolution satellite imagery is to tackle with mixed pixels. A mixed pixel is a pixel that contains multiple land-cover types, as compared to pure pixels that contain only one land-cover class. Traditional supervised and unsupervised classification techniques can only identify land-cover features at the pixel level and cannot effectively deal with mixed pixels. As a result, sub-pixel techniques need to be applied. To date, numerous sub-pixel classification methods have been developed for extracting impervious surface from remote sensing imagery (medium spatial resolution images),including linear spectral mixture analysis (LSMA), Regression Tree, artificial neural networks (ANN), and multiple regression (Civico and Hurd 1997; Wu and Murray 2003; Yang et al. 2003a, 2003b; Bauer et al. 2004; Lee and Lathrop 2005; Lu and Weng 2006). Object-based classification techniques have also been used for impervious surface estimation from remote sensing imagery, especially from high spatial resolution images. As the spatial resolution increases, the proportion of pure pixels increases and mixed pixels is reduced (Hsieh et al. 2001). Thus, the sub-pixel methods might not be appropriate. Object-based classification methods incorporate not only the color and tone of the pixels, but also other crucial characteristics such as shape, texture, and context, and thus can extract impervious surfaces with higher accuracy.

Weng (2012) provided a comprehensive review of remote sensing methods for estimating and mapping impervious surfaces in the urban areas. This chapter focuses on the examination of sub-pixel estimation techniques, including LSMA, ANNs, and fuzzy classifiers. A linear spectral mixture model is based on the assumption that each photon interacts with only one land-cover type on the ground before being reflected back to the sensor, and as a result, the mixed spectra can be modeled as a linear combination of the spectra of land-cover features weighted by the proportion of each feature within the instantaneous field of view (IFOV) (Singer and McCord 1979; Roberts et al. 1998; Small 2001). The spectra of these land-cover features are called end-members. End-member selection is a key step to LSMA, and end-members can be selected from image data themselves, spectra libraries, or reference spectra collected from field (Roberts et al. 1998; Small 2001). Selecting end-members directly from the image's feature space is relatively simple and thus has been employed by numerous previous studies. The results of LSMA are fraction images for end-members, in which the pixel values indicate the percentage of the end-member within that pixel (Small 2001). LSMA may not only provide a reasonable approximation for complex landscapes (Haglund 2000), but also contain several limitations (Foody et al. 1997). First, the linear assumption is not necessarily true. When scattered photons interact with multiple components, the mixture becomes nonlinear (Roberts et al. 1993; Gilabert et al. 2000). When the nonlinearity is significant, it should not be neglected (Roberts et al. 1993; Ray and Murray 1996). The difficulty of LSMA also comes from end-member selection. End-member selection is a challenge due to within-class spectral variability. The number of end-members is limited by the dimensionality of the image and the correlation between bands. Limited number of end-members reduces the capability of unmixing due to the image spectra being under sampled (Small 2001). The spectra of land-cover types can be very diverse. For example, there may be various degrees of bright/dark impervious surfaces, which are located in different locations within scatter plots. Thus, the selection of end-members is difficult that can utterly represent the spectra of specific land-cover classes.

Another approach for impervious surface estimation is ANNs. The neural networks mimic the functions of human brain and can learn through trial and error. The ANN can generate more accurate results, perform more rapidly, incorporate *a prior* knowledge in the calibration, and incorporate different types of data. There are many types of neural networks, and a commonly used algorithm is multilayer perceptron (MLP) neural network (Atkinson and Tatnall 1997; Foody et al. 1997). MLP has been used in land-use land-cover classification, change detection, and water properties estimation (Foody et al. 1997; Schiller and Doerffer 1999; Zhang and Foody 2001; Li and Yeh 2002; Corsini et al. 2003; Kavzoglu and Mather 2003). MLP has also been used in impervious surface estimation. Chormanski et al. (2008) conducted MLP to map the fractions of four major land-cover classes (impervious surfaces, vegetation, bare soil, and water/shade) with both high spatial resolution and medium-resolution imagery. Weng and Hu (2008) used LSMA and MLP neural network for impervious surface extraction at the sub-pixel level and compared the performance. Hu and Weng (2009) compared the performance of the MLP neural network and the self-organizing map (SOM) neural network for estimating impervious surfaces. In addition, Hu and Weng (2009) applied MLP to spectrally normalized images to extract impervious surfaces. Van de Voorde et al. (2011) employed an MLP-based supervised classification to extract impervious surfaces. Limitations of MLP are also obvious. First, it is a challenge to determine how many nodes are needed in each layer and how to design the number of the hidden layers. Secondly, the training of MLP requires both presence and absence data. However, in many cases, the absence data are not available.

SOM neural network has also been used for impervious surface estimation, although it has not been widely applied in other remote sensing applications. SOM were previously used for both supervised and unsupervised classifications. Ito and Omatu (1997) applied a SOM neural network and a *k*-nearest neighbor method on Landsat TM data. Moreover, Ito (1998) employed the SOM method to SAR data for image classification. Ji (2000) conducted a Kohonen self-organizing feature map (KSOFM) for land-use and land-cover classification. Lee and Lathrop (2006) combined SOM, learning vector quantization (LVQ), and Gaussian mixture model (GMM) to estimate the percentage of impervious surface coverage.

The third sub-pixel method for extracting impervious surfaces is for impervious surface estimation is fuzzy classification. Fuzzy classification is based on the fuzzy set theory that defines the strength of membership of a pixel to land-cover class by using fuzzy membership grade within a range from 0 to 100%. In general, the fuzzy membership grade is calculated by a fuzzy membership function. Different types of membership function have been developed, which include sigmoidal (S-shaped), J-shaped, and linear. Fuzzy classification has also been widely applied in land-use and land-cover classifications, vegetation classifications, change analysis, cloud cover classifications, and flooded area mapping (Fisher and Pathirana 1990; Foody and Cox 1994; Foody 1996, 1998; Zhang and Foody 1998, 2001; Mohan et al. 2000; Townsend 2000; Lee and Lathrop 2002; Amici et al. 2004; Tapia et al. 2005; Filippi and Jensen 2006; Ghosh et al. 2006; Okeke and Karnieli 2006a, 2006b; Tang et al. 2007). So far, the use of fuzzy classification in impervious surface estimation is limited. To date, Lee and Lathrop (2002) employed a supervised fuzzy c-means clustering (FCM) to extract impervious surface fractions from Landsat TM images. Hu and Weng (2011) compared the performance of fuzzy classification and LSMA in impervious surface extraction at sub-pixel level. Their results showed that fuzzy classification outperformed LSMA in terms of extraction accuracy.

4.3 Case Studies

4.3.1 Linear Spectral Mixture Analysis

The LSMA model can be expressed mathematically by the equation as follows:

$$R_b = \sum_{i=1}^{N} f_i R_{i,b} + e_b \tag{4.1}$$

where R_b is the reflectance for each band b, N is the number of end-members, f_i is the fraction of end-member i, $R_{i,b}$ is the reflectance of end-member i in band b, and e_b is the error for band b. For a fully constrained least-square unmixing solution, the following conditions are required to be satisfied:

$$\sum_{i=1}^{N} f_i = 1, \quad f_i \geq 0 \tag{4.2}$$

Model fitness is assessed by the root-mean-square error (RMSE). The mathematical expression of RMSE is shown as follows:

$$RMSE = \sqrt{\sum_{b=1}^{M} \frac{e_b^2}{M}} \tag{4.3}$$

where e_b is the un-modeled residual and M is the number of the bands.

A Hyperion image covering the City of Indianapolis, USA, was acquired in 12 April 2003 and used for the experiment of LSMA. The Hyperion image has 242 bands covering 400–2500 nm with a spatial resolution of 30 m. The image was georectified to a Universal Transverse Mercator coordinate system using nearest-neighbor resampling method. An RMSE of less than 0.2 pixels was obtained in geometric correction. End-members were selected at the vertexes of the triangles in the scatter plots (Figure 4.1). Four end-members were identified from the image plots, including high albedo, low albedo, soil, and vegetation.

The relationship between impervious surface, high albedo, and low albedo can be modeled in the following mathematical equation:

$$R_{imp,b} = f_{low} R_{low,b} + f_{high} R_{high,b} + e_b \tag{4.4}$$

where $R_{imp,b}$ is the spectra of impervious surfaces for band b, $R_{high,b}$ and $R_{low,b}$ are the spectra of high-albedo and low-albedo end-members, f_{low} and f_{high} are the fractions of low-albedo and high-albedo end-members, e_b is the un-modeled residual. The impervious surface fractions can be calculated by adding high albedo and low albedo together. Nevertheless, some low-albedo materials (e.g. water and shade) and high-albedo materials (e.g. cloud and sand) need to be removed because they tended to be confused with impervious surfaces spectrally. Figure 4.2 illustrates the unmixed results, including high albedo, vegetation, and low albedo. The brighter the pixel, the higher the percentage of the end-member within a particular pixel.

The impervious surface fractions were calculated using the Eq. (4.4) and illustrated in Figure 4.3. The high percentages of impervious surfaces are located in Central Business District (CBD) area, while the low percentages are within rural areas. Most residential areas had medium percentages (Figure 4.3b). The accuracy assessment result indicates that an RMSE of 17.2% and R^2 of 0.66 were achieved for the whole study area. However, RMSE for the residential area was highest (9.9%), followed by the rural area (16.5%) and commercial area (22.8%).

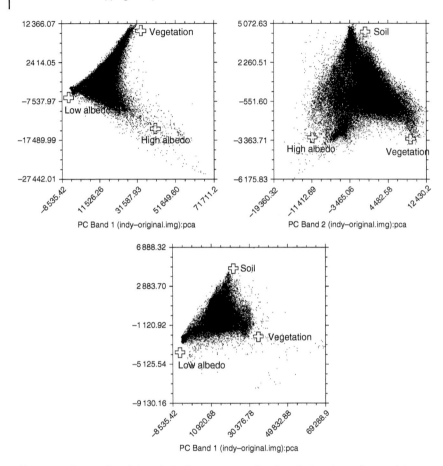

Figure 4.1 Scatterplot of the principal components showing the locations of potential endmembers. *Source:* after Weng et al. (2008). Reproduced with permission of Taylor & Francis.

4.3.2 The MLP Neural Network Method

An MLP neural network with a back-propagation learning algorithm was also implemented. The MLP classifier used the following algorithm to calculate the input that a single node *j* received:

$$net_j = \sum_i w_{ij} I_i \tag{4.5}$$

where net_j refers to the input that a single node *j* receives; w_{ij} represents the weights between node *i* and node *j*; and I_i is the output from node *i* of a sender layer (input or hidden layer). Output from a node *j* was calculated as follows:

$$O_j = f\left(net_j\right) \tag{4.6}$$

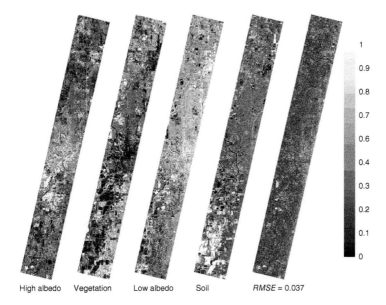

High albedo   Vegetation   Low albedo   Soil   *RMSE = 0.037*

Figure 4.2 Fractional (high albedo, low albedo, vegetation, and soil) and error images from spectral mixture analysis of the Hyperion image. *Source:* after Weng et al. (2008). Reproduced with permission of Taylor & Francis.

The function f usually is a nonlinear sigmoidal function. In the study, an input layer with 76 nodes corresponding to 76 selected Hyperion image bands (visible and near infrared [VNIR] and shortwave infrared [SWIR]), one output layer with four nodes corresponding to four land-cover classes, high albedo, low albedo, vegetation, and soil, were used. The number of the hidden layer nodes was calculated by the formula as follows:

$$N_h = INT \sqrt{N_i \times N_o} \tag{4.7}$$

where N_h is the number of hidden layer nodes, N_i is the number of input layer nodes, and N_o is the number of output layer nodes.

A Hyperion image covering the City of Atlanta, Georgia, USA, was acquired in 1 June 2003 and used for the experiment of LSMA. The fraction images of high albedo, low albedo, vegetation, and soil were illustrated in Figure 4.4. The impervious surface map is shown in Figure 4.5, which illustrates a similar spatial pattern as that shown in Figure 4.3. The accuracy assessment result indicates that an RMSE of 16.3% and R^2 of 0.75 were achieved for the whole study area.

4.3.3 Fuzzy Classification

Fuzzy sets are sets (classes) without sharp boundaries, and the transition from nonmembership to membership is gradual. A fuzzy set is measured by a fuzzy membership grade (possibilities) which ranges from 0.0 to 1.0, indicating from

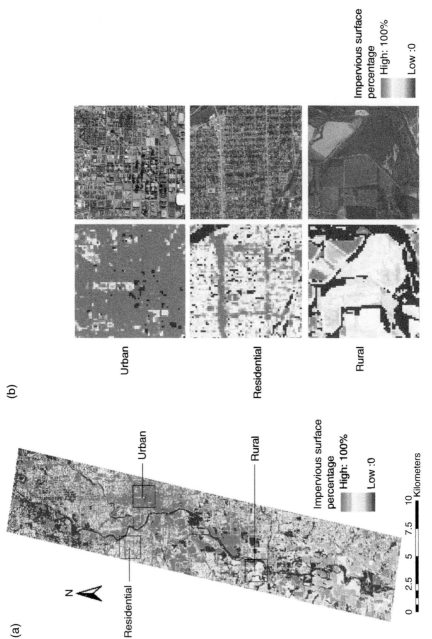

Figure 4.3 Fractional image of impervious surface of Indianapolis derived from linear spectral mixture analysis. (a) Entire image; (b) selected areas.

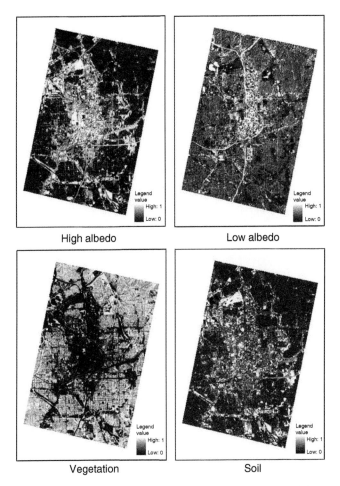

Figure 4.4 Fraction images of end members derived from multilayer perceptron (MLP) neural network. *Source:* Hu and Weng (2013). Reproduced with permission of Taylor & Francis.

a 0 to 100% membership. Fuzzy sets can provide better representation for geographical information. Both information classes and spectral classes can be represented as fuzzy sets. Spectral space can be divided into fuzzy sets without sharp boundaries, which can be mathematically expressed as

$$\forall x \in X$$
$$0 \le f_{F_i}(x) \le 1$$
$$\sum_{x \in X} f_{F_i}(x) > 0 \qquad (4.8)$$
$$\sum_{i=1}^{m} f_{F_i}(x) = 1$$

Figure 4.5 Fraction image of impervious surface derived from MLP neural network. *Source:* Hu and Weng (2013). Reproduced with permission of Taylor & Francis.

where $F_1, F_2, ..., F_m$ represent the spectral classes, X is the whole pixels, m is the number of the classes, x is the pixel measurement vector, and f_{F_i} is the membership function of the fuzzy set F_i $(1 \leq i \leq m)$ (Wang 1990).

To calculate the fuzzy representation for each spectral class, the probability measures of fuzzy events was applied. The mathematical expression can be defined as follows:

$$p(A) = \int_{\Omega} f_A(x) dP \tag{4.9}$$

where f_A is the membership function of the spectral class A $(0 \leq f_A(x) \leq 1)$, Ω is the spectral space, and $p(A)$ is the probability measure of the spectral class A (Wang 1990).

Then, fuzzy signatures can be extracted from original images using a chosen algorithm (Wang 1990). Fuzzy mean and fuzzy covariance can be calculated for each information class to yield fuzzy signatures. The mathematical expression of fuzzy mean is defined as

$$\mu_c^* = \frac{\sum_{i=1}^{n} f_c(x_i) x_i}{\sum_{i=1}^{n} f_c(x_i)} \tag{4.10}$$

where n is the total number of training pixel measurement vectors, x_i is a training pixel measurement vector $(1 \leq i \leq n)$, and f_c is the membership function of class c.

The mathematical expression of fuzzy covariance is defined as

$$\sum_c^* = \frac{\sum_{i=1}^{n} f_c(x_i)(x_i - \mu_c^*)(x_i - \mu_c^*)}{\sum_{i=1}^{n} f_c(x_i)} \tag{4.11}$$

where μ_c^* is the fuzzy mean, n is the total number of training pixel measurement vectors, x_i is a training pixel measurement vector $(1 \leq i \leq n)$, and f_c is the membership function of class c.

The membership function for class c can be expressed as

$$f_c(x) = \frac{p_c^*(x)}{\sum_{i=1}^{m} p_i^*(x)} \tag{4.12}$$

where

$$p_i^*(x) = \frac{1}{(2\pi)^{N/2} \left|\sum_i^*\right|^{1/2}} \cdot \exp\left[-\frac{1}{2}(x - \mu_i^*)^T \sum_i^{*-1}(x - \mu_i^*)\right] \tag{4.13}$$

N is the dimension of the pixel vectors, and m is the number of the classes $(1 \leq i \leq m)$.

Fuzzy signatures can then be inputted into a fuzzy classifier to calculate the fuzzy membership of each information class for each pixel. The membership value is determined by the standardized Euclidean distance of each pixel to the mean spectra of a signature on each band using a sigmoidal membership function. Although there are many types of membership function, such as sigmoidal, J-shaped, and linear, the sigmoidal membership function is the most commonly used in fuzzy set theory. The sigmoidal membership function is calculated as follows:

$$\mu = \cos^2 \alpha \qquad (4.14)$$

where μ is the sigmoidal membership function, in the case of a monotonically decreasing function:

$$\alpha = \frac{\left(x - point\,c\right)}{\left(point\,d - point\,c\right)} * \frac{pi}{2} \qquad (4.15)$$

when $x < point\,c$, $\mu = 1$; where, in the case of a monotonically increasing function,

$$\alpha = \frac{\left(x - point\,a\right)}{\left(point\,b - point\,a\right)} * \frac{pi}{2} \qquad (4.16)$$

when $x > point\,b$, $\mu = 1$. (Point a, b, c, and d refer to the control points.)

The fuzzy classifier provides fuzzy set membership images corresponding to each class as well as an image of classification uncertainty. Classification uncertainty indicates the degree to which no class clearly stands out above others in the assessment of fuzzy set membership of a pixel. It is calculated as follows:

$$Classification\ uncertainty = 1 - \frac{max - \left(sum/n\right)}{1 - \left(1/n\right)} \qquad (4.17)$$

where *max* is the maximum set membership value for the pixel; *sum* is the sum of the fuzzy set membership value for the pixel; and *n* is the number of classes (signatures) (Eastman 2006).

In the study conducted by Hu and Weng (2011), the fuzzy classification was conducted to extract impervious surfaces. Training data were manually selected from the original imagery. Training in the fuzzy classier was similar to training site selection in traditional supervised classification. The difference was that the training site needed to be homogeneous for conventional supervised classification, which was not required for the fuzzy classifier. For each image, 30 training samples for each land-cover class, including high albedo, low albedo, soil, and vegetation, were manually selected on the original imagery. Each sample had different size and contained a certain amount of training pixels (larger than 30 pixels). Four fuzzy membership images plus one classification uncertainty image were yielded. Four fuzzy membership images include a high-albedo image, a low-albedo image, a vegetation image, and a soil image. An impervious surface fraction image was then generated by adding two fuzzy membership images (high albedo and low albedo). Spectrally confused materials (e.g. water, shade, and dry soils) were removed to improve the quality of the impervious surface map using the method proposed by Wu and Murray (2003). Figure 4.6 illustrates the fuzzy membership images.

Although the fuzzy classification can yield better results of impervious surface estimation than LSMA, the quality of fuzzy signatures relies heavily on the identification of the fuzzy membership grade for all classed in each training site. Some important parameters need to be appropriately defined, such as z-score, which may significantly impact the accuracy of the final results. The fuzzy set membership was determined by the distance of a pixel to the signature means, and the z-score value was the distance where the fuzzy set membership value became zero. The fuzzy set membership at the signature means is one. If the distance increased, the fuzzy set membership would decrease until it became zero, where the z-score distance was reached. The z-score value was determined by the quality of the signature and the width of each class. If the signature was pure and the class width small, a small z-score should be selected; otherwise, a large z-score would be necessary. Compared to the fuzzy classifier, LSMA is a simpler model which is both time and computation efficient. Therefore, LSMA has recently become widely accepted as a sub-pixel classifier.

4.3.4 Kernel Fuzzy C-Means Clustering

The use of time series satellite imagery for land-cover classification and change detection has attracted increased attention in recent years (Zhu and Woodcock 2014). Temporal domain of satellite imagery has showed its advantages in resolving spectral confusion between classes with similar spectral characteristics (Bhandari et al. 2012; Schneider 2012). Landsat time series have been successfully applied to map dynamics of urban areas due to its long record of continuous measurement at effective spatial resolution and temporal frequency (Gao et al. 2012; Sexton et al. 2013; Li et al. 2015; Zhang and Weng 2016a). However, these methods focused on spectral differences or temporal consistency after classification. Little attention was paid to temporal data mining method to differentiate urban areas from other land cover using dense time series Landsat images.

Time series clustering has been shown to be effective in time series data mining (Liao 2005; Fu 2011). Kernel fuzzy C-means (KFCM), proposed by Zhang and Chen (2003), being introduced in this chapter, has shown potentials to provide a more robust signal-to-noise ratio and less sensitive to cluster shapes in comparison to other clustering algorithms (Du et al. 2005). Given time series data $X = \{x_1, x_2, ..., x_n\}$, $x_k \in R^d$ ($k = 1, 2, ..., n$), d was temporal dimension, and n was the number of samples. KFCM partitions X into c fuzzy subsets by minimizing the following objective function:

$$J_m(U,V) = 2\sum_{i=1}^{c} \sum_{k=1}^{n} u_{ik}^m \left(1 - K\left(x_k, v_i\right)\right) \tag{4.18}$$

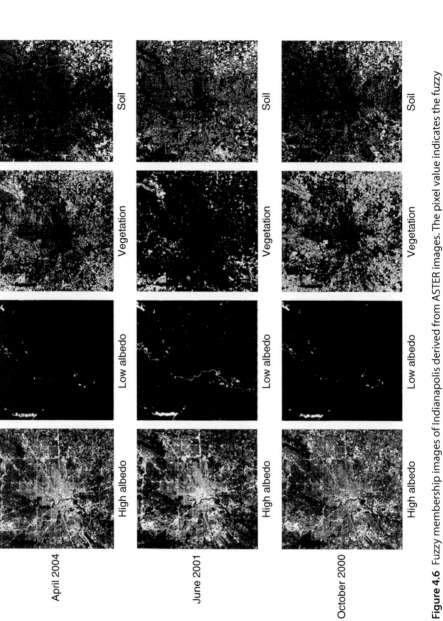

April 2004   High albedo   Low albedo   Vegetation   Soil

June 2001   High albedo   Low albedo   Vegetation   Soil

October 2000   High albedo   Low albedo   Vegetation   Soil

Figure 4.6 Fuzzy membership images of Indianapolis derived from ASTER images. The pixel value indicates the fuzzy membership of a class within that pixel. *Source:* Hu and Weng (2011). Reproduced with permission of Taylor & Francis.

where c is the number of clusters; v_i was ith cluster centroid; u_{ik} is the membership of x_k in class i, and $\Sigma_i u_{ik} = 1$; and $m \in [1, +\infty]$ is the weighting exponent determining the fuzziness of the clusters. $K(x_k, v_i)$ is the kernel function, aiming to map x_k from the input space X to a new space with higher dimensions. In this study, radial basis function (RBF) kernel was adopted:

$$K\left(x_k, v_i\right) = \exp\left(\frac{-\left\|x_k - v_i\right\|^2}{\sigma^2}\right) \tag{4.19}$$

where the parameter σ was computed by

$$\sigma = \frac{1}{c}\left(\sqrt{\frac{\sum_{i=1}^{n}\left\|x_i - m\right\|^2}{n}}\right) \tag{4.20}$$

In order to search for new clusters, objective function was minimized:

$$\min J_m\left(U, V\right) = 2\sum_{i=1}^{c}\sum_{k=1}^{n} u_{ik}^m\left(1 - K\left(x_k, v_i\right)\right) \tag{4.21}$$

$$s.t. \quad \sum_{i=1}^{c} u_{ik} = 1, \quad k = 1, 2, \ldots, n \tag{4.22}$$

Lagrange function converted the constrained objective as an unconstrained optimization model. By optimizing objective function, the membership u_{ik} and centroid v_i could be updated:

$$u_{ik} = \frac{\left(1 / \left(1 - K\left(x_k, v_i\right)\right)\right)^{1/(m-1)}}{\sum_{j=1}^{c}\left(1 / \left(1 - K\left(x_k, v_i\right)\right)\right)^{1/(m-1)}} \tag{4.23}$$

$$v_i = \frac{\sum_{k=1}^{n} u_{ik}^m K\left(x_k, v_i\right) x_k}{\sum_{k=1}^{n} u_{ik}^m K\left(x_k, v_i\right)} \tag{4.24}$$

Labeled time series samples were derived from stable time series, and remaining time series were as unlabeled samples. Given time series data X consisted of X_l and X_u, X_l was labeled samples and X_u was unlabeled samples. l and u indicated labeled or unlabeled data, respectively.

The whole process of semi-supervised KFCM algorithm was shown as follows: (i) Initialize the values of σ and u_{ik} using X_l and X_u. For X_l, the value of component u_{ik} was set to 1 if the data x_k was labeled with class i, and 0 otherwise. For X_u, positive random values within [0,1] were set to unlabeled data. The initial set of centroid v_i was calculated as

$$v_i^0 = \frac{\sum_{k=1}^{n'}\left(u_{ik}^l\right)^m x_k^l}{\sum_{k=1}^{n'}\left(u_{ik}^l\right)^m}$$

where n' was the number of labeled data. (ii) Update the membership u_{ik} in X_u and centroid v_i until the objective function was minimized.

Finally, inconsistent labeled pixels were mapped comparing the Land Surface Temperature (LST) L and Biophysical Composition Index (BCI) B clustering results. For those pixels, if the maximum membership $max(u_{ik})^L$ of the pixel k in L was higher than $max(u_{ik})^B$ in B, the pixel was labeled as the class with $max(u_{ik})^L$ in L, and vice versa. However, if the values were equal, the pixel was labeled as the class with $max(u_{ik})^L$.

In the study conducted by Zhang and Weng (2016b), we aimed at extracting urban areas from Landsat imagery by using a semi-supervised fuzzy time series clustering method through BCI (Deng and Wu 2012) and LST time series and applied the method to the Pearl River Delta, China, from 1990 to 2014. BCI and LST time series images were derived because of their strong correlation with urban areas (Zhang and Weng 2016a). BCI aimed to identify different urban biophysical compositions, which has been demonstrated to be effective in identifying the characteristics of impervious surfaces and vegetation and in distinguishing bare soil from impervious surfaces (Deng and Wu 2012). LST, as a significant parameter in urban environmental analysis, tended to be positively correlated with urban expansion (Yuan and Bauer 2007; Weng and Lu 2008). Figure 4.7 shows derived urban areas using time series fuzzy clustering.

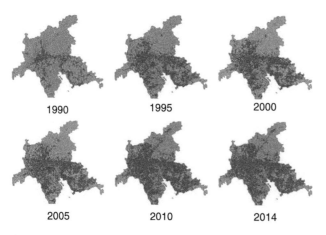

Figure 4.7 Annual urban areas from 1990 to 2014 in Pearl River Delta. The dark grey tone represented urban area. *Source:* modified after Zhang and Weng (2016b). Reproduced with permission of Taylor & Francis.

The accuracy annual clustering yielded from 78.23 to 91.32%, which showed the effectiveness of time series clustering method. However, the clustering accuracy varied yearly and fewer number of satellite images in a year could obscure the information of land-cover changes and reduce the separability of temporal features of urban areas from nonurban areas. Especially, the trend of vegetation phenology was weakened. Furthermore, the time series clustering suggested the value of imagery with cloud contamination or SLC-off data in identifying urban areas. Although cloud contamination and SLC-off data caused significant noise and resulted in incomplete time series, the method of gap filling and smoothing (Zhang and Weng 2016a) was helpful to solve the problem of missing data through enhancing temporal resolution of the time series.

4.4 Summary

This chapter provides a summary of the current research on urban impervious surface estimation and mapping. In addition, a case study was conducted to demonstrate the capability of two conventional methods (LSMA and MLP) for impervious surface estimation using Hyperion imagery. Satellite remote sensing provides a cost-effective and time-efficient way for impervious surface mapping. Medium spatial resolution imagery (e.g. Landsat TM, ETM+, and ASTER) has been utilized for large-area mapping, and high spatial resolution imagery (e.g. IKONOS and QuickBird), air photos, and Light Detection and Ranging (LiDAR) data for extracting urban features (e.g. roads and buildings). Numerous methods have been developed and applied in previous studies based on per-pixel, sub-pixel, and object-based algorithms. However, fewer studies have examined the spectral diversity of impervious surfaces. Hyperspectral imagery with rich spectral information is suitable for spectral analysis and should be extensively employed in future studies. The spatial and temporal variations of impervious surfaces are the two issues deserving more attention due to the diversity of urban environments and the changes caused by urbanization (Weng and Lu 2008). Another case study was illustrated to demonstrate time series data mining for long-term monitoring of imperious surfaces. The novel method of time series fuzzy clustering deserves future research to prove its effectiveness in diverse geographic settings of cities.

References

Amici, G., Dell'Acqua, F., Gamba, P., and Pulina, G. (2004). A comparison of fuzzy and neuro-fuzzy data fusion for flooded area mapping using SAR images. *International Journal of Remote Sensing* 25 (20): 4425–4430.

Arnold, C.L. and Gibbons, C.J. (1996). Impervious surface coverage: the emergence of a key environmental indicator. *Journal of the American Planning Association* 62: 243–258.

Atkinson, P.M. and Tatnall, A.R.L. (1997). Introduction: neural networks in remote sensing. *International Journal of Remote Sensing* 18: 699–709.

Bauer, M.E., Heinert, N.J., Doyle, J.K., and Yuan, F. (2004). Impervious surface mapping and change monitoring using Landsat remote sensing. *ASPRS Annual Conference*, Denver, CO.

Bhandari, S., Phinn, S., and Gill, T. (2012). Preparing Landsat image time series (LITS) for monitoring changes in vegetation phenology in Queensland, Australia. *Remote Sensing* 4 (6): 1856–1886.

Chormanski, J., Voorde, T.V.D., Roeck, T.D. et al. (2008). Improving distributed runoff prediction in urbanized catchments with remote sensing based estimates of impervious surface cover. *Sensors* 8: 910–932.

Civico, D.L. and Hurd, J.D. (1997). Impervious surface mapping for the state of Connecticut. *Proceedings of ASPRS/ACSM Annual Convention*, Seattle, Washington (7–10 April 1997), 3: 124–135.

Corsini, G., Diani, M., Grasso, R. et al. (2003). Radial basis function and multilayer perceptron neural networks for sea water optically active parameter estimation in case II waters: a comparison. *International Journal of Remote Sensing* 24 (20): 3917–3932.

Deng, C. and Wu, C. (2012). BCI: a biophysical composition index for remote sensing of urban environments. *Remote Sensing of Environment* 127: 247–259.

Du, W., Inoue, K., and Urahama, K. (2005). Robust kernel fuzzy clustering. In: *Fuzzy Systems and Knowledge Discovery*, 454–461. Berlin: Springer.

Eastman, J.R. (2006). *IDRISI Andes: Guide to GIS and Image Processing*. Worcester, MA: Clark University.

Filippi, A.M. and Jensen, J.R. (2006). Fuzzy learning vector quantization for hyperspectral coastal vegetation classification. *Remote Sensing of Environment* 100 (4): 512–530.

Fisher, P.F. and Pathirana, S. (1990). The evaluation of fuzzy membership of land cover classes in the suburban zone. *Remote Sensing of Environment* 34 (2): 121–132.

Flanagan, M. and Civco, D.L. (2001). Subpixel impervious surface mapping. *2001 ASPRS Annual Convention*, St. Louis, MO.

Foody, G.M. (1996). Approaches for the production and evaluation of fuzzy land cover classifications from remotely-sensed data. *International Journal of Remote Sensing* 17 (7): 1317–1340.

Foody, G.M. (1998). Sharpening fuzzy classification output to refine the representation of sub-pixel land cover distribution. *International Journal of Remote Sensing* 19 (13): 2593–2599.

Foody, G.M. and Cox, D.P. (1994). Sub-pixel land cover composition estimation using a linear mixture model and fuzzy membership functions. *International Journal of Remote Sensing* 15 (3): 619–631.

Foody, G.M., Lucas, R.M., Curran, P.J., and Honzak, M. (1997). Non-linear mixture modelling without end-members using an artificial neural network. *International Journal of Remote Sensing* 18: 937–953.

Fu, T.C. (2011). A review on time series data mining. *Engineering Applications of Artificial Intelligence* 24 (1): 164–181.

Gao, F., de Colstoun, E.B., Ma, R. et al. (2012). Mapping impervious surface expansion using medium-resolution satellite image time series: a case study in the Yangtze River Delta, China. *International Journal of Remote Sensing* 33 (24): 7609–7628.

Ghosh, A., Pal, N.R., and Das, J. (2006). A fuzzy rule based approach to cloud cover estimation. *Remote Sensing of Environment* 100 (4): 531–549.

Gilabert, M.A., Garcia-Haro, F.J., and Meli, J. (2000). A mixture modeling approach to estimate vegetation parameters for heterogeneous canopies in remote sensing. *Remote Sensing of Environment* 72 (3): 328–345.

Haglund, A. (2000). *Towards Soft Classification of Satellite Data: A Case Study Based Upon Resurs MSU-SK Satellite Data and Land Cover Classification Within the Baltic Sea Region*. Project report. Stockholm, Sweden: Royal Institute of Technology, Department of Geodesy and Photogrammetry.

Hsieh, P.-F., Lee, L.C., and Chen, N.-Y. (2001). Effect of spatial resolution on classification errors of pure and mixed pixels in remote sensing. *IEEE Transactions on Geoscience and Remote Sensing* 39: 2657–2663.

Hu, X. and Weng, Q. (2009). Estimating impervious surfaces from medium spatial resolution imagery using the self-organizing map and multi-layer perceptron neural networks. *Remote Sensing of Environment* 113 (10): 2089–2102.

Hu, X. and Weng, Q. (2011). Estimating impervious surfaces from medium spatial resolution imagery: a comparison between fuzzy classification and LSMA. *International Journal of Remote Sensing* 32 (20): 5645–5663.

Hu, X. and Weng, Q. (2013). Extraction of impervious surfaces from hyperspectral imagery: linear vs. nonlinear methods. In: *Remote Sensing of Natural Resources* (eds. G.X. Wang and Q. Weng), 141–153. Boca Raton, FL: CRC Press.

Ito, Y. (1998). Polarimetric SAR data classification using competitive neural networks. *International Journal of Remote Sensing* 19 (14): 2665–2684.

Ito, Y. and Omatu, S. (1997). Category classification method using a self-organizing neural network. *International Journal of Remote Sensing* 18 (4): 829–845.

Ji, C.Y. (2000). Land-use classification of remotely sensed data using Kohonen self-organizing feature map neural networks. *Photogrammetric Engineering & Remote Sensing* 66: 1451–1460.

Kavzoglu, T. and Mather, P.M. (2003). The use of backpropagating artificial neural networks in land cover classification. *International Journal of Remote Sensing* 24 (23): 4907–4938.

Lee, S. and Lathrop, R.G. (2002). Sub-pixel estimation of urban land cover intensity using fuzzy c-means clustering. *2002 ASPRS Annual Convention*, Washington, DC.

Lee, S. and Lathrop, R.G. (2005). Sub-pixel estimation of urban land cover components with linear mixture model analysis and Landsat thematic mapper imagery. *International Journal of Remote Sensing* 26 (22): 4885–4905.

Lee, S. and Lathrop, R.G. (2006). Subpixel analysis of Landsat ETM+ using self-organizing map (SOM) neural networks for urban land cover characterization. *IEEE Transactions on Geoscience and Remote Sensing* 44 (6): 1642–1654.

Li, X., Gong, P., and Liang, L. (2015). A 30-year (1984–2013) record of annual urban dynamics of Beijing City derived from Landsat data. *Remote Sensing of Environment* 166: 78–90.

Li, X. and Yeh, A.G.-O. (2002). Neural-network-based cellular automata for simulating multiple land use changes using GIS. *International Journal of Geographical Information Science* 16 (4): 323–343.

Liao, T.W. (2005). Clustering of time series data – a survey. *Pattern Recognition* 38 (11): 1857–1874.

Lu, D. and Weng, Q. (2006). Spectral mixture analysis of ASTER images for examining the relationship between urban thermal features and biophysical descriptors in Indianapolis, Indiana, USA. *Remote Sensing of Environment* 104 (2): 157–167.

Mohan, B.K., Madhavan, B.B., and Gupta, U.M.D. (2000). Integration of IRS-1A L2 data by fuzzy logic approaches for land use classification. *International Journal of Remote Sensing* 21 (8): 1709–1723.

Okeke, F. and Karnieli, A. (2006a). Methods for fuzzy classification and accuracy assessment of historical aerial photographs for vegetation change analyses. Part I: algorithm development. *International Journal of Remote Sensing* 27 (1/2): 153–176.

Okeke, F. and Karnieli, A. (2006b). Methods for fuzzy classification and accuracy assessment of historical aerial photographs for vegetation change analyses. Part II: practical application. *International Journal of Remote Sensing* 27 (9): 1825–1838.

Ray, T.W. and Murray, B.C. (1996). Nonlinear spectral mixing in desert vegetation. *Remote Sensing of Environment* 55: 59–64.

Roberts, D.A., Gardner, M., Church, R. et al. (1998). Mapping chaparral in the Santa Monica Mountains using multiple endmember spectral mixture models. *Remote Sensing of Environment* 65: 267–279.

Roberts, D.A., Smith, M.O., and Adams, J.B. (1993). Green vegetation, nonphotosynthetic vegetation, and soils in AVIRIS data. *Remote Sensing of Environment* 44 (2–3): 255–269.

Schiller, H. and Doerffer, R. (1999). Neural network for emulation of an inverse model operational derivation of Case II water properties from MERIS data. *International Journal of Remote Sensing* 20 (9): 1735–1746.

Schneider, A. (2012). Monitoring land cover change in urban and peri-urban areas using dense time stacks of Landsat satellite data and a data mining approach. *Remote Sensing of Environment* 124: 689–704.

Schueler, T. (1994). The importance of imperviousness. *Watershed Protection Techniques* 1 (3): 100–111.

Sexton, J.O., Urban, D.L., Donohue, M.J., and Song, C. (2013). Long-term land cover dynamics by multi-temporal classification across the Landsat-5 record. *Remote Sensing of Environment* 128: 246–258.

Singer, R.B. and McCord, T.B. (1979). Mars: large scale mixing of bright and dark surface materials and implications for analysis of spectral reflectance. In: *Proceedings of 10th Lunar and Planetary Science Conference*, 1835–1848. Washington, DC: American Geophysical Union.

Sleavin, W.J., Civco, D.L., Prisloe, S., and Giannotti, L. (2000). Measuring impervious surfaces for non-point source pollution modeling. *Proceedings of 2000 ASPRS Annual Convention*, Washington, DC (22–26 May 2000).

Small, C. (2001). Estimation of urban vegetation abundance by spectral mixture analysis. *International Journal of Remote Sensing* 22 (7): 1305–1334.

Stocker, J. (1998). *Methods for Measuring and Estimating Impervious Surface Coverage*. NEMO Technical Paper No.3. University of Connecticut, Haddam Cooperative Extension Center.

Tang, J., Wang, L., and Myint, S.W. (2007). Improving urban classification through fuzzy supervised classification and spectral mixture analysis. *International Journal of Remote Sensing* 28 (18): 4047–4063.

Tapia, R., Stein, A., and Bijker, W. (2005). Optimization of sampling schemes for vegetation mapping using fuzzy classification. *Remote Sensing of Environment* 99 (4): 425–433.

Townsend, P.A. (2000). A quantitative fuzzy approach to assess mapped vegetation classifications for ecological applications. *Remote Sensing of Environment* 72 (3): 253–267.

Van de Voorde, T., Jacquet, W., and Canters, F. (2011). Mapping form and function in urban areas: an approach based on urban metrics and continuous impervious surface data. *Landscape and Urban Planning* 102 (3): 143–155.

Wang, F. (1990). Fuzzy supervised classification of remote-sensing images. *IEEE Transactions on Geoscience and Remote Sensing* 28 (2): 194–201.

Weng, Q. (2012). Remote sensing of impervious surfaces in the urban areas: requirements, methods, and trends. *Remote Sensing of Environment* 117 (2): 34–49.

Weng, Q. and Hu, X. (2008). Medium spatial resolution satellite imagery for estimating and mapping urban impervious surfaces using LSMA and ANN. *IEEE Transactions on Geoscience and Remote Sensing* 46 (8): 2397–2406.

Weng, Q., Hu, X., and Lu, D. (2008). Extracting impervious surface from medium spatial resolution multispectral and hyperspectral imagery: a comparison. *International Journal of Remote Sensing* 29 (11): 3209–3232.

Weng, Q. and Lu, D. (2008). A sub-pixel analysis of urbanization effect on land surface temperature and its interplay with impervious surface and vegetation coverage in Indianapolis, United States. *International Journal of Applied Earth Observation and Geoinformation* 10 (1): 68–83.

Wu, C. and Murray, A.T. (2003). Estimating impervious surface distribution by spectral mixture analysis. *Remote Sensing of Environment* 84: 493–505.

Yang, L., George, X., Klaver, J.M., and Deal, B. (2003a). Urban land-cover change detection through sub-pixel imperviousness mapping using remotely sensed data. *Photogrammetric Engineering and Remote Sensing* 69 (9): 1003–1010.

Yang, L., Huang, C., Homer, C.G. et al. (2003b). An approach for mapping large-area impervious surfaces: synergistic use of Landsat-7 ETM+ and high spatial resolution imagery. *Canadian Journal of Remote Sensing* 29 (2): 230–240.

Yuan, F. and Bauer, M.E. (2007). Comparison of impervious surface area and normalized difference vegetation index as indicators of surface urban heat island effects in Landsat imagery. *Remote Sensing of Environment* 106 (3): 375–386.

Zhang, D.Q. and Chen, S.C. (2003). Kernel-based fuzzy and possibilistic c-means clustering. *Proceedings of the International Conference Artificial Neural Network*, Munich, Germany (June 2003), pp. 122–125.

Zhang, J. and Foody, G.M. (1998). A fuzzy classification of sub-urban land cover from remotely sensed imagery. *International Journal of Remote Sensing* 19 (14): 2721–2738.

Zhang, J. and Foody, G.M. (2001). Fully-fuzzy supervised classification of sub-urban land cover from remotely sensed imagery: statistical and artificial neural network approaches. *International Journal of Remote Sensing* 22 (4): 615–628.

Zhang, L. and Weng, Q. (2016a). Annual dynamics of impervious surface in the Pearl River Delta, China, from 1988 to 2013, using time series Landsat data. *ISPRS Journal of Photogrammetry and Remote Sensing* 113 (3): 86–96.

Zhang, L. and Weng, Q. (2016b). Assessment of urban growth in the Pearl River Delta, China, using time series Landsat imagery, Chapter 4. In: *Remote Sensing for Sustainability* (ed. Q. Weng), 45–60. Boca Raton, FL: CRC Press, Taylor & Francis Group.

Zhu, Z. and Woodcock, C.E. (2014). Continuous change detection and classification of land cover using all available Landsat data. *Remote Sensing of Environment* 144 (25): 152–171.

5

Land Surface Temperature Data Generation

5.1 Introduction

Land surface temperature (LST), as frequently referred to as the skin temperature of the Earth's surface and as derived from remotely sensed thermal infrared (TIR) data, is a key parameter in analyzing and modeling the surface energy balance (Trenberth 1992; Anderson et al. 2011), surface moisture and evapotranspiration (Gillies et al. 1997; Moran 2004; Carlson 2007), and climate change of various spatial scales (Jin et al. 2005; Weng 2009). LST and its spatial–temporal variations have long been foci of studies on surface urban heat island (UHI) (Oke 1982; Streutker 2003; Rajasekar and Weng 2009; Imhoff et al. 2010). Oke (1979) discriminated between the canopy layer UHI and the boundary layer UHI. The canopy layer UHI consists of air between the roughness elements, e.g. buildings and tree canopies, with an upper boundary just below roof level. Therefore, it relates closely to satellite-derived LST, although a precise transfer function between LST and the near ground air temperature is not yet available (Nichol 1994). In addition, LST is useful for the examination of heat-related health issues and the vulnerability of human beings to heat stress (Harlan et al. 2006; Lafortezza et al. 2009) and the outbreak and propagation of vector-borne diseases (Reisen et al. 2004; Liu and Weng 2009; Ruiz et al. 2010). Therefore, estimation of LST and assessing its variations are not only helpful to understand environmental and ecological processes, but also concerned with the well-being of humans.

Current satellite TIR data suitable for studying urban thermal environment or for solving urban environmental and health problems that are characterized by a high spatial variability – such as Landsat Thematic Mapper (TM), Enhanced Thematic Mapper Plus (ETM+), and Advanced Spaceborne Thermal Emission and Reflection Radiometer (ASTER) – have a much coarser temporal resolution than it is needed. Given the long repeat cycle of these satellites, their TIR data are not suited for UHI monitoring. Surface UHI is not only a

Techniques and Methods in Urban Remote Sensing, First Edition. Qihao Weng.
© 2020 by The Institute of Electrical and Electronics Engineers, Inc.
Published 2020 by John Wiley & Sons, Inc.

phenomenon of high spatial variability, but also of high temporal variability. For examining the health implications of UHI, such as heat-related epidemiological studies, routine LST estimation is essential (Liu and Weng 2012). Similarly, to assess the UHI impact on energy usage, LST measurements must match with simulated data of hourly energy consumption from urban buildings (Zhou et al. 2012). Thus, while some of the current satellite-borne TIR sensors can provide LST measurements at a reasonably high spatial resolution, their utilization in urban climate studies is restricted because of low temporal resolution and limited available nighttime image data (Stathopoulou and Cartalis 2009). No single satellite system currently provides TIR data of global coverage that combines both high spatial and temporal resolutions. For a list of major current satellite TIR imaging systems, please refer to Tomlinson et al. (2011). Due to technical constraints, these sensing systems reflect a tradeoff between temporal and spatial resolution such that the systems with high-spatial resolution possess low-temporal resolution, or vice versa.

It is, therefore, highly desirable to develop techniques to derive LST data of high spatial and temporal resolutions from available remotely sensed data. Existing techniques have been named differently, including image merging, image/data fusion, spatial sharpening, downscaling, disaggregation, etc., but basically fall into two categories. While the spatial thermal sharpening techniques aim at downscaling (disaggregating) radiometric surface temperature of a sensor to higher resolutions typically associated with its shorter wavebands (visible and near-infrared) (Liu and Moore 1998; Kustas et al. 2003; Pu et al. 2006; Dominguez et al. 2011), the temporal thermal sharpening techniques are developed to downscale TIR data from a coarser spatial-resolution but higher temporal-resolution sensor (typically associated with geostationary satellites) to generate highly temporally resolved LST diurnal cycles (Gottsche and Olesen 2001; Inamdar et al. 2008; Inamdar and French 2009; Bechtel et al. 2012; Zakšek and Oštir 2012). The former can produce TIR data on the order of 10^1–10^2 m in spatial resolution but is limited by temporal resolution. The latter, on the other hand, can generate TIR data of up to every 15 minutes in temporal resolution, but usually have very coarse spatial resolution (e.g. 1000×1000 m). The inability of existing techniques in producing proper spatial and temporal sampling of LST data for urban climate and environmental studies calls for development of new techniques in TIR data fusion.

Downscaling is the scaling process of converting remote sensing data from a low to a high spatial resolution. Thermal downscaling is also named thermal *sharpening or disaggregating*. Thermal downscaling typically requires preserving the radiometry of original TIR radiance or LST data for subsequent data analysis (Stathopoulou and Cartalis 2009). Therefore, it is important to understand the complexity and heterogeneity of thermal landscapes and key factors causing the spatial variability in LST. Optical and TIR data can provide complementary information about the Earth's surface but, due to instrumental reasons, TIR images are usually collected at coarser spatial resolution than do visible and near-infrared (NIR) bands on the same satellite platform. Various methods of thermal downscaling can be broadly grouped into physical and statistical approaches. Statistical downscaling techniques have been largely developed to disaggregate radiometric surface temperature of a sensor to higher resolutions associated with its shorter wavebands. Physical downscaling uses modulation methods, which take a thermal pixel as a block and distribute its thermal radiance into finer pixels corresponding to its shorter wavebands. Liu and Moore (1998) proposed the Pixel Block Intensity Modulation (PBIM) method, which was quantified by Landsat TM reflective bands, to adjust temperatures within the lower resolution pixel blocks of the TIR band based on topographic variations. Nichol (2009) suggested that the method by Liu and Moore is only suited for use within simple land-cover types where temperature variations are caused mainly by topography, but is not suitable for use in the urban areas where topography is mostly flat. Nichol (2009) proposed an alternative modulation method based on emissivity. Stathopoulou and Cartalis (2009) applied the PBIM method to downscale Advanced Very High-Resolution Radiometer (AVHRR) LST image data to that of TM band 6 by employing different scaling factors (effective emissivity, season-coincident Landsat LST data, or their combination). It is found that the spatial pattern of the downscaled AVHRR LST resembled reasonably well with time-coincident TM LST and the root-mean-square error (RMSE) yielded a range of 4.9–5.3 °C. It is worthy to note that Liu and Pu (2008) compared the modulation method with spectral unmixing of TIR radiance. The former disaggregated TIR radiance by using higher spatial resolution land-cover data as the distribution factor, while the latter employed spectral mixture analysis (SMA) to decompose mixed TIR pixels into multiple isothermal components. However, the spatial details within mixed pixels remained unresolved in the decomposed component temperatures (Gillespie 1992). Moreover, the isothermal assumption that underpins various modulation methods for retrieving component temperature or emissivity may not be valid. This is especially true within urban landscapes where component surfaces are often seen smaller than the Instantaneous Field of View (IFOV) of satellite sensors, resulting in a mixture of different temperature components.

While these thermal downscaling methods provide useful means to improve the resolution of TIR data of a sensor or LST to higher spatial resolutions, they do not enhance the temporal resolution of the sensor simultaneously. Gao et al. (2006) developed a data fusion technique that allows improving spatial resolution and temporal coverage at the same time. The technique was named "Spatial and Temporal Adaptive Reflectance Fusion Model," or for short, STARFM, that blends Landsat and Moderate Resolution Imaging Spectroradiometer (MODIS) data to generate synthetic Landsat-like daily surface reflectance. The basic assumption is that surface reflectance at a predicted date may be estimated by a weighted sum of the spectrally similar neighborhood information from both Landsat and MODIS reflectance at observed dates (close to the predicted date). Essentially, STARFM integrates daily information from MODIS with periodic Landsat data to interpolate surface reflectance at the Landsat resolution of 30 m on a daily basis. This data fusion approach has received a lot of attention lately, because it can provide successful monitoring of seasonal changes in vegetation cover (Gao et al. 2006; Hilker et al. 2009b) and larger changes in land use (Hansen et al. 2008; Potapov et al. 2008). The STARFM approach was later adjusted and revised for specific applications under different conditions. Hilker et al. (2009a) developed Spatial Temporal Adaptive Algorithm for mapping Reflectance Change (STAARCH) model based on the STARFM approach for mapping disturbance events between two input dates. The forest changes are mapped using the dense time series of MODIS imagery. Moreover, the STAARCH allows selecting an optimal MODIS–Landsat image pair from one of two inputs for making a prediction (Hilker et al. 2009a). Zhu et al. (2010) developed an enhanced STARFM (ESTARFM) approach for application in a heterogeneous area by introducing a conversion coefficient to the fusion model, which represented the ratio of change between the MODIS pixels and ETM+ end-members. The ESTARFM approach provides a solution for the heterogeneous (mixed) pixels, but it still cannot accurately predict short-term, transient changes not recorded in any of the bracketing (observed) fine-resolution images (Zhu et al. 2010).

Although STARFM was originally designed to fuse shortwave reflectance fields from MODIS and Landsat to create daily reflectance and vegetation index maps, it appears to hold great utility for high-resolution thermal mapping too (Anderson et al. 2011; Liu and Weng 2012). Anderson et al. (2012) used STARFM to predict evapotranspiration at 30-m resolution and compared it with flux estimation derived from spatially disaggregated ALEXI (Atmosphere-Land Exchange Inverse) model. They found that the prediction yielded errors on the order of 10%. Liu and Weng (2012) applied the STARFM model to simulate ASTER-like land surface reflectance and LST images for Los Angeles for five dates in the five epidemiological weeks in summer 2007 and used the simulated data to assess the environmental conditions of the WNV (West Nile Virus) outbreak. The mean absolute difference between the observed and the simulated surface reflectance was found less than 0.2 and the LST residual less than 1 °C for all images. However, the application of the STARFM and its variants for LST prediction is immature in terms of methodology. Many critical issues have not been solved, especially in terms of thermal landscape heterogeneity, land-cover change, and vegetation phenology. Moreover, LST of a landscape patch may be affected by the surrounding materials seriously (Oke 1982). In this chapter, a new fusion algorithm for TIR data is introduced to predict daily LST at 120-m resolution by blending Landsat TM and MODIS LST data, i.e. Spatiotemporal *A*daptive *D*ata *F*usion *A*lgorithm for *T*emperature mapping (SADFAT) (Weng et al. 2014).

Further, to reconstruct a long-term LST data set for a specific region, it is necessary to develop an algorithm that can transcend the techniques of thermal sharpening and LST interpolation under cloudy conditions to produce LSTs of both high spatial and temporal resolutions. The development of the spatial–temporal fusion algorithms (Gao et al. 2006; Liu and Weng 2012; Huang et al. 2013; Weng et al. 2014; Wu et al. 2015) has provided a foundation for long-term LST reconstruction. Despite all these progresses, existing fusion algorithms are still subject to several key limitations and cannot directly be used for generating a consistent, long-term LST data set. The first limitation is that LSTs under cloudy conditions cannot be interpolated if input images are cloud contaminated, which is common for areas experiencing frequent cloud coverage. In addition, uncertainties remain in selecting the best imagery pairs as the inputs for predictions. Thus, the accuracy of the data fusion algorithms (e.g. STARFM and SADFAT) for deriving LSTs must be fully assessed. The third limitation is that these algorithms are not effective in generating LSTs for areas where disturbance events, such as deforestation, forest degradation, desertification, and other land-cover and land-use changes, occur (Hilker et al. 2009b; Julien and Sobrino 2012), since the corresponding LST variations are not stationary over time. Finally, the interannual trend within LST variations cannot be captured by these data fusion algorithms. The last issue does not pose a big challenge for predicting LSTs over a short time period; however, the maximum annual trend change may reach as high as 0.34 K (Julien and Sobrino 2012). Therefore, it is highly desirable to develop a new technique that can overcome the limitations discussed above and to generate consistent, long-term LSTs.

Consistent time series LSTs are of prime importance for assessing climate change of different scales (Jin and Dickinson 2002; Jin et al. 2005; Sun et al. 2006). A long-term LST data set of high quality can benefit analyses of impact of urbanization on thermal characteristics. Therefore, in this chapter, an algorithm will be introduced that allows reconstructing historical LST measurements at daily interval based solely on irregularly spaced Landsat imagery. Instead of blending data among different satellite sensors, this algorithm takes advantage of unevenly distributed time series Landsat imagery and goes through the modules of Data filtEr, temporaL segmentation, periodic and Trend modeling, and GAussian (DELTA).

5.2 Generating Daily Land Surface Temperature by Data Fusion

5.2.1 Study Area and Data

The study area covers the majority of the Los Angeles County, California, USA, a small portion of Simi Valley in Ventura County and a small part of Orange County (Figure 5.1). This area consists of various land covers, including water,

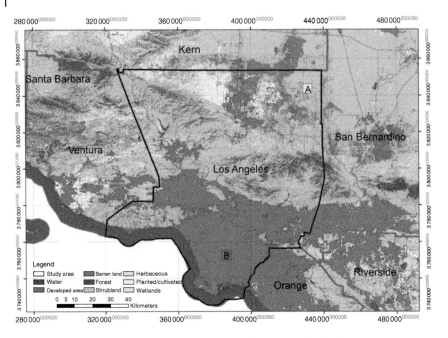

Figure 5.1 The study area Los Angeles County, California, USA. The background image shows land-cover types derived from the National Land-Cover Database (NLCD) 2006. *Source:* Weng et al. (2014). Reproduced with permission of Elsevier.

developed urban, barren land, forest, shrub land, herbaceous, planted/cultivated, and wetlands as identified by the 2006 The National Land-Cover Database (NLCD) database. The primary mountain ranges are Santa Monica Mountains and the San Gabriel Mountains in the southwestern and southeastern part of Los Angeles County, respectively. The valleys are largely the population centers and compose a large percentage of the urban areas. The area possesses a Subtropical–Mediterranean climate with a dry summers and moist winters. The average high air temperature is 29 °C in August and 20 °C in January based on the weather records from the Downtown University of Southern California campus. According to a study by the UHI group, the urbanization of city has negatively affected the urban community, such as the increased energy use, impaired air quality, and the aggravation of heat-related and respiratory illness (https://heatisland.lbl.gov/coolscience/urban-heat-islands). Thus, the generation of daily LST data will be conducive to monitoring the dynamic patterns of LST and to investigating the UHI effect on energy use and environmental and public health.

Landsat TM images of Path 41, Row 36, acquired on 24 June, 10 July, 27 August, 28 September, 14 October, 30 October, and 15 November 2005 were used as reference data to estimate daily LST images at 120-m resolution (Table 5.1). The corresponding daily MODIS LST (MOD11A1) and reflectance

Table 5.1 Characteristics of Landsat and MODIS land surface temperature (LST) data used in the study.

Date	Landsat overpass time (H : M)	MODIS overpass time (H : M)	Landsat minimum	Landsat median	Landsat maximum	MODIS minimum	MODIS median	MODIS maximum	Correlation
24 June 2005	10 : 15	10 : 06	259.51	313.60	333.75	250.72	310.54	319.24	0.83
10 July 2005	10 : 16	10 : 06	283.38	309.27	331.37	250.00	306.46	317.80	0.67
27 August 2005	10 : 16	10 : 06	278.19	318.53	341.76	250.02	315.04	322.02	0.77
28 September 2005	10 : 16	10 : 06	263.60	310.56	329.32	252.66	307.48	314.44	0.79
14 October 2005	10 : 16	10 : 06	268.78	310.79	334.94	254.20	308.46	319.38	0.63
30 October 2005	10 : 16	10 : 06	257.50	299.33	320.01	250.02	297.62	306.48	0.66
15 November 2005	10 : 16	10 : 06	261.07	299.12	319.16	250.16	298.00	304.82	0.75

Note: The overpassing time recorded is the local time in the format of hour and minute, and the unit for the minimum, median, and maximum LST values is Kelvin (K).

data (MOD09GA) were obtained through the Level 1 and Atmosphere Archive and Distribution System website data portal available at https://ladsweb. modaps.eosdis.nasa.gov. These MODIS LST products were selected and used as another set of references for the daily LST mapping. Before the implementation of the fusion method, both the Landsat TM and MODIS data were registered to the same coordinate system and resampled to the same spatial resolution (120 m). A shared cloud mask was created for both Landsat and MODIS images to remove cloudy pixels from the computation.

5.2.2   LST Retrieval from Landsat TM Imagery

The accuracy of LST computation for Landsat TM image is crucial in implementing the proposed algorithm. Sobrino et al. (2004) investigated and compared three methods to derive LST over an agricultural area in Spain with land surface emissivity (LSE) estimated from the visible and near infrared bands. The retrieved results disclosed that the error of the single channel method developed by Jimenez-Munoz and Sobrino (2003) in retrieving LST is below 1 K. Therefore, the generalized single channel method was utilized for LST estimation in this study.

The main characteristic of the generalized single channel method is that in situ radio soundings or effective mean atmospheric temperature values are not required compared to other single channel methods. More importantly, specific atmospheric functions for Landsat TM 6 were obtained using the Institute for Genomic Research (TIGR) database and simulations from MODTRAN 3.5. The following equations show how to implement the single channel method for TM TIR data.

$$T_s = \gamma\left[\varepsilon^{-1}\left(\psi_1 L_{sensor} + \psi_2\right) + \psi_3\right] + \delta \tag{5.1}$$

with

$$\gamma = \left(\frac{c_2}{T_{sensor}^2}\left[\frac{\lambda^4}{c_1}L_{sensor} + \lambda^{-1}\right]\right)^{-1} \tag{5.2a}$$

$$\delta = -\gamma L_{sensor} + T_{sensor} \tag{5.2b}$$

where L_{sensor} is the at-satellite radiance, T_{sensor} is the at-sensor brightness temperature, λ is the effective wavelength for TM sensor (11.475 μm), and c_1 and c_2 are the constants. The atmospheric functions are defined based on the water vapor content:

$$\begin{bmatrix} \psi_1 \\ \psi_2 \\ \psi_3 \end{bmatrix} = \begin{bmatrix} 0.147\,14 & -0.155\,83 & 1.123\,4 \\ -1.183\,6 & -0.376\,07 & -0.528\,94 \\ -0.045\,54 & 1.8719 & -0.390\,71 \end{bmatrix} \begin{bmatrix} \omega^2 \\ \omega \\ 1 \end{bmatrix} \tag{5.3}$$

where ω is the water vapor content, which can be obtained from the satellite images or in situ device (Jimenez-Munoz and Sobrino 2003). Since MODIS can provide data of water vapor content, this research adopted the ratio method using the atmospheric water channels (Kaufman and Gao 1992) to provide the input for the atmospheric function.

Another parameter assumed to be known in the single channel method is LSE. According to Sobrino and Raissouni (2000), it is possible to acquire LSE data from normalized difference vegetation index (NDVI) values for the areas comprise soil, vegetation, and mixed soil/vegetation components. The problem of the NDVI threshold method, as pointed out by Sobrino et al. (2008), is the lack of continuity emissivity values at soil and vegetation thresholds. Therefore, a simplified version of the NDVI threshold method was employed in this study for the derivation of LSE data. The values of NDVI for soil and vegetation were obtained from the NDVI histogram of each image. However, the NDVI threshold method is not suitable to derive emissivity values for the urban areas. Therefore, the emissivity data product from the ASTER Global Emissivity Database 3.0 (Hulley et al. 2008; Hulley and Hook 2009) was utilized for the urban areas identified by the NLCD 2006. The mean emissivity value of Bands 13 and 14 was selected for use to match the spectral range of TM Band 6. The original emissivity product was delivered in 1° by 1° tiles with geographic projection (WGS84) at 100 m spatial resolution. It was later mosaicked and resampled to 120 m using Universal Transverse Mercator (UTM) projection system to match the Landsat LST data. It should also be noted that the study area contains open water and inland lakes, which are not suitable for the direct use of the threshold method either. As a result, water areas were first extracted based on their spectral characteristics because water bodies have a low value for the infrared reflectance and NDVI (Jensen 2005). The effective emissivity value was calculated for water bodies by using the ASTER spectral library 2.0 and the TM spectral response filter. The Kirchhoff's law was applied to convert hemispherical reflectance to surface emissivity (Nicodemus 1965).

5.2.3 Theoretical Basis of the TIR Data Fusion Model

The STARFM algorithm was initially designed to predict surface reflectance and is based on the assumption that MODIS and Landsat surface reflectance are highly consistent (Gao et al. 2006; Masek et al. 2006). For homogeneous pixels, as long as this assumption holds true for the thermal regime, the STARFM procedure can be applied to TIR image data either at radiance or at LST level. Since our improved method must handle the heterogeneous characteristics of LSTs and radiance can be spatially aggregated using linear form, this research focuses on data fusion at the radiance level. For homogenous pixels, remotely sensed TIR data from different sensors at a close acquisition time should be comparable and correlated after radiometric calibration, geometric rectification, and atmospheric correction. Nevertheless, such factors as acquisition time, bandwidth, orbit parameters, geo-location errors, effective pixel coverage, and spectral response function can introduce some system biases into the subsequent analysis. Assuming that MODIS radiance images have been resampled to the same spatial resolution of Landsat radiance images, the following discussion focuses on how to derive daily radiance images at 120-m resolution. For the convenience, the MODIS pixel will be described as M pixel, and the Landsat pixel will be simply described as L pixel.

For a homogeneous M pixel covered only by one land-cover type, the radiance difference between the resampled M pixel and the L pixel results from the system biases and should be stable in a short period. Thus, the relationship between the observations of radiance from the two sensors for the homogeneous pixels can be expressed as

$$R_L(x,y,t) = a * R_M(x,y,t) + b \tag{5.4}$$

where R defines the radiance, (x, y) represents a given location, t is the acquisition date, and a, b are the coefficients for relative adjustment needed for the Landsat and MODIS radiance pixels. Therefore, suppose there is one pair of Landsat and MODIS image acquired at t_0, and another MODIS image acquired at t_p, and the land cover and sensor calibration does not change during the period between t_0 and t_p, then Eq. (5.1) can have two instances:

$$R_L(x,y,t_0) = a * R_M(x,y,t_0) + b \tag{5.5}$$

$$R_L(x,y,t_p) = a * R_M(x,y,t_p) + b \tag{5.6}$$

There is a possibility that the relationship between MODIS and Landsat LST may vary from day to day depends on weather and surface moisture conditions. Table 5.1 shows that the relationship remained stable as indicated by correlation coefficient (CC) (Masek et al. 2006). Thus, the inference of Eqs. (5.5) and (5.6) is reasonable.

From Eqs. (5.5) and (5.6), the following expression can be derived:

$$R_L(x,y,t_p) = R_L(x,y,t_0) + a * \left[R_M(x,y,t_p) - R_M(x,y,t_0) \right] \tag{5.7}$$

Equation (5.7) states that for a homogenous pixel at t_p, its L radiance equals to the sum of L radiance at t_0 and the scaled difference of M radiance between t_0 and t_p. The coefficient a can be calculated from the system biases given the stability of the two sensors. If two pairs of L image and M image can be obtained, the coefficients can also be determined by the regression of the L radiance with M radiance at t_1 and t_2. However, it should be noted that Eq. (5.7) is only valid for the non-changing surfaces if the relationship between MODIS and Landsat LST remains stable.

However, a large proportion of the pixels from medium- and coarse-resolution imagery contains more than one land-cover type, i.e. they are mixed pixels. According to linear SMA (LSMA) theory, the radiance of a mixed pixel can be defined as

$$R = \sum_{i=1}^{N} f_i R_i + \varepsilon \tag{5.8}$$

where R represents the radiance received by the satellite sensor, N is the number of end-member, f_i denotes the fraction of each land-cover component, and ε is the residual. Suppose that each L pixel can be regard as one end-member of an M pixel, then the radiance of the M pixel at t_1 and t_2 can be described as Eqs. (5.9) and (5.10), according to Eqs. (5.4) and (5.8):

$$R_M(t_1) = \sum_{i=1}^{N} f_i \left(\frac{1}{a} R_L(t_1) - \frac{b}{a} \right) \tag{5.9}$$

$$R_M(t_2) = \sum_{i=1}^{N} f_i \left(\frac{1}{a} R_L(t_2) - \frac{b}{a} \right) \tag{5.10}$$

It should be noted that in Eqs. (5.9) and (5.10), the coefficients a and b remain stable and the fraction of each L pixel end-member does not vary, either. Therefore, the radiance change of an M pixel from t_1 to t_2 can be computed as

$$R_M(t_2) - R_M(t_1) = \frac{1}{a} \sum_{i=1}^{N} f_i \left(R_{iL}(t_2) - R_{iL}(t_1) \right) \tag{5.11}$$

The temporal variability of LST shows a strong diurnality (Sabins 1997) and seasonality (Weng et al. 2008). The LST seasonal change can be modeled using the annual temperature cycle (ATC) approximated by a sinusoidal function (Bechtel 2012):

$$LST = MAST + YAST * \sin(w * d + \theta) \tag{5.12}$$

where $MAST$ is the mean annual surface temperature, $YAST$ is the yearly amplitude surface temperature, w is the angular frequency, d is the day of year (DOY) relative to the equinox, and θ is the phase shift. Since spectral radiance is related to LST by the Plank's law, the radiance change of an L pixel from time t_1 to t_2 can be quantified as

$$R_{iL}(t_2) - R_{iL}(t_1) = 2c \cos\left(\theta_i + w \frac{d_1 + d_2}{2} \right) \sin w \frac{d_2 - d_1}{2} = C \cos\left(\theta_i + w\overline{d} \right) \tag{5.13}$$

where θ is the phase shift, or heat lag, c is the amplitude of the radiance variation, C is the constant, and $\bar{d}$ is the mean acquisition date, d_1 and d_2 are the parameters input to the algorithm. Incorporating the ATC model, Eq. (5.11) can be rewritten as

$$R_M\left(t_2\right) - R_M\left(t_1\right) = \frac{2c \sin w\left(\left(d_2 - d_1\right)/2\right)}{a} \sum_{i=1}^{N} f_i \cos\left(\theta_i + w\bar{d}\right) = \frac{C}{a} \sum_{i=1}^{N} f_i \cos\left(\theta_i + w\bar{d}\right)$$

(5.14)

If the radiances of the kth L pixel at date t_1 and t_2 are known, Eq. (5.12) has an instance as

$$R_{kL}\left(t_2\right) - R_{kL}\left(t_1\right) = 2c \cos\left(\theta_k + w\bar{d}\right) \sin w \frac{d_2 - d_1}{2}$$

(5.15)

By combining Eq. (5.14) with Eq. (5.15), Eq. (5.16) can be obtained:

$$\frac{R_{kL}\left(t_2\right) - R_{kL}\left(t_1\right)}{R_M\left(t_2\right) - R_M\left(t_1\right)} = \frac{\cos\left(\theta_k + w\bar{d}\right)}{\frac{1}{a} \sum_{i=1}^{N} f_i \cos\left(\theta_i + w\bar{d}\right)} = h_k$$

(5.16)

Since θ reflects the phase shift of a pixel and is associated with the thermal properties of land surface materials, it can be regarded as constant as long as the land cover does not change in the observational period. Therefore, the ratio of the radiance change of kth L pixel to that of the corresponding M pixel is constant for a certain L pixel. Here, h_k is called the conversion coefficient for the purpose of consistency (Zhu et al. 2010).

Based on Eq. (5.16), if one pair of L and M radiance image at t_0 and another M radiance image at t_p are available, the L radiance image at t_p can be predicted using the following formula:

$$R_L\left(x,y,t_p\right) = R_L\left(x,y,t_0\right) + h\left(x,y\right) * \left[R_M\left(x,y,t_p\right) - R_M\left(x,y,t_0\right)\right]$$

(5.17)

Apparently, Eqs. (5.7) and (5.16) only utilize information from a single pixel to infer the L radiance. By introducing additional information from neighboring spectrally similar pixels, the solution to Eq. (5.17) can be determined uniquely, and a moving window (Gao et al. 2006) can be employed to compute the radiance of the central pixel. Therefore, assuming w is the moving window size, the predicted L pixel radiance can be computed as

$$R_L\left(x_{w/2},y_{w/2},t_p\right) = R_L\left(x_{w/2},y_{w/2},t_0\right) + \sum_{i=1}^{N} W_i * h_i * \left[R_M\left(x_i,y_i,t_p\right) - R_M\left(x_i,y_i,t_0\right)\right]$$

(5.18)

where W_i is the weight of a neighboring similar pixel, and N is the number of the spectrally similar pixel. The computed radiance can then be converted to LST using the Planck's law. It deserves to note that the use of ATC parameters ignores the influence of synoptic and surface conditions on the variations of the thermal radiance; however, since Eq. (5.13) has coped with the change of radiance, it is reasonable to assume the stability of the synoptic and surface conditions.

Figure 5.2 illustrates the procedure of SADFAT, which contains five steps. The inputs for the algorithm are two pairs of L and M images at t_1 and t_2, respectively, and one M image at the prediction date t_p. At the first step, all the images should be registered to the same coordinate system and atmospherically calibrated and corrected to the surface radiance. Secondly, the two L images are used to search for the spectrally similar pixels according to the predefined principles. The third step is to compute the combined weight for each similar pixel. At the fourth step, the conversion coefficients are determined by the regression analysis. Finally, the M images at t_p and the calculated

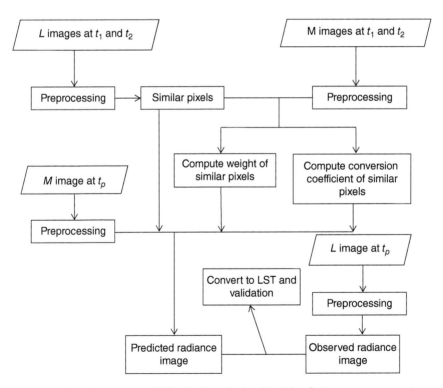

Figure 5.2 The Spatiotemporal Adaptive Data Fusion Algorithm for Temperature mapping (SADFAT) procedure for predicting the daily land surface temperature (LST) image and the validation. *Source:* Weng et al. (2014). Reproduced with permission of Elsevier.

conversion coefficients are employed to obtain the predicted L radiance image. The computed radiance image will then be converted to LST using the Planck's law. The predicted LST images are validated against the real TM image data using the coefficient of correlation, mean difference, and mean absolute difference values.

5.2.4 Implementation of the Spatiotemporal Adaptive Data Fusion Algorithm for Temperature Mapping

The implementation of Spatiotemporal Adaptive Data Fusion Algorithm for Temperature Mapping (SADFAT) requires preprocessing of Landsat and MODIS data, selection of spectrally similar pixels, and computation of the conversion coefficient. The calculation of weights for spectrally similar pixels involves weighing the contribution of the neighboring pixels to the computation of a central pixel. The conversion coefficient reflects the combined changes of LST for Landsat and MODIS from one date to another. Using a local moving window, neighboring spectrally similar pixels were included for the computation of the LST for a central pixel with the temporal weights of the two dates. Below are the detailed descriptions for the implementation of SADFAT.

5.2.4.1 Data Preprocessing and Selection of Spectrally Similar Pixels

Before the application of SADFAT, both MODIS and Landsat data need to be preprocessed geometrically to the same pixel size and radiometrically to ensure the data accuracy. In this study, Landsat level 1T product was employed and atmospherically calibrated (Bands 1–5 and 7) using Landsat Ecosystem Disturbance Adaptive Processing System (LEDAPS) (Masek et al. 2006). The LST retrieval for Landsat TM was based on the single channel method (Jimenez-Munoz and Sobrino 2003). MODIS daily LST and surface reflectance data were re-projected to the Landsat coordinate system using the MODIS Reprojection Tools (MRTs).

Similar pixels provide needed spectral and spatial information for the LST computation of a central pixel within the local moving window. In the original STARFM and the enhanced STARFM models (Gao et al. 2006; Zhu et al. 2010), two methods were used to obtain spectrally similar pixels. The unsupervised classification can be applied to Landsat images to identify the pixels belonging to the same cluster as the central pixel. Another method is to define a threshold of difference between the central pixel and the neighboring pixels, through which similar pixels can be identified. Differences among the spectrally similar pixels were computed based on the standard deviation of fine-resolution images and the number of classes used (Gao et al. 2006). A larger number of classes mean a stricter condition for selecting spectrally similar pixels. Since this study utilized multiple bands, i.e. Landsat red, NIR, and TIR bands, to search for similar pixels, it is reasonable to adopt the threshold method. The threshold method can avoid global misclassifications which would have an adverse impact on the calculation (Gao et al. 2006). In searching for the similar pixels, both Landsat images were used to provide the effective similar pixels. If only one Landsat image was used for the identification of similar pixels, it may result in inconsistency with the actual surface conditions, e.g. land cover may change from one date to another date. In addition, it is possible that a certain pixel did not have any spectrally similar pixel. In such case, the weight for the central pixel would be set to 1. It is also possible that all of the similar pixels cannot provide better prediction than the central pixel. Therefore, additional filtering was applied to the selected candidates to remove the poor quality pixels (Gao et al. 2006). First, according to the QA layer in Landsat and MODIS data (surface reflectance and LST product), poor quality data were removed. Second, neighboring similar pixels were filtered out if they cannot provide better reflective and thermal information than the central pixel.

5.2.4.2 Computation of Weight

The weight defined the contribution of neighboring pixels to the calculation of a central pixel. It was determined by the location of the similar pixels and the reflective and thermal similarity between the fine and coarse resolution data. Higher reflective and thermal similarity and shorter distance between a central and a neighboring pixel yielded a higher weight in the computation. This research computed the CC to determine the reflective and thermal similarity between Landsat and the corresponding MODIS pixels. The CC can be computed as

$$R_i = \frac{E\left[\left(L_i - E\left(L_i\right)\right)\left(M_i - E\left(M_i\right)\right)\right]}{\sqrt{D\left(L_i\right)}\sqrt{D\left(M_i\right)}} \tag{5.19a}$$

with

$$L_i = \left[L_i\left(x,y,t_1,B_1\right),\ldots,L_i\left(x,y,t_1,B_n\right),\ L_i\left(x,y,t_2,B_1\right),\ldots,L_i\left(x,y,t_2,B_n\right)\right] \tag{5.19b}$$

$$M_i = \left[M_i\left(x,y,t_1,B_1\right),\ldots,M_i\left(x,y,t_1,B_n\right),\ M_i\left(x,y,t_2,B_1\right),\ldots,M_i\left(x,y,t_2,B_n\right)\right] \tag{5.19c}$$

where R_i is the combined reflective and thermal CC between L and M images for the ith pixel. L_i and M_i are the collection of similar pixels from Band 3, Band 4, and Band 6 for Landsat and its corresponding bands for MODIS at t_1 and t_2. E and D are the values of expectation and variance, respectively. The value of R ranged from -1 to 1, and a higher R meant a higher reflective and thermal similarity. The reasons to include both the reflective and thermal bands of different dates were (i) LST varied with land-cover type (Weng et al. 2004). The inclusion of additional spectral bands ensured that the selection and computation only occurred for the same land cover; (ii) both pairs of L and M images were contained in the vectors to provide a more accurate calculation of similarity given that land cover may change over the time.

The location of similar pixels also impacted their contributions to a central pixel. The distance between the central pixel and the neighboring ith pixel can be calculated as

$$d_i = 1 + \frac{\sqrt{\left(x_{w/2} - x_i\right)^2 + \left(y_{w/2} - y_i\right)^2}}{\left(w/2\right)} \tag{5.20}$$

where d_i is the computed distance and ranges from 1 to $2^{0.5}$, $(x_{w/2}, y_{w/2})$ is the spatial position of the central pixel, and (x_i, y_i) is the position of the ith neighboring similar pixel. With the location of the similar pixels and the reflective and thermal similarity, a combined weight, CW, can be calculated as

$$CW = \left(1 - R_i\right) * d_i \tag{5.21}$$

A spectrally similar pixel with a larger CW value would have a less weight in the calculation. Therefore, the combined weight needs to be normalized based on the inverse:

$$W_i = \frac{1/CW_i}{\sum_{i=1}^{N} 1/CW_i} \tag{5.22}$$

W_i is the final combined weight for the ith similar pixel. The range of W ranged from 0 to 1, and the total value of each W_i would be 1. When there were p similar pixels among all the similar pixels whose corresponding coarse resolution pixels were pure ($R = 1$), the weight for the similar p pixels would be set to $1/p$ and other pixels to 0. That is to say, only homogeneous pixels were used for the calculation of the weight.

5.2.4.3 Conversion Coefficient and Computation of LST for Central Pixel

The conversion coefficient defines the relationship between L and M radiance changes (Zhu et al. 2010). Since neighboring similar pixels are introduced, it is practical to compute the conversion coefficient using the regression analysis within each local moving window. Regression analysis is reasonable, because, based on the theoretical basis of SADFAT, similar pixels within the same coarse resolution pixel have the same conversion coefficient. Due to the additional spatial filter, it is suitable to calculate the conversion coefficients using all the selected similar pixels rather than using only similar pixels within the same coarse resolution. Considering the potential geometrical error involved in the data preprocessing, we decided to compute the conversion coefficient for all the selected similar pixels. As a special case, if there were no similar pixels or a linear regression model cannot be built with the defined statistical significance, the central pixel can still provide the conversion coefficient although this may introduce some errors.

According to Eq. (5.18), L radiance image at t_p can be predicted based on the weighted scaled M radiance changes and L radiance image either at t_1 or t_2. An accurate radiance image can be obtained by using the weighted combination of the two predicted radiance images based on t_1 and t_2, while the temporal weights of the two images may be given by the temporal changes in coarse resolution radiance images. Eq. (5.23) shows how to calculate the temporal weight:

$$T_k = \frac{1 / \left(\sum_{i=1}^{w} \sum_{j=1}^{w} M\left(x_i, y_j, t_k, B\right) - \sum_{i=1}^{w} \sum_{j=1}^{w} M\left(x_i, y_j, t_p, B\right) \right)}{\sum_{k=t_1, t_2} 1 / \left(\sum_{i=1}^{w} \sum_{j=1}^{w} M\left(x_i, y_j, t_k, B\right) - \sum_{i=1}^{w} \sum_{j=1}^{w} M\left(x_i, y_j, t_p, B\right) \right)}$$

$$(5.23)$$

where $M(x, y, t, B)$ is the resampled radiance at time t. Therefore, the final predicted radiance image can be calculated as follows:

$$L\left(x_{w/2}, y_{w/2}, t_p\right) = T_{t_1} * L_{t_1}\left(x_{w/2}, y_{w/2}, t_p\right) + T_{t_2} * L_{t_2}\left(x_{w/2}, y_{w/2}, t_p\right) \quad (5.24)$$

where L_{t_1} is the predicted radiance image using L image at t_1 as the base image, L_{t_2} is the predicted radiance image using L image at t_2 as the base image. With the final radiance image, LST image can be computed.

5.2.5 Results of Data Fusion

The accuracy of SADFAT was evaluated through validating the predicted LSTs against the observed LSTs obtained from the Landsat TIR data. In this study, five pairs of Landsat and MODIS images were used for accuracy assessment. We selected CC, mean difference, and mean absolute difference to serve as indicators to evaluate the accuracy. Two pairs of MODIS and Landsat images, acquired on 24 June and 15 November 2005, respectively, were used as the first inputs to the fusion model (Figure 5.3). The white areas in the LST images were masked-out cloudy areas according to the QA layer of the MODIS data. From Figure 5.3, it is apparent that LST spatial patterns in MODIS and Landsat images were substantially different in the two dates. Although the minimum temperatures of MODIS and Landsat images seemed similar, the maximum

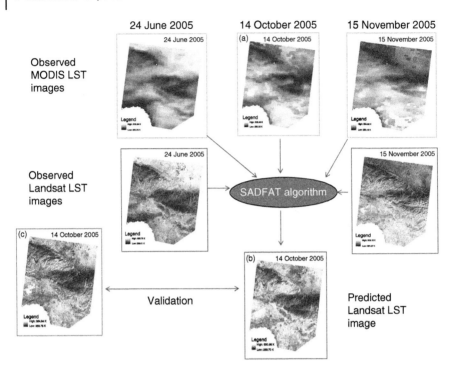

Figure 5.3 Illustration of predicting Landsat LST image based on a pair of observed MODIS LST and a pair of observed Landsat LST images. *Source:* after Weng et al. (2014). Reproduced with permission of Elsevier.

temperatures differed in about 15° in either date (Table 5.1). Considering the land-cover characteristics of the study area, the size of the searching window was set to 3 MODIS pixels (i.e. 25 Landsat TM pixels), and the number of land-cover types was set to 5. The second input for the fusion model was the MODIS LST image used to predict the corresponding Landsat LST image at the same date (t_p). Figure 5.4 showed the results of prediction in comparison with the inputs: (A) MODIS LST images (model inputs); (B) predicted Landsat LST images; and (C) observed Landsat LST images on 10 July, 27 August, 28 September, 14 October, and 30 October 2005, respectively. For all the five predictions, the predicted LST images (B) and the observed LST images (C) matched well in terms of the overall spatial patterns of LST. The predicted LST images contained the vast majority of the spatial details found in the observed images, including such surface features as roads, urban streets, and lakes. Moreover, LST variations with terrain were also preserved and the unique LST patterns in the mountainous region can also be discerned.

Figure 5.4 displays the scatter plot between the predicted and observed LSTs for each prediction date. The data points fell close to the diagonal line in each

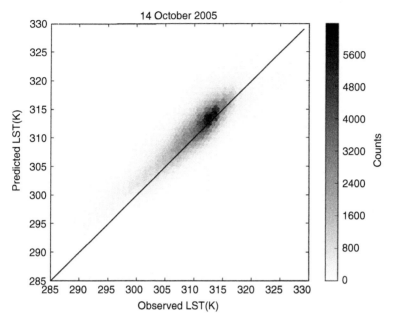

Figure 5.4 The scatter plots between the observed and the predicted LSTs on 14 October 2005. *Source:* Weng et al. (2014). Reproduced with permission of Elsevier.

panel, indicating that the predictions were all in good agreement with the observations. To quantify the prediction accuracy, CC, mean difference, and mean absolute difference were computed. Results show that the values of mean difference and mean absolute difference between the predicted and the observed LSTs were quiet small, whereas the CC values were found to be all greater than 0.90, except for the one on 14 October 2005. This is because that the correlation between MODIS and Landsat LST on that date was the weakest (0.63), as seen in Table 5.1. Because all the predictions shared the same pairs of Landsat and MODIS LST images as the model inputs, a comparison of their accuracies was also feasible. The mean absolute difference ranged from 1.25 K to approximately 2.0 K. Since the selected images reflected well the phenological change from June to November, SADFAT proved to be effective to account for the temporal variations of LST. Compared with previous studies using the original STARFM method (Liu and Weng 2012) or with a revised filter considering surrounding pixels to predict LST (Huang et al. 2013), the improvement of LST prediction in this study was due largely to the inclusion of the ATC model and LSMA into the fusion algorithm to calculate the conversion coefficient. The ATC model can delineate the trend of annual mean temperature variations while ignoring the specific daily LST changes (Bechtel 2012; Weng and Fu 2014a) and thus facilitated the prediction of LST temporal changes between the two dates. However, it should be noted in the scatter plots that there were some pixels showing large differences between the predicted and observed LSTs. These discrepancies presented a major limitation of SATFAT that associated with the assumption on the stability of phase shift parameter (θ) in the ATC model. While it may be largely true that the land cover and other surface conditions did not change over the study period, however, if land covers or other surface conditions did change, it could lead to larger errors in the prediction (Weng et al. 2014).

5.2.6 Summary

This research proposed a data fusion model, SADFAT, for predicting LSTs at high temporal frequency and at the medium spatial resolution (120 m) using the combined data sets of MODIS and Landsat. SADFAT presented several improvements over its precedents. The most significant improvement was to incorporate an ATC model to characterize the annual variations of LST, based on which the conversion coefficients were computed. The use of a conversion coefficient allowed relating the thermal radiance change of a mixed pixel at the coarse resolution to that of a fine resolution pixel. In this way, the prediction with a regression model can well be justified. In contrast with the ESTARFM (Zhu et al. 2010) that used a linear model to predict the change in surface reflectance, SADFAT employed a nonlinear model to approximate annual change of LST. In addition, since SADFAT aims at prediction of LSTs, both the reflective (Landsat band 3 and 4) and TIR bands were utilized for searching for similar pixels. The inclusion of the TIR band allowed for the obtainment of more accurate similar pixels, because reflective and thermal information were complementary. Two-pair images used for the selection ensured that selected similar pixels possessed the same spectral and thermal trajectories. In the computation of the weights for selected similar pixels, the CC of the pixel-wise reflective and thermal vector between MODIS and Landsat were utilized to measure the thermal "similarity." The higher similarity would provide a larger weight in the final calculation for the central pixel. With the proposed model SADFAT, it is possible to predict a radiance image using one Landsat reference image, either at time t_1 or t_2. To reduce the uncertainty, the final predicted images in this study were computed based on the two predicted images weighted by the temporal change in radiance at the coarse resolution level. Another merit of SADFAT is to employ LSMA to address the issue of thermal landscape heterogeneity. This issue is especially important when SADFAT is applied to urban landscapes, where mixed pixels often prevail in medium- or coarse-resolution satellite imagery (Lu and Weng 2004, Weng and Lu 2009). To ensure a consistent comparison between MODIS and Landsat data, LST values were first converted to radiances at the Landsat effective thermal wavelength. Then, LSMA linked the radiation relationship between Landsat and MODIS

TIR information. Thus, the temporal change in radiance can be included in the fusion model.

However, it should be noted that SADFAT contains a few limitations. An assumption for SADFAT was that the phase shift parameter would keep constant. Therefore, SADFAT did not have the ability to predict LST changes that were not reflected in the MODIS and/or Landsat pixels. The use of an ATC model can well approximate the seasonal cycles of LST, but it cannot delineate the specific daily weather and surface conditions that may also affect LST temporal variation, leading to uncertainty in the modeling. Considering the difference in the overpassing time between Landsat-5 and MODIS sensors was small (less than 10 minutes) for the study area, this study did not correct for the difference in diurnal temperature change between the two sensors. Since their orbital parameters were equal, the viewing (near-nadir) and solar geometries of MODIS were close to those of the corresponding Landsat acquisition (Gao et al. 2006). Nevertheless, the proposed algorithm should be applicable to Landsat-like sensors, such as ASTER, to enhance their temporal frequency, or to any pair of satellite sensors that parallel the relationship between MODIS and Landsat. One important feature of SADFAT is to establish a linkage in radiance change over time between MODIS and Landsat; therefore, as long as radiance changes between paired satellite sensors can be modeled by LSMA and ATC, SADFAT will be practically useful for thermal sharpening, given satellite overpassing, orbital parameters, and viewing geometry being considered. In addition, several parameters must be carefully set, e.g. the window size and the number of land-cover classes. Different study areas may require setting up different parameter values, and it is necessary to perform a sensitivity analysis of these parameters before the modeling. The mismatching of Landsat and MODIS pixel was neglected in this study. The variation of MODIS pixel footprint especially at off-nadir viewing may cause some errors and need to be cautioned. Further research is also needed to fill up the missing values caused by clouds. The problem was not so serious in this study since images were carefully selected to avoid cloud contamination. An improved fusion model that can resolve the cloud contamination issue would be useful for areas experiencing more cloudy skies than Los Angeles.

5.3 Reconstructing Consistent LSTs at Landsat Resolution

5.3.1 Study Area and Data

The study area consists of both metropolitan and rural areas of Beijing. The metropolis, located in the northern tip of the roughly triangular North China Plain, has 14 urban and suburban districts and 2 rural counties (Figure 5.5). Beijing experiences elevation decrease from the northwest to the southwest

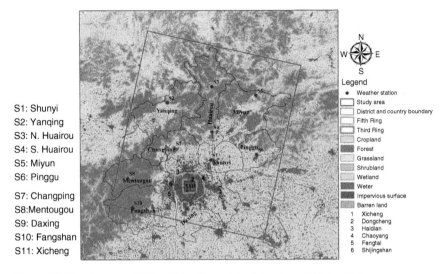

S1: Shunyi
S2: Yanqing
S3: N. Huairou
S4: S. Huairou
S5: Miyun
S6: Pinggu

S7: Changping
S8: Mentougou
S9: Daxing
S10: Fangshan
S11: Xicheng

Figure 5.5 The study area Beijing, China. It consists of six urban districts – Xicheng, Dongcheng, Haidian, Chaoyang, Fengtai, and Shijingshan; eight suburban districts – Changping, Daxing, Fangshan, Huairou, Mengtougou, Pinggu, Shunyi, and Tongzhou; and two rural counties – Miyun and Yanqing. The figure also shows the land-cover types identified by the GlobalLand30 in 2010 and 11 weather stations (named from S1 to S11) for the study area. *Source:* Fu and Weng (2016). Reproduced with permission of Elsevier.

with the mountains in the north and northwest shielding the city from the encroaching desert steppes. This region of China exhibits a typical temperate continental climate generally characterized by hot and wet summers and dry and cold winters. The study area covers more than 95% of the Beijing metropolis, captured by the Landsat scene of path/row 123/32. The global land-cover mapping project (GlobalLand30) (Chen et al. 2015) identifies eight land covers including croplands, forest, grassland, shrubland, wetland, water, impervious surface, and barren land in the study area for the baseline of year 2010. Since the 1980s, Beijing underwent rapid urban growth. The urban area of Beijing increased from $183.84\,km^2$ in 1973 to $1209.97\,km^2$ in 2005 with an annual expansion rate of built-up area at $32.07\,km^2$ (Mu et al. 2007). The population reached 21.51 million in 2014 and the average population density was 1311 persons/km (Beijing Municipal Statistical Bureau 2014). The intensive urbanization in the past decades has also caused a series of environmental issues, such as haze pollution, extreme rainstorms, and water contamination. It has been reported that surface temperature and UHI intensity in Beijing increased at the rate of 0.25 and 0.31 °C per decade, respectively, after 1981 (Lin and Yu 2005). Therefore, generation of a time series LST data set can help understand the impact of urban growth on environment and public health and is also of significance for sustainable urban development in Beijing.

All L1T Landsat images available for the study area were downloaded through the United States Geological Survey (USGS) Center Science Processing Architecture (ESPA) On Demand Interface. The data sets included surface and top-of-atmosphere (TOA) reflectance and brightness temperature processed by the Landsat Ecosystem Disturbance Adaptive Processing System (Masek et al. 2006). According to the metadata, images with cloud percentage less than 90% were selected and a total of 512 images were finally collected. A further screening procedure *Tmask* based on the multi-temporal images were adopted to remove the outliers possibly caused by cloud, cloud shadow, and snow contaminations (Zhu and Woodcock 2014a). Then, all the images were resampled to 120 m and subset to the study area. The retrieval of LSTs from brightness temperature was accomplished by using the single channel algorithm (Jimenez-Munoz and Sobrino 2003) since its reported computation error was close to 1.5 K for water vapor from 0.5 to 3 g/cm^2. The input of water vapor was derived from the National Centers for Environmental Prediction (NCEP) Reanalysis data (Kalnay et al. 1996). LSE values were first calculated from both the classification-based method (Snyder et al. 1998) and the NDVI threshold method (Sobrino et al. 2008). The comparison of the emissivity values computed from the GlobalLand30 with those derived from the NDVI data showed the mean difference of 0.01. According to Jiménez-Muñoz and Sobrino (2006), the 0.01 emissivity difference may produce the LST retrieval error less than 0.4 K. Thus, the NDVI threshold method was adopted because of its feasibility to derive time series emissivity data set. In addition, surface temperature measurements from 11 weather stations in the years of 2008, 2009, and 2010 were collected and compared with the corresponding satellite-derived LSTs to assess the accuracies of satellite-derived LSTs and LSTs reconstructed by the DELTA algorithm.

5.3.2 The DELTA Algorithm

The DELTA algorithm consisted of four modules: (i) data filter, (ii) temporal segmentation, (iii) periodic and trend modeling, and (iv) Gaussian process (GP). Each module was applied on the pixel-by-pixel basis. Figure 5.6 shows the flowchart of the DELTA algorithm to generate the daily LSTs at the Landsat spatial resolution. The data filter module removed outliers and constrained surface reflectance and LSTs within the set range. Temporal segmentation aimed at dividing time series LSTs into relatively stationary segments to reduce the effect of disturbance events on LST reconstructions. In the phase of periodic and trend modeling, the unevenly distributed time series LSTs were fitted by a parametric model consisting of ATC and trend variations. Then, the residuals between the LST observations and model predictions were analyzed by a nonparametric Bayesian technique, Gaussian process regression (GPR). Finally, LSTs at the daily interval were reconstructed by

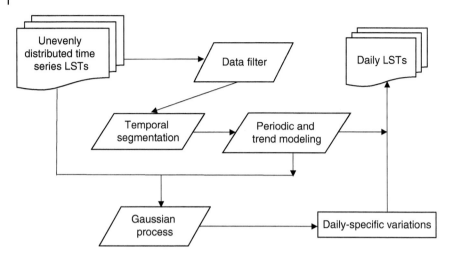

Figure 5.6 The flowchart of the algorithm DELTA consisting of four modules: data filter, temporal segmentation, periodic and trend modeling, and Gaussian process. The final daily LSTs reflected annual temperature cycle, interannual trend, and daily-specific variations. *Source:* Fu and Weng (2016). Reproduced with permission of Elsevier.

adding up the variations of ATC and trend component, and daily-specific anomalies inferred by the GPR.

5.3.2.1 Data Filter

The data filter ensured that only valid surface reflectance and LST measurements over time were utilized. First, the filter screened out the pixels which had the values of surface reflectance beyond the range of from 0 to 10 000 (the reflectance data downloaded was first scaled by a factor of 10 000) and LSTs outside the range from 250 to 340 K to exclude extreme outliers. In addition, quality flags from both the metadata and the *Tmask* were used to ensure that only clear-sky pixels were selected and input into the DELTA.

5.3.2.2 Temporal Segmentation

The temporal segmentation procedure was based on the consideration that disturbance events, such as deforestation, desertification, and urbanization, can induce nonstationary LST variations over time. The division of the whole time series into a sequence of discrete segments was conducted because different land covers and surface conditions may generate different temporal characteristics in LSTs. The treatment of individual data segments thus eased the LST reconstruction without impacting the nonstationary variations.

The segmentation method was adapted from the Continuous Change Detection and Classification (Zhu and Woodcock 2014b) algorithm and was applied only to band 2, band 3, band 4, band 5, and band 7 to reduce the computation time. First, it identified land-cover changes by using the time series additive model (Eq. 5.25).

$$\hat{P}(i,d) = a_i + b_{1i}\cos(\omega d) + b_{2i}\sin(\omega d) + c_i d \qquad (5.25)$$

where $\hat{P}(i,d)$ is the value predicted for surface reflectance at Julian date d, ω is the angular frequency, a_i, b_i, and c_i are the coefficients for the mean value, intra- and interannual changes, respectively. The change was flagged if the normalized difference exceeded the predefined thresholds as showed in Eq. (5.26) in three consecutive days.

$$\frac{\left|P(i,d) - \hat{P}(i,d)\right|}{3 \times RMSE} > 1 \qquad (5.26)$$

Given that land-cover change may happen in the first and last observations during the process of the model initialization, another three conditions (Eq. 5.27) were listed below to identify abnormal observations for the first and last observations of the model initialization and the observations between them.

$$\frac{1}{k}\sum_{i=1}^{k}\frac{\left|c_i(d)\right|}{3\times\left(RMSE_i / T_{model}\right)} > 1 \ \text{or} \ \frac{1}{k}\sum_{i=1}^{k}\frac{\left|P(i,d_1) - \hat{P}(i,d_1)\right|}{3\times RMSE_i} > 1$$

$$\text{or} \ \frac{1}{k}\sum_{i=1}^{k}\frac{\left|P(i,d_n) - \hat{P}(i,d_n)\right|}{3\times RMSE_i} > 1 \qquad (5.27)$$

With d_1 and d_n the Julian dates for the first and last observation during model initialization, T_{model} is the total time used for the model initialization.

5.3.2.3 Periodic and Trend Modeling

In this phase, the parametric model (Eq. 5.25) consisted of two sinusoidal functions, a mean value and a linear trend. The two sinusoidal functions could characterize the ATC, which was part of the fluctuation attributed to the Earth's changing position over the course of the year (Thomson 1995; Bechtel 2012; Weng and Fu 2014a). ATC parameters were effective only if satellite observations were cyclic stationary. The mean value and the linear trend functions included in the parametric model can capture the overall and the interannual LST changes over the study period. Although previous studies (Huang et al. 2013; Weng et al. 2014; Wu et al. 2015) have proven that the data fusion algorithms can predict LSTs at high spatial and temporal resolutions, inter-year LST changes have so far been neglected. The inter-year LST change may become apparently obvious when long-term LST (e.g. 20 years) observations were analyzed (Julien and Sobrino 2012). Given the factors above, the periodic and trend modeling procedure was applied only to within individual homogeneous segments to avoid the impacts of land-cover changes.

5.3.2.4 Gaussian Process Regression

The GP module was included because the ATC and interannual model may not capture all the LST variations. Weng and Fu (2014a) employed an ATC model to characterize the landscape thermal patterns and revealed the overall mean RMSE of 7.4 K and the overall mean RMSE of 4.3 K by using annual, semiannual, and the trend models to fit LSTs (Fu and Weng 2015). Since the model residuals (differences between the model predictions and satellite observations) may be regarded to relate to daily-specific variations associated with weathers, GPR was employed to make inference of the residuals at the daily basis.

The GPR can be interpreted as a distribution over and inference occurring in the space of function from the function-space view (Rasmussen and Williams 2006). It has received much attention in the field of machine learning and can provide the Bayesian approach to establish the relationship between input and output variables with kernels in the form of

$$\hat{f}(x) = \sum_{i=1}^{n} \alpha_i k(x_i, x) \tag{5.28}$$

where α is the kernel used to assign weight, and k is the kernel function evaluating the covariance (similarity) between the input observations x_i (i = 1, ..., n) and the test data x. For the current study, based on a trial and error test, the kernel covariance function of "Materniso" was adopted (Eq. 5.29).

$$k(r) = \sigma_s^2 \frac{2^{1-v}}{\mathcal{T}(v)} \left(\sqrt{2v} \frac{r}{l} \right)^v K_v \left(\sqrt{2v} \frac{r}{l} \right) + \delta * \sigma_n^2 \tag{5.29}$$

where K_v is a modified Bessel function; $\mathcal{T}$ is the gamma function; r is the distance between x and x_i; v and l are the positive parameters; σ_s is the signal standard deviation; σ_n is the noise standard deviation; and δ is the Kronecker's symbol. The kernel was parameterized collectively by a hyperparameter $\theta = [\sigma_s, v, l, \sigma_n]$. Now suppose that the observed variable was formed by noisy observations of the true underlying function $y = f(x) + \varepsilon$ and the noise was additive independently identically Gaussian distributed with zero mean and variance σ_m, under the prior assumption, the joint distribution of the observed target values and the function values at the test locations were given by Eq. (5.30).

$$\begin{pmatrix} y \\ f_* \end{pmatrix} \sim \mathcal{N} \left(0, \begin{bmatrix} k(x,x) + \delta_n^2 I & k(x,x_*) \\ k(x,x_*) & k(x_*,x_*) \end{bmatrix} \right) \tag{5.30}$$

For the prediction purposes, the GPR was employed by computing the posterior distribution over the unknown values f_* with the hyperparameters θ typically selected by maximizing Type-II Maximum Likelihood, using the marginal likelihood of the observations. Therefore, the predictive mean $\overline{f_*}$ and variance $v\left(\overline{f_*}\right)$ can be estimated using Eq. (5.31).

$$\overline{f_*} = k_*^T \left(k + \delta_n^2 I\right)^{-1} y, \quad v\left(\overline{f_*}\right) = k\left(x_*, x_*\right) k_*^T \left(k + \delta_n^2 I\right)^{-1} k_* \tag{5.31}$$

Based on the four modules, daily LSTs at 120-m resolution were reconstructed by adding up the ATC, trend, and daily-specific variations. Given that the segmentation procedure may produce segments with less than six pixel dates, these pixel dates were merged with other longer segments to ease the computation.

5.3.3 Results

Accuracy assessment was performed to compare satellite-derived LSTs with in situ LSTs from 11 weather stations. Since the ground measurements were recorded hourly, the in situ LSTs did not match the satellite overpassing times (approximately 10 : 30 a.m. local time). This study utilized linear interpolation method to estimate reference LSTs at the satellite acquisition time based on ground measurements at 10 and 11 a.m.

Satellite-derived LSTs that matched the locations of weather stations were extracted and compared with the corresponding field LST measurements from 2008 to 2010 (Figure 5.7). Validation results show that the CC ranged

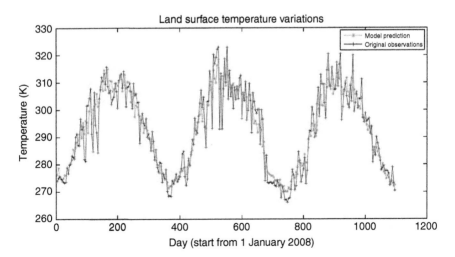

Figure 5.7 Daily temperature difference between values estimated from DELTA and from weather stations S1–S11, 2008–2010. *Source:* reproduced with permission of Elsevier.

Table 5.2 Comparisons between satellite-derived and weather-station LSTs.

Station name	Correlation coefficient	Mean error (K)	Mean absolute error (K)
S1	0.94	−0.9	2.4
S2	0.94	−0.8	2.7
S3	0.89	0.4	2.4
S4	0.98	−2.5	2.8
S5	0.96	−0.1	1.8
S6	0.92	−1.0	2.3
S7	0.98	−0.6	2.5
S8	0.80	−0.3	2.5
S9	0.95	−0.8	2.4
S10	0.87	−0.6	1.8
S11	0.92	1.4	2.1

Note: S1–S11 are weather stations in the study area. The distributions of the 11 weather stations can be seen in Figure 5.5.

from 0.80 to 0.98 (Table 5.2), indicating that the good agreement between the satellite-derived LSTs and in situ LSTs was achieved. However, the validation also indicates that in general LSTs provided by satellite sensors were less than ground measurements. In this study, satellite LSTs were retrieved from the TIR sensors by the single channel algorithm with input of water vapor from the NCEP Reanalysis data set. The dependence of water vapor for the single-channel algorithm resulted from the fact that water vapor is the mainly absorber in the TIR region. The water vapor data used for LST computation came from the linear interpolation between the two time points that were close to the satellite overpassing time. According to the pattern of diurnal variations of water vapor (Dai et al. 2002), this study may have used a smaller water vapor value. Consequently, it is reasonable that satellite-derived LSTs showed smaller values than field measurements. In addition, Table 5.2 shows that the mean absolute error (MAE) between satellite-derived and weather-station LSTs is at least 1.8 K, larger than the error of 1.5 K reported by Jimenez-Munoz and Sobrino (2003). The difference between the two studies may be attributed to different methods in collecting field temperatures. For the current study, temperature measurements were recorded hourly by using thermometers positioned on the ground surface. The in situ LSTs by Jimenez-Munoz and Sobrino (2003) were obtained by combining soil and vegetation temperatures based on the fractions of bare soil and vegetation inside the field of view of the field radiometer. The field measurements by Jimenez-Munoz and Sobrino (2003) were more accurate than LSTs from the weather stations especially for mixed pixels. However, the absolute mean error (ME) of 1.8–2.8 K (Table 5.2) among the 11 weather stations suggests that the accuracy of satellite-derived LSTs was in the acceptable range.

To assess the effectiveness of DELTA to generate LSTs, daily in situ LSTs collected from the 11 weather stations were compared against the reconstructed daily LSTs in year 2008. LSTs reconstructed at the same date with the satellite pixels were excluded before the comparison. The CC, ME, and MAE were computed as accuracy measures. It is found that MAE ranged from 3.4 to 3.5 K. Since satellite-derived LSTs had an average retrieval bias of 2.3 K, it is expected that reconstructed daily LSTs possessed a larger error. The relatively even MAE among 11 weather stations indicates that the DELTA algorithm performed well. This result should be mainly attributed to the inclusion of GPR to estimate daily-specific LST variations associated with weathers. Without GPR, the DELTA algorithm could only predict the harmonic and inter-year variations within individual stationary segments. For example, the modeling performance for the ATC, interannual trend, and the daily specific variations reduced the LST prediction error to 1.9 K. Although the Bayesian model was powerful in learning the daily specific variations, the GPR technique was still influenced by the Landsat revisiting time and poor atmospheric conditions. The prediction errors arose since these factors limited the images that could be used to provide the input of weather-related LST variations. As a result, it is impossible for the GPR to predict all the daily-specific LST variations accurately. Overall, based on the in situ LSTs from the 11 weather stations, the ME of reconstructed daily LSTs at the Landsat resolution was 3.5 K.

One of interesting applications of consistent long-term LSTs generated by the DELTA algorithm is to characterize and quantify the spatiotemporal dynamics of UHI. Here, LST maps of the 15 day of each month in the year 2000 were produced to reveal changes in the spatial and temporal patterns (Figure 5.8). Furthermore, LST maps were generated for each 15 August from 1984 to 2011 to quantify changes in the UHI intensity. The UHI effect refers to the increased temperatures of the dense urban areas with respect to their rural or suburban surroundings. However, there was no consensus about what constitute the representative "urban" and "rural" surfaces (Oke et al. 1991). For the current study, the urban and rural area were identified based on the land-cover map. In addition, an appropriate rural reference should be from the same type of topographical setting without major elevation change (Weng et al. 2011). Therefore, it was assumed the region within the Third Ring was the urban area and the non-mountain region that was outside of the Fifth Ring constituted the rural area. The UHI intensity was calculated as the temperature difference between the two regions. It shows mean LSTs in both urban and rural regions and the UHI intensity from 1984 to 2011. Despite of temporal fluctuations, the mean temperatures in the two regions exhibited an increase trend. The UHI intensity ranged from 3.3 K (in the years 2007 and 2008) to 5.3 K (in the year 2005).

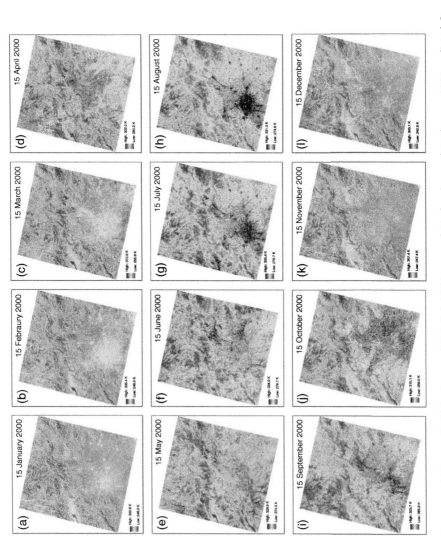

Figure 5.8 LST maps (A–L) for the year 2000. Only LST images of 15 of each month in the year 2000 are listed. *Source:* reproduced with permission of Elsevier.

Furthermore, the mean UHI intensity was 4.2 K from 1998 to 2011, 0.4 K higher than the mean value from 1984 to 1997, suggesting an intensified UHI in Beijing. Further analysis are warranted to correlate the temporal UHI evolution with the urban expansion history to explain the underlying the mechanisms.

5.3.4 Summary

This chapter presents the DELTA algorithm to reconstruct consistent, daily LSTs at Landsat resolution based solely on Landsat imagery. The DELTA algorithm possessed some advantages compared with the algorithms by blending Landsat and MODIS imagery to generate LSTs (Gao et al. 2006; Huang et al. 2013; Weng et al. 2014; Wu et al. 2015). First, DELTA did not require the input of MODIS images. The requirement of MODIS data input limited the prediction of LSTs before 2000 when MODIS data were not available. Second, the utilization of temporal segmentation (Zhu and Woodcock 2014b) in DELTA reduced the impact of land-cover changes on LST prediction over time. If land-cover changes occurred at some points, the corresponding LSTs would be nonstationary leading to the changes of ATC over the time. The assumption to use only one annual frequency to characterize the intra-annual thermal variations was only reasonable to be applied within the stationary (homogenous) segments. Based on in situ measurements from the weather stations, results showed that retrieval errors of satellite-derived LSTs ranged from 1.8 to 2.8 K with the average error of 2.3 K. The DELTA algorithm was able to predict daily LSTs at the average bias of 3.5 K. The LST maps of 15 August from 1984 to 2011 showed that the mean LST in both urban and rural regions increased and the UHI intensity in Beijing ranged from 3.3 to 5.3 K. However, it is suggested that the DELTA algorithm was still influenced by the Landsat revisit frequency and poor atmospheric conditions. A possible way to improve the algorithm is to resort to weather-station LSTs for inferencing daily-specific LST variations associated with different weather patterns. In addition, given that DELTA algorithm required intensive computation, parallel distribution of the algorithm among powerful computer clusters is necessary.

References

Anderson, M.C., Allen, R.G., Morse, A., and Kustas, W.P. (2012). Use of Landsat thermal imagery in monitoring evapotranspiration and managing water resources. *Remote Sensing of Environment* 122: 50–65.

Anderson, M.C., Kustas, W.P., Norman, J.M. et al. (2011). Mapping daily evapotranspiration at field to continental scales using geostationary and polar orbiting satellite imagery. *Hydrology and Earth System Sciences* 15 (1): 223–239.

Bechtel, B. (2012). Robustness of annual cycle parameters to characterize the urban thermal landscapes. *IEEE Geoscience and Remote Sensing Letters* 9 (5): 876–880.

Bechtel, B., Zakšek, K., and Hoshyaripour, G. (2012). Downscaling land surface temperature in an urban area: a case study for Hamburg, Germany. *Remote Sensing* 4 (10): 3184–3200.

Beijing Municipal Statistical Bureau (2014). http://tjj.beijing.gov.cn/nj/main/2014-en/index.htm (accessed 4 July 2019).

Carlson, T. (2007). An overview of the "triangle method" for estimating surface evapotranspiration and soil moisture from satellite imagery. *Sensors* 7 (8): 1612–1629.

Chen, J., Chen, J., Liao, A. et al. (2015). Global land cover mapping at 30m resolution: a POK-based operational approach. *ISPRS Journal of Photogrammetry and Remote Sensing* 103: 7–27.

Dai, A., Wang, J., Ware, R.H., and Van Hove, T. (2002). Diurnal variation in water vapor over North America and its implications for sampling errors in radiosonde humidity. *Journal of Geophysical Research: Atmospheres* 107: 11–14.

Dominguez, A., Kleissl, J., Luvall, J.C., and Rickman, D.L. (2011). High-resolution urban thermal sharpener (HUTS). *Remote Sensing of Environment* 115 (7): 1772–1780.

Fu, P. and Weng, Q. (2015). Temporal dynamics of land surface temperature from Landsat TIR time series images. *IEEE Geoscience and Remote Sensing Letters* 12 (10): 2175–2179.

Fu, P. and Weng, Q. (2016). Consistent land surface temperature data generation from irregularly spaced Landsat imagery. *Remote Sensing of Environment* 184 (10): 175–187.

Gao, F., Masek, J., Schwaller, M., and Hall, F. (2006). On the blending of the Landsat and MODIS surface reflectance: predicting daily Landsat surface reflectance. *IEEE Transactions on Geoscience and Remote Sensing* 44 (8): 2207–2218.

Gillespie, A.R. (1992). Spectral mixture analysis of multispectral thermal infrared images. *Remote Sensing of Environment* 42 (2): 137–145.

Gillies, R.R., Carlson, T.N., Cui, J. et al. (1997). A verification of the "triangle" method for obtaining surface soil water content and energy fluxes from remote measurements of the normalized difference vegetation index (NDVI) and surface radiant temperature. *International Journal of Remote Sensing* 18 (15): 3145–3166.

Gottsche, F.M. and Olesen, F.S. (2001). Modelling of diurnal cycles of brightness temperature extracted from METEOSAT data. *Remote Sensing of Environment* 76 (3): 337–348.

Hansen, M.C., Roy, D.P., Lindquist, E. et al. (2008). A method for integrating MODIS and Landsat data for systematic monitoring of forest cover and change in the Congo Basin. *Remote Sensing of Environment* 112 (5): 2495–2513.

Harlan, S.L., Brazel, A.J., Prashad, L. et al. (2006). Neighborhood microclimates and vulnerability to heat stress. *Social Science and Medicine* 63 (11): 2847–2863.

Hilker, T., Wulder, M.A., Coops, N.C. et al. (2009a). A new data fusion model for high spatial- and temporal-resolution mapping of forest disturbance based on Landsat and MODIS. *Remote Sensing of Environment* 113 (8): 1613–1627.

Hilker, T., Wulder, M.A., Coops, N.C. et al. (2009b). Generation of dense time series synthetic Landsat data through data blending with MODIS using a spatial and temporal adaptive reflectance fusion model. *Remote Sensing of Environment* 113 (9): 1988–1999.

Huang, B., Wang, J., Song, H. et al. (2013). Generating high spatiotemporal resolution land surface temperature for urban heat island monitoring. *IEEE Geoscience and Remote Sensing Letters* 10 (5): 1011–1015.

Hulley, G.C. and Hook, S.J. (2009). The North American ASTER Land Surface Emissivity Database (NAALSED) version 2.0. *Remote Sensing of Environment* 113 (9): 1967–1975.

Hulley, G.C., Hook, S.J., and Baldridge, A.M. (2008). ASTER land surface emissivity database of California and Nevada. *Geophysical Research Letters* 35 (13): L13401.

Imhoff, M.L., Zhang, P., Wolfe, R.E., and Bounoua, L. (2010). Remote sensing of the urban heat island effect across biomes in the continental USA. *Remote Sensing of Environment* 114 (3): 504–513.

Inamdar, A.K. and French, A. (2009). Disaggregation of GOES land surface temperatures using surface emissivity. *Geophysical Research Letters* 36 (2): L02408. https://doi.org/10.1029/2008GL036544.

Inamdar, A.K., French, A., Hook, S. et al. (2008). Land surface temperature retrieval at high spatial and temporal resolutions over the southwestern United States. *Journal of Geophysical Research: Atmospheres* 113: D07107. https://doi.org/10.1029/2007JD009048.

Jensen, J.R. (2005). *Introductory Digital Image Processing: A Remote Sensing Perspective*, 3e. Upper Saddle River: Prentice Hall.

Jimenez-Munoz, J.C. and Sobrino, J.A. (2003). A generalized single-channel method for retrieving land surface temperature from remote sensing data. *Journal of Geophysical Research: Atmospheres* 108 (D22): 4688. https://doi.org/10.1029/2003JD003480.

Jiménez-Muñoz, J.C. and Sobrino, J.A. (2006). Error sources on the land surface temperature retrieved from thermal infrared single channel remote sensing data. *International Journal of Remote Sensing* 27 (5): 999–1014.

Jin, M.L. and Dickinson, R.E. (2002). New observational evidence for global warming from satellite. *Geophysical Research Letters* 29 (10) https://doi.org/10.1029/2001GL013833.

Jin, M.L., Dickinson, R.E., and Zhang, D.L. (2005). The footprint of urban areas on global climate as characterized by MODIS. *Journal of Climate* 18 (10): 1551–1565.

Julien, Y. and Sobrino, J.A. (2012). Correcting AVHRR long term data record V3 estimated LST from orbital drift effects. *Remote Sensing of Environment* 123: 207–219.

Kalnay, E., Kanamitsu, M., Kistler, R. et al. (1996). The NCEP/NCAR 40-year reanalysis project. *Bulletin of the American Meteorological Society* 77: 437–471.

Kaufman, Y.J. and Gao, B.C. (1992). Remote-sensing of water-vapor in the near IR from EOS/MODIS. *IEEE Transactions on Geoscience and Remote Sensing* 30 (5): 871–884.

Kustas, W.P., Norman, J.M., Anderson, M.C., and French, A.N. (2003). Estimating subpixel surface temperatures and energy fluxes from the vegetation index-radiometric temperature relationship. *Remote Sensing of Environment* 85 (4): 429–440.

Lafortezza, R., Carrus, G., Sanesi, G., and Davies, C. (2009). Benefits and well-being perceived by people visiting green spaces in periods of heat stress. *Urban Forestry and Urban Greening* 8 (2): 97–108.

Lin, X. and Yu, S. (2005). Interdecadal changes of temperature in the Beijing region and its heat island effect. *Chinese Journal of Geophysics* 48: 39–45.

Liu, D.S. and Pu, R.L. (2008). Downscaling thermal infrared radiance for subpixel land surface temperature retrieval. *Sensors* 8 (4): 2695–2706.

Liu, H. and Weng, Q. (2009). An examination of the effect of landscape pattern, land surface temperature, and socioeconomic conditions on WNV dissemination in Chicago. *Environmental Monitoring and Assessment* 159 (1–4): 143–161.

Liu, H. and Weng, Q. (2012). Enhancing temporal resolution of satellite imagery for public health studies: a case study of West Nile virus outbreak in Los Angeles in 2007. *Remote Sensing of Environment* 117: 57–71.

Liu, J.G. and Moore, J.M. (1998). Pixel block intensity modulation: adding spatial detail to TM band 6 thermal imagery. *International Journal of Remote Sensing* 19 (13): 2477–2491.

Lu, D. and Weng, Q. (2004). Spectral mixture analysis of the urban landscape in Indianapolis with Landsat ETM+ imagery. *Photogrammetric Engineering and Remote Sensing* 70 (9): 1053–1062.

Masek, J.G., Vermote, E.F., Saleous, N.E. et al. (2006). A Landsat surface reflectance dataset for North America, 1990–2000. *IEEE Geoscience and Remote Sensing Letters* 3 (1): 68–72.

Moran, M.S. (2004). Thermal infrared measurement as an indicator of plant ecosystem health. *Thermal Remote Sensing in Land Surface Processes*: 257–257.

Mu, F.-Y., Zhang, Z.-X., Chi, Y.-B. et al. (2007). Dynamic monitoring of built-up area in Beijing during 1973–2005 based on multi-original remote sensed images. *Journal of Remote Sensing* 11: 257.

Nichol, J. (1994). A GIS-based approach to microclimate monitoring in Singapore's high-rise housing estates. *Photogrammetric Engineering and Remote Sensing* 60 (10): 1225–1232.

Nichol, J. (2009). An emissivity modulation method for spatial enhancement of thermal satellite images in urban heat island analysis. *Photogrammetric Engineering and Remote Sensing* 75 (5): 547–556.

Nicodemus, F.E. (1965). Directional reflectance and emissivity of an opaque surface. *Applied Optics* 4 (7): 767–773.

Oke, T.R. (1979). *Technical Note No. 169: Review of Urban Climatology* (43 p). Geneva, Switzerland: World Meteorological Organization.

Oke, T.R. (1982). The energetic basis of the urban heat island. *Quarterly Journal of the Royal Meteorological Society* 108 (455): 1–24.

Oke, T.R., Johnson, G.T., Steyn, D.G., and Watson, I.D. (1991). Simulation of surface urban heat islands under "ideal" conditions at night part 2: diagnosis of causation. *Boundary-Layer Meteorology* 56 (4): 339–358.

Potapov, P., Hansen, M.C., Stehman, S.V. et al. (2008). Combining MODIS and Landsat imagery to estimate and map boreal forest cover loss. *Remote Sensing of Environment* 112 (9): 3708–3719.

Pu, R.L., Gong, P., Michishita, R., and Sasagawa, T. (2006). Assessment of multi-resolution and multi-sensor data for urban surface temperature retrieval. *Remote Sensing of Environment* 104 (2): 211–225.

Rajasekar, U. and Weng, Q.H. (2009). Spatio-temporal modelling and analysis of urban heat islands by using Landsat TM and ETM plus imagery. *International Journal of Remote Sensing* 30 (13): 3531–3548.

Rasmussen, C.E. and Williams, C.K.I. (2006). *Gaussian Processes for Machine Learning (Adaptive Computation and Machine Learning)*. The MIT Press.

Reisen, W., Lothrop, H., Chiles, R. et al. (2004). West Nile virus in California. *Emerging Infectious Diseases* 10 (8): 1369–1378.

Ruiz, M.O., Chaves, L.F., Hamer, G.L. et al. (2010). Local impact of temperature and precipitation on West Nile virus infection in Culex species mosquitoes in Northeast Illinois, USA. *Parasites and Vectors* 3: 19.

Sabins, F.F. (1997). *Remote Sensing: Principles and Interpretation*, 3e. New York: W. H. Freeman and Co.

Snyder, W.C., Wan, Z., Zhang, Y., and Feng, Y.Z. (1998). Classification-based emissivity for land surface temperature measurement from space. *International Journal of Remote Sensing* 19 (14): 2753–2774.

Sobrino, J.A., Jimenez-Munoz, J.C., and Paolini, L. (2004). Land surface temperature retrieval from LANDSAT TM 5. *Remote Sensing of Environment* 90 (4): 434–440.

Sobrino, J.A., Jimenez-Munoz, J.C., Soria, G. et al. (2008). Land surface emissivity retrieval from different VNIR and TIR sensors. *IEEE Transactions on Geoscience and Remote Sensing* 46 (2): 316–327.

Sobrino, J.A. and Raissouni, N. (2000). Toward remote sensing methods for land cover dynamic monitoring: application to Morocco. *International Journal of Remote Sensing* 21 (2): 353–366.

Stathopoulou, M. and Cartalis, C. (2009). Downscaling AVHRR land surface temperatures for improved surface urban heat island intensity estimation. *Remote Sensing of Environment* 113 (12): 2592–2605.

Streutker, D.R. (2003). Satellite-measured growth of the urban heat island of Houston, Texas. *Remote Sensing of Environment* 85 (3): 282–289.

Sun, D.L., Pinker, R.T., and Kafatos, M. (2006). Diurnal temperature range over the United States: a satellite view. *Geophysical Research Letters* 33 (5) https://doi.org/10.1029/2005GL024780.

Thomson, D.J. (1995). The seasons, global temperature, and precession. *Science* 268: 59–68.

Tomlinson, C.J., Chapman, L., Thornes, J.E., and Baker, C.J. (2011). Remote sensing land surface temperature for meteorology and climatology: a review. *Meteorological Applications* 18 (3): 296–306.

Trenberth, K.E. (1992). *Climate System Modeling*. Cambridge, UK: Cambridge University Press.

Weng, Q. (2009). Thermal infrared remote sensing for urban climate and environmental studies: methods, applications, and trends. *ISPRS Journal of Photogrammetry and Remote Sensing* 64 (4): 335–344.

Weng, Q. and Fu, P. (2014a). Modeling annual parameters of clear-sky land surface temperature variations and evaluating the impact of cloud cover using time series of Landsat TIR data. *Remote Sensing of Environment* 140: 267–278.

Weng, Q., Fu, P., and Gao, F. (2014). Generating daily land surface temperature at Landsat resolution by fusing Landsat and MODIS data. *Remote Sensing of Environment* 145: 55–67.

Weng, Q., Liu, H., Liang, B.Q., and Lu, D.S. (2008). The spatial variations of urban land surface temperatures: pertinent factors, zoning effect, and seasonal variability. *IEEE Journal of Selected Topics in Applied Earth Observations and Remote Sensing* 1 (2): 154–166.

Weng, Q. and Lu, D. (2009). Landscape as a continuum: an examination of the urban landscape structures and dynamics of Indianapolis city, 1991–2000. *International Journal of Remote Sensing* 30 (10): 2547–2577.

Weng, Q., Lu, D.S., and Schubring, J. (2004). Estimation of land surface temperature-vegetation abundance relationship for urban heat island studies. *Remote Sensing of Environment* 89 (4): 467–483.

Weng, Q., Rajasekar, U., and Hu, X. (2011). Modeling urban heat islands with multi-temporal ASTER images. *IEEE Transactions on Geoscience and Remote Sensing* 49 (10): 4080–4089.

Wu, P., Shen, H., Zhang, L., and Göttsche, F.-M. (2015). Integrated fusion of multi-scale polar-orbiting and geostationary satellite observations for the mapping of high spatial and temporal resolution land surface temperature. *Remote Sensing of Environment* 156: 169–181.

Zakšek, K. and Oštir, K. (2012). Downscaling land surface temperature for urban heat island diurnal cycle analysis. *Remote Sensing of Environment* 117: 114–124.

Zhou, Y.Y., Weng, Q., Gurney, K.R. et al. (2012). Estimation of the relationship between remotely sensed anthropogenic heat discharge and building energy use. *ISPRS Journal of Photogrammetry and Remote Sensing* 67: 65–72.

Zhu, X.L., Chen, J., Gao, F. et al. (2010). An enhanced spatial and temporal adaptive reflectance fusion model for complex heterogeneous regions. *Remote Sensing of Environment* 114 (11): 2610–2623.

Zhu, Z. and Woodcock, C.E. (2014a). Automated cloud, cloud shadow, and snow detection in multitemporal Landsat data: an algorithm designed specifically for monitoring land cover change. *Remote Sensing of Environment* 152: 217–234.

Zhu, Z. and Woodcock, C.E. (2014b). Continuous change detection and classification of land cover using all available Landsat data. *Remote Sensing of Environment* 144: 152–171.

6

Urban Heat Islands Modeling and Analysis

6.1 Introduction

Land surface temperature (LST) and emissivity data derived from satellite thermal infrared (TIR) imagery have been used in urban climate studies mainly for analyzing LST patterns and its relationship with surface characteristics, assessing the urban heat island (UHI), and relating LSTs to surface energy fluxes for characterizing landscape properties, patterns, and processes (Weng 2009). Permanent meteorological station data and moving observations using air temperatures cannot provide a synchronized view of temperature over a city. Only remotely sensed TIR data can provide a continuous and simultaneous view of a whole city, which is of prime importance for detailed investigation of urban surface climate. However, it should be noted that LSTs and air temperatures can be very different, especially in summer daytime with clear skies and high solar loading. LSTs show much larger spatial variability associated with the properties of the surface, whereas air temperatures are much less variable due to mixing of heat by the atmosphere. Among much of the existing literature in remote sensing, there is confusion between LST patterns and UHIs. Nichol (1996) suggested that the concept of a "satellite derived" UHI is largely an artifact of low spatial resolution imagery used, and the term "surface temperature patterns" is more meaningful than surface heat island. However, more and more researchers have accepted the term of surface UHI (Streutker 2002, 2003; Tran et al. 2006; Yuan and Bauer 2007).

The UHI was first defined in terms of air temperatures to represent the difference between warmer urban compared to cooler rural areas, but sufficient care must be taken to defining the urban and rural surfaces chosen for the comparison (Oke et al. 1991). The definition of UHI has since been adapted to looking at spatial patterns of surface temperature using remote sensing (Rao 1972); however, there were different opinions about what constitute the representative "urban" and "rural" surfaces. An appropriate rural reference would be

Techniques and Methods in Urban Remote Sensing, First Edition. Qihao Weng.
© 2020 by The Institute of Electrical and Electronics Engineers, Inc.
Published 2020 by John Wiley & Sons, Inc.

from a representative land-cover type typical of nonurban surroundings of a city and should be from the same type of topographical setting without major elevation change. Stewart and Oke (2009) recently developed a classification scheme of urban land-cover zones which can be used to select representative surfaces for studies of surface and air temperature heat islands. When imaged at high spatial resolution, the apparent UHI for surface temperatures may potentially be very large since extremely hot and cold individual surfaces (pixels) may be seen and measured. Therefore, what needs to be considered is what an appropriate scale is for assessing the UHI based on LSTs. It is of great scientific significance to investigate how satellite-derived LSTs can be utilized to characterize UHI phenomenon and to derive UHI parameters. Only a few studies have focused on the modeling of UHI by deriving its parameters, such as magnitude/intensity, center, spatial extent, shape, and orientation, from LST measurements (e.g. Streutker 2002, 2003; Rajasekar and Weng 2009a, 2009b; Imhoff et al. 2010; Keramitsoglou et al. 2011). In spite of these research efforts, characterizing and modeling UHIs across various spatial scales in the urban areas, especially estimation of UHI intensity/magnitude based on multi-temporal and multi-resolution TIR imagery remains a promising research direction given the increased interest among the urban climate and the environment science communities.

In this chapter, two methods for characterizing and modeling UHIs using remotely sensed LST data are introduced. A kernel convolution modeling method for two-dimensional LST imagery will be introduced to characterize and model the UHI in Indianapolis, USA, as a Gaussian process (Weng et al. 2011). The main contribution of this method lies in that UHIs can be examined as a scale-dependent process by changing the smoothing kernel parameter. Furthermore, an object-based image analysis procedure will be introduced to extract hot spots from LST maps with a case study of Athens, Greece (Keramitsoglou et al. 2011). The spatial and thermal attributes associated with these objects (hot spots) are then calculated and used for the analyses of the intensity, position, and spatial extent of UHIs.

6.2 Characterizing UHIs Using a Convolution Model

6.2.1 Study Area and Data

The City of Indianapolis, located in Marion County (Figure 6.1), Indiana has been chosen as the area of study. With over 0.8 million population, the city is the 12th largest one in the nation. Situated in the middle of the country, Indianapolis possesses several other advantages that make it an appropriate choice. It has a single central city, and other large urban areas in the vicinity have not influenced its growth. The city is located on a flat plain and is

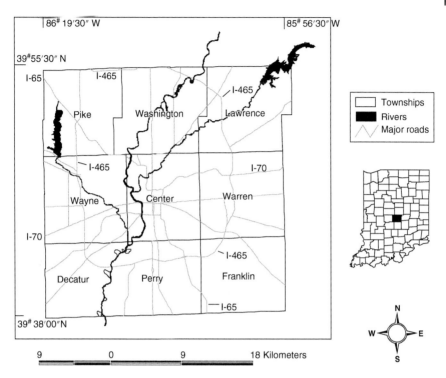

Figure 6.1 Study area – City of Indianapolis, Indiana, USA.

relatively symmetrical, having possibilities of expansion in all directions. Like most American cities, Indianapolis is increasing in population and in area. It is one of the three major cities in the Midwest having an annual population growth rate over 5%. The areal expansion is through encroachment into the adjacent agricultural and nonurban land. As a result, a new metropolitan area of nine counties (Marion, Hamilton, Johnson, Madison, Hancock, Shelby, Boone, Hendricks, and Morgan) is forming, that comprise approximately 9100 km² in area and 1.6 million population in 2000. Certain decision-making forces have encouraged some sectors of the Metropolitan Indianapolis (especially towards north, northeast, northwest, south, and more recently towards southwest) to expand faster than others. Detecting its urban expansion and the relationship to UHI development is significant to mitigate its effect and also provides important insights into the city's future urban planning and environmental management.

The primary remote sensing data used in this research include Advanced Spaceborne Thermal Emission and Reflection Radiometer (ASTER) (Fujisada 1998) and US Geological Survey's Digital Raster Graphic (DRG) imagery. ASTER images were acquired on 3 October 2000 (12 : 00 : 51 eastern standard time), 16 June 2001 (11 : 55 : 29), 5 April 2004 (11 : 46 : 39), and 13 October 2006 (11 : 40 : 01). Although we were approved by Jet Propulsion Laboratory (JPL) NASA to acquire four images annually for the study area, clouds and other weather conditions have prevented us to get a quality image every season. These selected images are the best available ones, which we intend to use for examination of the seasonal variability of urban landscape and the effect of urban growth on the UHI. The local times of satellite overpasses were all between 11 : 40 a.m. and 12 : 01 p.m. Satellite detection of UHIs using TIR sensors has demonstrated that the UHI intensity reaches the greatest in the daytime and in the warm season and least at the nighttime – the opposite to the UHI results based on the measurement of air temperatures (Roth et al. 1989).

ASTER data, including the level-1B registered radiance at the sensor and the level-2 surface reflectance, kinetic temperature, and emissivity data, were obtained. The theoretical basis and computation algorithm for surface reflectance products have been reported by Thome et al. (1999). The algorithm for converting ASTER thermal infrared measurements to LSTs has been reported by Gillespie et al. (1998) and the ASTER Temperature/Emissivity Working Group (TEWG) (1999), with which surface kinetic temperature is determined by applying Planck's Law using the emissivity values from the Temperature-Emissivity Separation (TES) algorithm. LST values calculated using this procedure are expected to have an absolute accuracy of 1–4 K and relative accuracy of 0.3 K, and surface emissivity values an absolute accuracy of 0.05–0.1 and relative accuracy of 0.005 (TEWG (1999)).

Separated single bands of visible and near-infrared (VNIR), short-wave infrared (SWIR), and TIR data were first stacked into one set. An image-to-image registration was conducted. The ASTER images were georectified to Universal Transverse Mercator (UTM) projection with NAD27 Clarke 1866 Zone 16, by using 1 : 24 000 DRG maps as the reference data. Approximately, 40–50 ground control points were chosen for each image. The images were resampled to the spatial resolution of 15 m with the nearest-neighbor resampling algorithm. The root-mean-square error (RMSE) for the geocorrection was all less than 0.3 pixel. An improved image-based dark object subtraction model has proved effective and was applied to reduce the atmospheric effects (Chavez 1996; Lu et al. 2003).

6.2.2 Kernel Convolution

A kernel convolution model implementing Gaussian bivariate function was employed to characterize LST pattern. Higdon (2002) explained this process convolution for a single-dimensional process and made suggestions for its extension to two or three dimensions. The kernel convolution results in smoothened data surfaces showing the spatial patterns of LST, GV, and IS across scales. We have written a computer program for this task using the language *R*. In order to understand how the kernel convolution works, we first need to know the definitions of kernel and convolution and then the process convolution that was extended in this study to model LST, GV, and IS as a two-dimensional Gaussian process.

Kernel smoothing refers to a general class of techniques for nonparametric estimation of functions. The kernel is a smooth positive function $k(x, h)$ which peaks at zero and decreases monotonically as x increases in size. The smoothing parameter h controls the width of the kernel function and hence the degree of smoothing applied to the data (Bowman and Azzalini 1997). One can define the kernel as a function k that for all $x, h \in X$ satisfies (Taylor and Cristianini 2004):

$$k(x,h) = (\phi(x), \phi(h)) \tag{6.1}$$

where ϕ is a mapping from X to a (inner product) feature space F.

$$\phi : x \rightarrow \phi(x) \in F \tag{6.2}$$

The degree of smoothing (i.e. the extent of standard deviation of the Gaussian distribution) to be performed by the kernel is a function defined by its bandwidth h. The value of h increases as the degree of smoothing increases and vice versa.

The convolution of f and g could be written as $f * g$. It is defined as the integral of the product of the two functions after one is reversed and shifted.

$$(f * g)(t) = \int f(\tau) g(t - \tau) dt \tag{6.3}$$

If X and Y are two independent variables with probability densities f and g, then the probability density of the sum $X + Y$ is given by the convolution $f * g$. For discrete functions, one can use a discrete version of the convolution. It is given below by Eq. (6.4):

$$(f * g)(m) = \sum_{n} f(n) g(m - n) \tag{6.4}$$

A Gaussian process over R^d is to take the independent and identically distributed Gaussian random variables on a lattice in R^d and convolve them with a kernel. The process involves successive increase in the density of the lattice by a factor of two in each dimension and reduces the variance of the variates by a factor of $2d$, leading to a continuous Gaussian white noise process over R^d. A Gaussian white noise is a white noise with normal distribution. The convolution of this process can be equivalently defined by using some covariogram in R^d. The process of convolution gives very similar results to defining a process by the covariogram.

Let $y_{(1,1)}, \ldots, y_{(i,j)}$ (where y is a two-dimensional matrix of $(1,1), \ldots, (i,j)$) be data recorded over the two-dimensional spatial locations $s_{(1,1)}, \ldots, s_{(i,j)}$ in S, a spatial process $z = (z_{(1,1)}, \ldots, z_{(i,j)})^T$, and Gaussian white noise $\varepsilon = (\varepsilon_{(1,1)}, \ldots, \varepsilon_{(i,j)})^T$ with variance σ_ε^2. The spatial method represents the data as the sum of an overall mean μ.

$$y = s + z + \varepsilon \tag{6.5}$$

where the elements of z are the restriction of the spatial process $z(s)$ to the two-dimensional data locations $s_{(1,1)}, \ldots, s_{(i,j)}$. $z(s)$ is defined to be a mean zero Gaussian process. Rather than specifying $z(s)$ through its covariance function, it is determined by the latent process $x(s)$ and the smoothing kernel $k(s)$. The latent process $x(s)$ is restricted to be nonzero at the two-dimensional spatial sites $\omega_{(1,1)}, \ldots, \omega_{(a,b)}$, also in S and define $x = (x_{(1,1)}, \ldots, x_{(a,b)})^T$ where $x\omega_p = x(\omega_p)$ and $p = (1,1), \ldots, (a,b)$. Each x_p is then modeled as independent variable and draws from an $N(0, \sigma_\varepsilon^2)$ distribution. The resulting continuous Gaussian process is then

$$Z(S) = \sum_{p=(1,1)}^{(a,b)} x_j k(s - \omega_p) \tag{6.6}$$

where $k(s - \omega_p)$ is a kernel centered at ω_p. This gives a linear model:

$$y = \mu l_{(i,j)} + Kx + \varepsilon \tag{6.7}$$

where $l_{(i,j)}$ is the (i,j)th vector of l's, the elements of K are given by

$$K_{pq} = k(s_p - \omega_q) x_q \tag{6.8}$$

$$x \sim N(0, \sigma_x^2 l_{(a,b)}) \text{ and} \tag{6.9}$$

$$\varepsilon \sim N(0, \sigma_x^2 l_{(i,j)}) \tag{6.10}$$

The smoothing kernel or the parameter, as described in the Eq. (6.8), defines the degree of smoothing and is a crucial parameter in the kernel convolution modeling. The degree of smoothing is inversely proportional to the smoothing kernel. In other words, as the value of the smoothing kernel decreases, the degree of smoothing over the spatial domain increases. When the degree of smoothing reached the maximum value of one, a kernel-convoluted image was formed, whose values were equivalent to the mean of the original image. When the degree of smoothing was minimum (i.e. zero smoothing), the final kernel-convoluted image would be the same as that of the original image.

6.2.3 UHI Intensity Across Spatial Scales

By changing the smoothing parameter from 0.05 to 1.0, with an incremental rate of 0.05, the kernel convolution modeling resulted in 20 convoluted maps for each LST image. Figure 6.2 shows the convoluted LST maps with the smoothing parameter ranging from 0.05 to 0.9 on 3 October 2000. We further computed the magnitude (intensity) of UHI by subtracting the minimum temperature value from the maximum temperature value of each convoluted image. Table 6.1 shows the UHI intensity value at each smoothing level. An examination of data distribution pinpointed two breaks at the smoothing level of 0.2 and 0.5. When the smoothing parameter ranged from 0.05 to 0.2, the UHI intensity mostly reached 6 K, peaking around 14 K. The thermal landscape exhibited scattered hot spots (Figure 6.2). This type of thermal map would be suitable for UHI studies at the microscale. The term "surface temperature patterns" is more meaningful here than surface UHI according to Nichol (1996). The information on microscale LST patterns may be useful for zoning considerations in the process of city planning and environmental management at the relevant geographical scale. For the purpose of comparison, Table 6.1 also lists the minimum–maximum temperature differences for the original LST images, which have values from 15.7 to 18.3 K. Streutker (2002) suggested that UHI

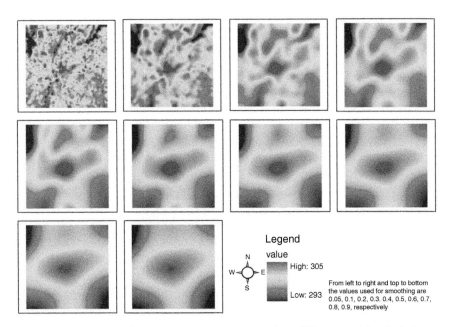

Figure 6.2 Urban heat island (UHI) pattern in Indianapolis at different spatial scales in 3 October 2000. *Source:* after Weng et al. (2011).

Table 6.1 Urban heat island (UHI) intensity values at each smoothing parameter level.

Smoothing parameter	UHI intensity (K)			
	3 October 2000	16 June 2001	5 April 2004	13 October 2006
None	18.3	15.7	16.3	15.8
0.05	12.5	9.6	11.4	14.1
0.1	9.0	8.2	8.5	6.9
0.15	7.5	7.3	7.0	5.5
0.2	6.6	6.6	6.2	5.0
0.25	6.0	6.0	5.5	4.6
0.3	5.5	5.6	5.1	4.3
0.35	5.0	5.1	4.7	3.9
0.4	4.5	4.6	4.2	3.6
0.45	4.1	4.2	3.8	3.3
0.5	3.6	3.8	3.4	3.0
0.55	3.3	3.5	3.1	2.8
0.6	2.9	3.3	2.8	2.6
0.65	2.7	3.0	2.5	2.3
0.7	2.4	2.8	2.3	2.1
0.75	2.2	2.6	2.1	2.0
0.8	2.0	2.5	2.0	1.8
0.85	1.9	2.3	1.8	1.6
0.9	1.7	2.2	1.7	1.5
0.95	1.6	2.0	1.6	1.4
1.0	1.5	1.9	1.4	1.3

Source: Weng et al. (2011). Reproduced with permission of IEEE.

magnitude computed by using single-pixel comparison method was strongly influenced by localized extreme temperatures and tended to be several times higher than the Gaussian magnitude. When the smoothing parameter ranged from 0.25 to 0.5, the UHI intensity yielded a value from 3 to 6 K, subject to the seasonality. Two to four hot spots can be identified from each LST image, with the largest hot spot in the city center corresponding to its central business district. These hot spots can be directly linked to the biophysical structure of the city and the structure of the urban atmosphere and are thus suitable for studies at the mesoscale (i.e. the city scale). These types of maps would provide valuable information for city planning, especially for transportation, large infrastructures, and clustered development (e.g. leap-frog development). Figure 6.2 shows that the hot spots identified at this scale appeared to associate with major infrastructures and urban clusters in the city (e.g. major transportation network, CBD, commercial/industrial cluster, etc.). By setting the smoothing parameter in the range of 0.55–1.0, the kernel convolution modeling yielded consistently single-hot spot thermal landscape pattern. This kind of spatial pattern is essentially in agreement with the conventional definition of UHI, which reveals the difference between the urban and the rural temperatures of a specific city. It should be noted that the conventional definition of UHI being referred to here is the air temperature heat island, which has shown the greatest intensity at night (Oke 1982). Most UHI intensity values computed in this study ranged from 1.5 to 3.5 K, despite of overtime changes in the UHI center position, spatial extent, and shape. The studies on such UHI pattern and its dynamics would contribute to understanding of urban climate. More importantly, the modeling at this macroscale would allow for concurrent comparison of UHIs across nearby cities in the region. The information provided by such studies would be beneficial for regional planning and watershed management of a city region (i.e. an urban cluster).

By comparing the convoluted LST maps of different dates at the same smoothing levels, it is possible to observe the changes in the urban thermal landscape pattern over the time. The changes in the thermal pattern reflected largely the changes in "surface" factors, including the thermal properties of surface materials, the composition and layout of land-use and land-cover (LULC) types, and anthropogenic effects related to the existence of automobiles, air conditioning units, industries, and air pollution. As these images were acquired in different seasons (spring summer and fall) spanning several years (2000–2006), the variation in the surface factors embraced the effects of seasonal change and long-term change due to urban sprawl. The UHI intensity was greatest in the summer (16 June 2001) as far as the macro- and mesoscales were concerned. This observation is consistent with the findings in previous literature (Roth et al. 1989). However, at the microscale, the UHI intensity was strongly influenced by pixels with extreme LST values. At all smoothing levels, the UHI intensity was lowest on 13 October 2006. The only exception was observed at the smoothing parameter 0.05, when the UHI intensity was detected to be the greatest on 13 October 2006. This anomaly can be explained by the coexistence of the extreme low LST (291 K) and the extreme high LST (305 K) on that date.

6.2.4 Summary

A key issue in thermal remote sensing of urban areas is how to use pixel-based LST measurements to characterize and quantify UHIs observed at various spatial scales. This study has employed a kernel convolution modeling method to characterize the UHI in Indianapolis as a two-dimensional Gaussian process. The modeling results allowed us to analyze several key UHI parameters, including the intensity, center, spatial extent, and orientation, but this study focused on the UHI intensity only.

In this study, we utilized the smoothing kernel to characterize the spatial nonstationarity in the urban thermal landscape. This characterization and the linkage allow us to examine UHI as a scale-dependent process. Oke (2008) suggested that there were often several linked UHIs in one city, and that each distinguished from one another mainly by *scales* imposed by the biophysical structure of a given city and the structure of the urban atmosphere. Each UHI is therefore required measurement arrays appropriate to the scale (Oke 2008). Most previous literature failed to account for the existence of multiple UHIs in a city (Streutker 2002; Imhoff et al. 2010). By changing the smoothing parameter, this study was able to characterize multiple UHIs of various sizes (spatial extent) in Indianapolis. Further, we categorized the smoothing parameters into three groups. Each group was found suitable to study the urban thermal landscape at a particular spatial scale, i.e. the microscale, mesoscale, and regional scale, such that the identified scales can be matched to various applications in urban planning (and environmental management). Further studies are necessary to reconcile studies of the UHI scale (e.g. this study) and the operational scale of LST (Schmid 1988; Voogt and Oke 1998; Weng et al. 2004; Liu and Weng 2009) in order to better understand urban thermal landscapes and dynamics from the remote sensing perspective.

6.3 Object-Based Extraction of Hot Spots

6.3.1 Study Area and Data Sets

The study area is the Athens Greater Area, the capital and largest city of Greece. According to the recent census paper of Eurostat (http://epp.eurostat. ec.europa.eu), the Athens Larger Urban Zone (LUZ) is the eighth most populated LUZ in the European Union with a population of about 4 000 000. The area lies at the southeasternmost edge of the Greek mainland (Figure 6.3). Athens sprawls across a central basin bound by four mountains and the Saronic Gulf in the southwest. The basin is bisected by a series of small hills. These specific topographic characteristics make Athens an example of a coastal city located in very complex terrain, where sea-breeze and heat-island circulations interact (Dandou et al. 2009) making it especially interesting for UHI and thermal pattern studies. In terms of climate, Athens enjoys a prolonged warm and dry period during the year with July and August being the hottest and driest months. The normal value of the summer (from June to August) daily maximum temperature at Athens is 31.6 °C (wrt 1961–1990 period; Founda and

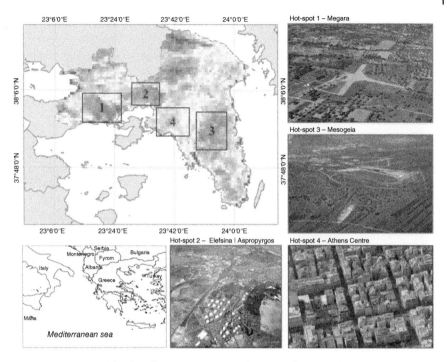

Figure 6.3 A typical land surface temperature (LST) map of Greater Athens Area, Greece, during summer. Numbers 1–4 denote the four characteristic surface temperature patterns (hot spots) of the study area, namely 1: Megara, 2: Elefsina and Aspropyrgos, 3: Mesogeia, and 4: City Centre. A representative subarea of each hot spot is also provided. *Source:* Keramitsoglou et al. (2011). Reproduced with permission of Elsevier.

Giannakopoulos 2009). In summertime, the city suffers from heat wave events. Particularly in June 2007, a severe heat wave lasted for seven days reaching temperatures as high as 46 °C. In the daytime, the LST patterns of Athens Greater Area yield the presence of two dominant and one weaker hot spots as shown in Figure 6.3. The two dominant hot spots occur at the Municipalities of Megara (denoted by rectangular "1" in the figure) and Elefsina and Aspropyrgos ("2") while the weaker one occurs at the area of Mesogeia ("3"). During nighttime, a typical UHI is developed related clearly to the urban and industrial zone city center ("4").

A nine-year time series of MODIS summer images was analyzed. Specifically, 1085 daytime MODIS-Terra LST images and 1956 nighttime MODIS-Terra and Aqua LST images were produced and archived as part of the requirements of UHIs and Urban Thermography project (21913/08/I-LG) funded by the European Space Agency. In total, more than 3000 raster LST maps of 1-km spatial resolution were tested and processed using the algorithm described in this paper.

The original MODIS data used in the present study were procured at no charge from the Warehouse Inventory Search Tool (WIST web site) repository. Firstly, the MODIS onboard Terra satellite was considered and all the available scenes covering the area of interest were selected. The subset of the daytime scenes with a central viewing angle larger than 45° was identified and removed, due to the well-documented directional dependence of the LST (Lagouarde and Irvine 2008).

The selected TERRA images consisted of one day- and one nighttime scene, although a number of days (about two days every six) had no available data sets due to the viewing angle constraint. Consequently, MODIS Aqua images were selected so as to ensure that at least one daytime scene is available per day and one to two nighttime scenes are also available considering that the UHI effect is more evident during the night. At that stage no cloud coverage limitation was set, nevertheless, this was taken into consideration further in the processing chain.

6.3.2 Methodology

The complete procedure includes gap filling, filtering, global thresholding, generation of pixel regions, and extracting features from regions. The methodology implemented for the calculation of the LST maps is based on the Split Window Technique (SWT) from Jiménez-Muñoz and Sobrino (2008). LST maps were preprocessed. The input data were the MODIS LST products, as described above. The LST maps were in Hierarchical Data Format, commonly abbreviated HDF. The data preprocessing included masking out not valid pixels based on a corresponding flag to assure that only cloud-free, not obliquely viewed (>40°) land pixels would be used further in the analysis. Once the valid images and pixels were selected, the problem of missing (flagged) pixels had to be tackled by applying an averaging technique on invalid pixels, where their LST value was actually substituted by the average LST value of their adjacent valid pixels. This procedure – based on a kernel convolution method suggested by Rajasekar and Weng (2009a) – can be applied in an iterative manner deploying subsequent passes through the invalid pixels of the image. In our case, a single pass was selected as the procedure's critical step since complete restoration would probably produce erroneous patterns within the image. The procedure adopted was necessary in order to facilitate the process of automated information extraction from a large set of images. The resulting image was subsequently filtered using a similar averaging technique as the one presented in the gap filling problem. Filtering was used for smoothing pixel LST so that extreme values were appropriately throttled.

The next step of the preprocessing was the extraction of hot-spot objects through global image thresholding. In the case of global thresholding, the window addressed was the entire image. All pixels whose LST was higher than the suburban LST plus a predefined threshold value (let us say 6 °C) were registered as potential hot pixels. Therefore, the definition of the suburban area, which would serve as the reference area for the analysis, was an important issue. The area selected was the municipality of Thrakomakedones in Attica. The municipality is located at the foot of Parnitha Mountain. It is a sparsely built up municipality near a forested area. For each individual image, the procedure evaluated the mean LST for the pixels corresponding to this municipality and hereafter we will refer to this temperature as RLST (Reference Land Surface Temperature). It is stressed that each LST image had its own RLST value. Following the calculation of the reference LST, each pixel LST of the image was compared to RLST.

An object-based image analysis procedure was developed to extract hot spots. According to this procedure, hot pixel image groups were extracted using a global thresholding algorithmic procedure. Separation of hot pixel groups was automatically performed through appropriate segmenting of the initial image by partitioning the pixels whose LST values were above a threshold value plus the average LST of reference area pixels. For these pixels, segmentation was performed by determining the k-groups (Pal and Pal 1993). This type of groups were characterized by a certain proximity property for the pixels of the group, that is to say, each pixel in the group had at least one pixel neighbor that was close to the former by a distance of k pixels in any direction. This distance metric adopted was actually the norm of the digital space processed (image). In this way, depending on the value of k, smaller groups could be merged to bigger ones. The k pixel distance parameter was set to one (each pixel had at least one neighbor at a distance of one pixel). An algorithm of linear computational complexity was used to separate groups obtained. According to this algorithm, each pixel was checked for its neighbors, which in turn were checked for theirs in a recursive way. The grouped pixels were then treated as different regions (objects). Following that, several features related to these objects, which represent the city hot spots, were extracted.

The hot spots represented as objects underwent a series of procedures to extract the necessary information for the area. As a first step, a procedure counted the number of pixels that belonged to each object. If the object area was smaller than $10 \, km^2$ (i.e. number of pixels was less than 10), then the object was not further studied. Only the objects whose weighted centroid belonged to a predefined area of interest were taken into consideration. Once the objects of interest were extracted, a number of parameters per object were calculated and appropriately stored in a database. These parameters included the satellite sensor as well as temporal, spatial (weighted centroid location, extent in square kilometer), and thermal information (e.g. minimum, maximum LST). In total, 1237 daytime hot spots were identified with a 6 °C threshold value and 358 nighttime hot spots with a 4 °C threshold.

6.3.3 General Thermal Pattern in the Greater Athens

Four areas of interest can be defined in the majority of the LST maps for the Athens Greater Area. In particular, there are three daytime hot spots (hot spots No. 1, 2, 3; Figure 6.3 and Table 6.2) and one nighttime (hot spot No. 4; Figure 6.3). On the daytime images (most of which were acquired before 12 : 00 Local time), surface temperature spatial patterns revealed that the center of Athens did not exhibit the highest LST; this is referred to as urban heat sink or negative heat island. It can be simply considered as a brief stage in the development of the UHI that occurs during the morning and midday period due to differences in the urban–rural warming rates (Oke 1982). This behavior is well known for Athens, as already reported in Stathopoulou and Cartalis (2007b). Therefore, in the daytime, the LST patterns of Athens Greater Area yielded the presence of two dominant and one weaker hot spots, as presented in Figure 6.3.

The typical daytime pattern in 3D can be also observed in Figure 6.4a. In this figure, the LST images were plotted in a wire frame model to appreciate visually the thermal variations as "height" (z-axis) and color. Visual analysis of the hot spots (especially the two dominant ones, hot spots Nos. 1 and 2) showed that there was an LST discrepancy of at least 4–6 °C from the suburban areas. They appeared in almost every morning image of June to August; the phenomenon was weaker in May and September. These hot spots were clearly identifiable by photointerpretation of satellite images on all spatial resolutions (from

Table 6.2 Fitted Maximum LST of the four hot spots of Athens Greater Area per year with the corresponding date when it occurred.

	Hot-spot							
	1 (daytime)		2 (daytime)		3 (daytime)		4 (nighttime)	
Year	Date	Max LST	Date	Max LST	Date	Max LST	Date	Max LST
2000	7 July	50.2	9 July	50.3	10 July	47.9	19 July	26.4
2001	13 July	49.5	15 July	49.0	12 July	46.3	1 August	27.8
2002	11 July	47.9	16 July	49.1	16 July	46.8	30 July	26.1
2003	17 July	48.0	18 July	48.2	19 July	45.5	29 July	25.8
2004	16 July	48.7	18 July	49.4	15 July	47.0	2 August	26.5
2005	17 July	48.3	16 July	48.0	18 July	45.9	29 July	25.9
2006	13 July	46.6	12 July	47.4	12 July	45.3	24 July	26.1
2007	12 July	49.2	14 July	48.4	15 July	46.9	1 August	27.4
2008	12 July	50.4	17 July	49.1	10 July	47.4	27 July	26.2

Source: Keramitsoglou et al. (2011). Reproduced with permission of Elsevier.

Landsat Thematic Mapper [TM] to Meteosat Second Generation Spinning Enhanced Visible and Infrared Imager [MSG-Seviri], not shown here). It should be noted that the time of MODIS acquisitions allowed studying of the phenomenon at certain time (around 09 : 00 UTC; midday 12 : 00 local summertime). Another interesting feature of Figure 6.4a is the presence of

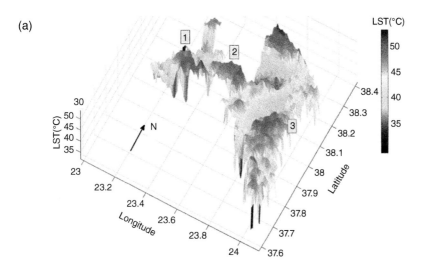

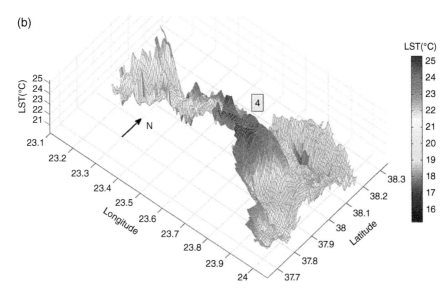

Figure 6.4 Typical 3D LST patterns in Athens Greater area using MODIS. (a) Daytime hot spots No. 1, 2, and 3. (b) Nighttime hot spot No. 4. *Source:* Keramitsoglou et al. (2011). Reproduced with permission of Elsevier.

cooler pixels along the coastline, which may be partially caused by sea breeze. In general, the thermal environment of Athens during daytime depends on the combined influence of the area topography and surface cover characteristics. The open plain of Mesogeia (hot spot No. 3, Figure 6.3) is mainly covered with sparse low vegetation (particularly olive trees and vineyards) and bare soil (Athens International Airport "El. Venizelos" is also located at this area) while Aspropyrgos and Elefsina (hot spot No. 2 of Figure 6.3) are mainly industrial zones. These plains become warm faster than densely built-up urban areas which are extensively covered by materials of high thermal inertia such as concrete and asphalt. The influence of topography is also evident on the thermal pattern of the study area, as higher altitudes exhibit lower LST.

During the nighttime, the thermal pattern of Athens was inverted as higher surface temperatures were related with the residential urban zones rather than the different urban use zones and rural areas. There was one dominant hot spot affecting a number of central municipalities, as described in Table 6.1. A 3D representation of a typical nighttime LST image is shown in Figure 6.4b. At that time of day, cooling or warming of a surface is determined by its thermal characteristics. Thus, during night hours, the continuous urban fabric is a few degrees warmer than rural areas, due to the lower thermal inertia of the soil compared to concrete. In contrast, the morning hot spots of the city have faded out in the night and appear to be cooler than the continuous urban fabric owning to the fact that in industrial and agricultural areas usually extended open spaces of bare soil cover most of the area. The presence of a heat wave increases LST and makes less definable the hot spots.

6.3.4 Time Series of Maximum LST

The high volume of data allowed the extraction of statistics and the investigation of possible correlations and trends. In this paper, we present the results of the analysis including the behavior of the maximum LST of the different hot spots over time, the frequency of occurrence of the thermal intensity (discrepancy between hot spot maximum and RLST), and the correlation between the thermal intensity and the size of the hot spots.

The maximum LST of the hot spots (hereafter referred to as maxLST) was analyzed by plotting the maxLST acquired between 08 : 55 and 10 : 00 UTC against date (1 May–30 September) for every year separately and fitting a quadratic polynomial to the data. The reason behind the selection of a certain time window was to eliminate the diurnal variation signal from the analysis. In any case, this specific window was the one with most of the daytime MODIS acquisitions ensuring minimum loss of data. The results per year and hot spot are presented in Table 6.2, while for all years are shown in Figure 6.5 (day = 0 is the 1 May and day = 152 is the 30 September; the year was irrelevant). The plot clearly shows a consistent seasonal variation with increasing maxLST from the beginning until approximately middle of the season and then the gradual decrease of temperature until the end of the season (end of September). Figure 6.5a shows the described seasonal trend of Megara hot spot (No. 1) with a maximum of 48.6 °C on 14 July, after when the maxLST starts decreasing. The same behavior was observed at Elefsina-Aspropyrgos hot spot (No. 2); in this case, the maximum was observed on 16 July and was of the same value. The third slightly weaker hot spot of Mesogeia (No. 3) was consistently cooler (between 1 and 3 °C; Table 2) compared to the other two dominant hot spots. As a result, the maximum of the overall seasonal trend (Figure 6.5a) was 2 °C cooler and was observed on 15 July.

For the nighttime images, a 4 °C threshold was selected for the city center, as it was more realistic and increased the number of objects and subsequently

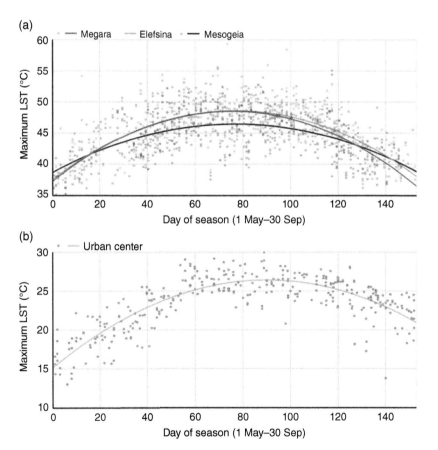

Figure 6.5 Maximum LST acquired at (a) the three daytime hot spots between 08 : 55 and 10 : 00 UTC during the summer months (May to September) from 2000 to 2009 as a function of day of season, and (b) Athens City Centre hot spot (No. 4) at nighttime for the same period. *Source:* Keramitsoglou et al. (2011). Reproduced with permission of Elsevier.

statistical samples. The maxLST for the city center (Figure 6.5b) follows a similar seasonal pattern. As expected, the maxLST was considerably lower than the morning hot spots. The quadratic line fitted to all the available maxLST nighttime data versus day of the season peaked on 29 July, two weeks later than the daytime surface patterns. The same behavior was observed for all years (Table 2). By definition, this was the UHI pattern of Athens city. Given the special interest in UHI, an additional plot depicting the yearly average behavior of nighttime UHI was generated.

6.3.5 Summary

The main innovation of this research is that the thermal hot spots were extracted and treated as objects. This allowed for the calculation of several features related to the hot spots (e.g. the area extent, the maximum, the mean, and the minimum LST) from the original LST maps. Other recent studies (Streutker 2003; Rajasekar and Weng 2009b) had dealt with modeled approximations (e.g. fast Fourier transformation; FFT) of the temperature patterns. For instance, Streutker (2003) approximated the area extent of the UHIs by calculating the longitudinal and latitudinal extent assuming an ellipsoid footprint. Additionally, our method was used to extract the attributes and to present them as a series of functions rather than a series of images, as suggested by Rajasekar and Weng (2009a). The thermal features retained their original values circumventing modeling and smoothing out the maximum values of the original data. The proposed methodology can be considered appropriate for both coastal cities, where the discontinuity of sea–land interface would cause complications in the modeling approach, as well as inland cities. The methodology can be customized to cover other geographical areas. Nevertheless, sufficient knowledge of a city is required to efficiently define the areas of interest as well as the reference rural/suburban area. Information on the city's land cover and topography is also important for the interpretation of the patterns. However, once the area of interest is set, the algorithm processes thousands of LST images in a few seconds without any manual interference. As comparison of thermal patterns across different cities is currently hampered by the lack of an appropriate methodology to extract the patterns and characterize them, future research should examine the application of the method to different cities and comparison of their UHIs behavior.

References

ASTER Temperature/Emissivity Working Group (1999). *Temperature/Emissivity Separation Algorithm Theoretical Basis Document*, Version 2.4. JPL, California Institute of Technology. https://unit.aist.go.jp/igg/rs-rg/ASTERSciWeb_AIST/en/documnts/pdf/2b0304.pdf (accessed 30 November 2009).

Bowman, A.W. and Azzalini, A. (1997). *Applied Smoothing Techniques for Data Analysis*, Oxford Statistical Science Series. Oxford, New York: Clarendon Press.

Chavez, P.S. Jr. (1996). Image-based atmospheric corrections – revisited and improved. *Photogrammetric Engineering & Remote Sensing* 62: 1025–1036.

Dandou, A., Tombrou, M., and Soulakellis, N. (2009). The influence of the city of Athens on the evolution of the sea-breeze front. *Boundary-Layer Meteorology* 131: 35–51.

Founda, D. and Giannakopoulos, C. (2009). The exceptionally hot summer of 2007 in Athens, Greece – a typical summer in the future climate? *Global and Planetary Change* 67: 227–236.

Fujisada, H. (1998). ASTER level-1 data processing algorithm. *IEEE Transactions on Geoscience and Remote Sensing* 36: 1101–1112.

Gillespie, A., Rokugawa, S., Matsunaga, T. et al. (1998). A temperature and emissivity separation algorithm for advanced spaceborne thermal emission and reflection radiometer (ASTER) images. *IEEE Transactions on Geoscience and Remote Sensing* 36: 1113–1126.

Higdon, D. (2002). Space and space–time modeling using process convolution. In: *Quantitative Methods for Current Environmental Issues* (eds. C.W. Anderson, V. Barnett, P.C. Chatwin and A.H. El-Shaarawi), 37–56. London: Springer-Verlag.

Imhoff, M.L., Zhang, P., Wolfe, R.E., and Bounoua, L. (2010). Remote sensing of the urban heat island effect across biomes in the continental USA. *Remote Sensing of Environment* 114 (3): 504–513.

Jiménez-Muñoz, J.C. and Sobrino, J.A. (2008). Split-window coefficients for land surface temperature retrieval from low-resolution thermal infrared sensors. *IEEE Geoscience and Remote Sensing Letters* 5: 806–809.

Keramitsoglou, I., Kiranoudis, C.T., Ceriola, G. et al. (2011). Extraction and analysis of urban surface temperature patterns in Greater Athens, Greece, using MODIS imagery. *Remote Sensing of Environment* 115 (12): 3080–3090.

Lagouarde, J.-P. and Irvine, M. (2008). Directional anisotropy in thermal infrared measurements over Toulouse city center during the CAPITOUL measurement campaigns: first results. *Meteorology and Atmospheric Physics* 102: 173–185.

Liu, H. and Weng, Q. (2009). Scaling-up effect on the relationship between landscape pattern and land surface temperature. *Photogrammetric Engineering & Remote Sensing* 75 (3): 291–304.

Lu, D., Moran, E., and Batistella, M. (2003). Linear mixture model applied to Amazônian vegetation classification. *Remote Sensing of Environment* 87: 456–469.

Nichol, J.E. (1996). High-resolution surface temperature patterns related to urban morphology in a tropical city: a satellite-based study. *Journal of Applied Meteorology* 35 (1): 135–146.

Oke, T.R. (1982). The energetic basis of the urban heat island. *Quarterly Journal of the Royal Meteorological Society* 108: 1–24.

Oke, T.R. (2008). The continuing quest to understand urban heat islands. Presentation at the 2nd Workshop on Earth Observation for Urban Planning and Management, Hong Kong Polytechnic University, China, 20–21 May 2008.

Oke, T.R., Johnson, G.T., Steyn, D.G., and Watson, I.D. (1991). Simulation of surface urban heat islands under "ideal" conditions at night. Part 2: diagnosis of causation. *Boundary-Layer Meteorology* 56 (4): 339–358.

Pal, N.R. and Pal, S.K. (1993). A review on image segmentation techniques. *Pattern Recognition* 26: 1277–1294.

Rajasekar, U. and Weng, Q. (2009a). Urban heat island monitoring and analysis by data mining of MODIS imageries. *ISPRS Journal of Photogrammetry and Remote Sensing* 64 (1): 86–96.

Rajasekar, U. and Weng, Q. (2009b). Spatio-temporal modeling and analysis of urban heat islands by using Landsat TM and ETM+ imagery. *International Journal of Remote Sensing* 30 (13): 3531–3548.

Rao, P.K. (1972). Remote sensing of urban heat islands from an environmental satellite. *Bulletin of the American Meteorological Society* 53: 647–648.

Roth, M., Oke, T.R., and Emery, W.J. (1989). Satellite derived urban heat islands from three coastal cities and the utilisation of such data in urban climatology. *International Journal of Remote Sensing* 10: 1699–1720.

Schmid, H.P. (1988). Spatial scales of sensible heat flux variability: representativeness of flux measurements and surface layer structure over suburban terrain. PhD thesis. The University of British Columbia, Vancouver, BC.

Stathopoulou, M. and Cartalis, C. (2007b). Use of Satellite Remote Sensing in Support of Urban Heat Island Studies. *Advances in Building Energy Research* 1: 203–212.

Stewart, I. and Oke, T. (2009). A new classification system for urban climate sites. *Bulletin of the American Meteorological Society* 90 (7): 922–923.

Streutker, D.R. (2002). A remote sensing study of the urban heat island of Houston, Texas. *International Journal of Remote Sensing* 23: 2595–2608.

Streutker, D.R. (2003). Satellite-measured growth of the urban heat island of Houston, Texas. *Remote Sensing of Environment* 85: 282–289.

Taylor, J.S. and Cristianini, N. (2004). *Kernel Methods for Pattern Analysis*. Cambridge, UK: Cambridge University Press.

Thome, K., Biggar, S., and Takashima, T. (1999). Algorithm theoretical basis document for ASTER level 2B1-surface radiance and ASTER level 2B5-surface reflectance, p. 45. https://lpdaac.usgs.gov/documents/74/AST_05_ATBD.pdf (accessed 4 July 2019).

Tran, H., Uchihama, D., Ochi, S., and Yasuoka, Y. (2006). Assessment with satellite data of the urban heat island effects in Asian mega cities. *International Journal of Applied Earth Observation and Geoinformation* 8 (2006): 34–48.

Voogt, J.A. and Oke, T.R. (1998). Effects of urban surface geometry on remotely sensed surface temperature. *International Journal of Remote Sensing* 19: 895–920.

Weng, Q. (2009). Thermal infrared remote sensing for urban climate and environmental studies: methods, applications, and trends. *ISPRS Journal of Photogrammetry and Remote Sensing* 64 (4): 335–344.

Weng, Q., Lu, D., and Schubring, J. (2004). Estimation of land surface temperature-vegetation abundance relationship for urban heat island studies. *Remote Sensing of Environment* 89: 467–483.

Weng, Q., Rajasekar, U., and Hu, X. (2011). Modeling urban heat islands with multi-temporal ASTER images. *IEEE Transactions on Geosciences and Remote Sensing* 49 (10): 4080–4089.

Yuan, F. and Bauer, M.E. (2007). Comparison of impervious surface area and normalized difference vegetation index as indicators of surface urban heat island effects in Landsat imagery. *Remote Sensing of Environment* 106 (3): 375–386.

7

Estimation of Urban Surface Energy Fluxes

7.1   Introduction

The seasonal and spatial variability of surface heat fluxes is crucial to the understanding of urban heat island (UHI) phenomenon and dynamics. To estimate energy fluxes, remote sensing-derived biophysical variables need to be integrated with surface atmospheric parameters measured in meteorological stations or *in situ* field measurements. Most of previous UHI studies have been conducted at the mesoscale using an energy budget approach, which separates the energy flow into measurable interrelated components for simulation modeling. Numerous methods have been proposed and applied, such as *in situ* field measurement, the one-source surface energy balance algorithm (SEBAL), and the two-source energy balance (TSEB) algorithm (Bastiaanssen et al. 1998a, 1998b; Oke et al. 1999; Kato and Yamaguchi 2005, 2007; Timmermans et al. 2007). Oke et al. (1999) obtained the net radiation, sensible, and latent heat flux densities at the roof level by direct measurement. Bastiaanssen et al. (1998a, 1998b) developed the SEBAL algorithm to independently calculate heat fluxes. Kato and Yamaguchi (2005) developed a method to quantify anthropogenic heat discharge as the residue of the energy balance model and further applied the method to estimate storage heat flux (Kato and Yamaguchi 2007). Timmermans et al. (2007) compared SEBAL and TSEB and found that they both contained advantages and disadvantages. The major difference between SEBAL and TSEB lies in whether soil and vegetation are treated separately. For TSEB, heat fluxes are calculated separately for soil and vegetation, while for SEBAL, they are calculated as a composite. A large portion of the Earth's land surfaces are only partially vegetated. Hence, a two-source model can generally estimate the surface energy balance with higher accuracy than a one-source model, especially when the two sources show very different radiometric behavior and atmospheric coupling (Timmermans et al. 2007). The existing literature suggests that there has been a great deal of effort in assessing urban surface energy balance from both observations (Arnfield 2003; Grimmond 2006) and

Techniques and Methods in Urban Remote Sensing, First Edition. Qihao Weng.
© 2020 by The Institute of Electrical and Electronics Engineers, Inc.
Published 2020 by John Wiley & Sons, Inc.

from modeling (Martilli 2007). However, satellite-based assessment of urban energy budgets has received the attention of scientific community less than it deserves, in particular, modeling energy fluxes at the intra-urban scale, in spite of the existence of a few interesting studies (Kato and Yamaguchi 2005, 2007; Rigo and Parlow 2007; Frey et al. 2011).

Because the morphological characteristics of urban areas that modify urban climate occur at the microscale, it is a great challenge to conduct field studies of energy budget and to understand fully the UHI mechanism without a good understanding of urban morphology. Numerous remote sensing techniques have been developed in support of effective urban sensing of morphological characteristics, detection and monitoring of discrete land-use/cover types, and estimation of biophysical variables (Jensen and Cowen 1999; Weng 2012). In urban climate studies, remotely sensed thermal infrared (TIR) imagery has been used to measure land surface temperature (LST) and emissivity (Weng 2009). These measurements provide essential data for analyzing urban thermal landscape pattern and its relationship with surface biophysical characteristics, assessing UHI effect, and relating LST with surface heat fluxes for character-izing landscape properties, patterns, and processes (Quattrochi and Luvall 1999). Through the combined use of TIR and optical sensing data, remote sensing imagery can be used to estimate key surface parameters necessary for estimating surface energy fluxes, which are related to the soil–vegetation sys-tem and surface soil moisture, radiation forcing components, and indicators of the surface response to them (i.e. LST) (Schmugge et al. 1998). Previous studies have focused on the methods for estimating variables related to the soil–vegetation system and radiation forcing components, but few studies have been done to estimate the surface atmospheric parameters with validation from ground-based meteorological or surface radiation instruments (Schmugge et al. 1998). Estimates of surface energy fluxes using remote sensing have been validated with *in situ* data, but for the most part, have been limited to studies of agricultural or forested landscapes at relatively small spatial scales. For example, Humes et al. (2000) estimated spatially distributed surface energy fluxes over two small instrumented watersheds in Oklahoma with airborne and satellite-borne data. The model estimates of surface energy fluxes com-pared well with ground-based measurements of surface flux. Wang and Liang (2009) compared Moderate Resolution Imaging Spectroradiometer (MODIS) and Advanced Spaceborne Thermal Emission and Reflection Radiometer (ASTER) data LST and emissivity products with longwave radiation observa-tions during 2000–2007 derived from Surface Radiation Budget Network (SURFRAD) sites. They found that ASTER LSTs had an average bias of 0.1 °C, while MODIS LSTs had an average bias of –0.2 °C. Examples of other studies that relate remotely sensed surface–atmosphere fluxes (primarily evapotran-spiration from instrumented agricultural sites) are provided by Czajkowski et al. (2000), Kustas et al. (2004), and French et al. (2010). There is a paucity of studies, however, that have related estimates of surface–atmosphere fluxes derived from satellites with ground-based routine meteorological data in urban areas. Therefore, there is a critical need for integrating the information derived from remote sensing imagery and that from the network of weather stations and/or *in situ* field measurements for estimating urban surface energy fluxes. The urban thermal remote sensing literature is rapidly increasing, but that works on urban surface energy estimations has not been extended. The majority of previous work (Carlson et al. 1981; Zhang et al. 1998; Hafner and Kidder 1999; Chrysoulakis 2003; Kato and Yamaguchi 2005, 2007) combined satellite measurements with meteorological data to estimate and analyze the spatial patterns of urban heat fluxes based on a one-source surface energy model.

Human-induced energy discharge has an important impact on the urban environment in terms of surface energy balance. Quantification of anthropogenic heat discharge and its spatial pattern is important to improve the understanding of human impacts on the urban environment. The methods used for estimating anthropogenic heat discharge can be grouped into three major categories: inventory methods, micrometeorologically based energy budget closure methods, and building energy modeling methods (Sailor 2011). For example, Lee et al. estimated anthropogenic heat emission in the Gyeong-In region of Korea in 2002 based on the energy consumption statistics data using an inventory approach (Lee et al. 2009). Kato and Yamaguchi separated the contribution of anthropogenic heat discharge and heat radiation due to solar input to the sensible heat flux by using an energy balance method (Kato and Yamaguchi 2005). Heiple and Sailor (2008) estimated the hourly energy consumption from residential and commercial buildings at 100-m spatial resolution in Houston, Texas, using the building energy modeling method. Each of these methods has its strengths and limitations. Estimation of anthropogenic heat emissions using inventory approach is based on the real energy consumption data at the spatial scale of utility service territories (Klysik 1996; Taha 1997; Ichinose et al. 1999; Sailor and Lu 2004; Lee et al. 2009). It is difficult to quantify the spatial distribution of energy consumption at fine spatial and temporal scales. Spatial and temporal downscaling can only be achieved when additional data such as land use are employed. Energy balance method has been applied to estimate energy fluxes in urban areas from site and neighborhood scale to city scale (Oke 1988; Oke et al. 1999; Masson et al. 2002; Pigeon et al. 2007, 2008; Belan et al. 2009). High spatial resolution anthropogenic heat discharge can be estimated based on micrometeorologically based energy budget closure method by the combined use of remote sensing and meteorological data (Schmugge et al. 1998; Bastiaanssen et al. 1998a, 1998b; French et al. 2005; Kato and Yamaguchi 2005; Kato et al. 2008; van der Kwast et al. 2009). However, as the estimation of anthropogenic heat discharge is based on the residual of other components in the surface energy balance, each component introduces uncertainties and propagates errors towards the final estimation of anthropogenic heat flux. By integrating the geospatial data and simulated temporal pattern of energy consumption for representative buildings, energy consumption in all buildings can be estimated using the building energy modeling method (Ichinose et al. 1999; Heiple and Sailor 2008; Zhou and Gurney 2010; Zhou et al. 2012). However, the representative buildings for each category may not be easily identified and the use of the representative buildings may introduce bias in the estimation of spatial and temporal patterns of energy use for some buildings as these buildings may have different behaviors of energy consumption in terms of magnitude and temporal pattern compared to the representative buildings.

In this chapter, we first applied an analytical protocol, based on the TSEB algorithm, to estimate urban surface heat fluxes by combined use of remotely sensed data and routine measurements of a weather station. This method was applied to four Terra's ASTER images of Indianapolis, Indiana, United States, to assess seasonal spatial pattern in the surface energy balance and intra-urban variability of energy fluxes (Weng et al. 2014). Furthermore, anthropogenic heat discharge and energy use from residential and commercial buildings were investigated using two different methods in Indianapolis (Zhou et al. 2012). First, anthropogenic heat discharge was estimated using a surface energy balance method. Then energy use from residential and commercial buildings was calculated using a building energy modeling method. Finally, the relationship between remotely sensed anthropogenic heat discharge and building energy use was examined across multiple spatial scales.

7.2 Data and Methodology

7.2.1 Study Area and Data Sets

Indianapolis is selected as the study area. Rapid urbanization is increasing its built-up area through encroachment into the adjacent agricultural, forest, and other nonurban lands. Indianapolis experiences a continental climate. The average July high is 30 °C, with the low being 16 °C. January highs average 1 °C, and lows −8 °C. Humidity varies, depending on the position of weather fronts and prevailing winds. Winters may be rather long and cold, with significant snowstorms blowing in from the Great Lakes region. Wind chills can reach −30 °C, with no natural features like mountains to protect the area from the onslaught of arctic Canadian air. Spring and fall bring pleasant air temperatures and the occurrence of many thunderstorms. The city's average annual precipitation is 41 in. (1040 mm), while snowfall varies from about 20 to 30 in. (500–760 mm) per year (https://climate.agry.purdue.edu/climate/index.asp).

The ground meteorological data needed to calculate heat fluxes include solar radiation, air temperature, relative humidity, air pressure, and wind speed. The solar radiation data was acquired from the National Climatic Data Center

(NCDC), which supplied hourly modeled solar radiation data along with hourly meteorological data between 1991 and 2005 from the National Solar Radiation Database (NSRDB). An earlier version of the database was also available for the Indianapolis Airport station covering 1961–1990. Since the shortwave radiation changes with date and time, we selected the shortwave radiation value modeled for the same day as the image acquisition date and at the closest matching time between the NSRDB measurement and the satellite overpass. The data for air temperature, relative humidity, air pressure, and wind speed were obtained from Indiana State Climate Office and NCDC. The stations reporting to NCDC provided hourly measurements for these routine parameters in addition to precipitation and other weather conditions. The meteorological data measured at the time closest to each satellite overpass were used in this study. The meteorological parameters were assumed constant throughout the study area. Since the study area is flat, air temperature was not corrected for differences in elevation. The precipitation records were checked prior to each satellite overpass to understand the moisture conditions over land surfaces. The meteorological data used in the study are shown in Table 7.1.

7.2.2 Remote Sensing Data Acquisition and Preprocessing

ASTER images covering the City of Indianapolis, acquired on 3 October 2000 (12 : 00 : 51 eastern standard time, the fall image), 16 June 2001 (11 : 55 : 29, the summer image), 5 April 2004 (11 : 46 : 39, the spring image), and 6 February 2006 (11 : 45 : 36, the winter image), were used to study seasonal changes in surface heat fluxes. The ASTER data products including surface kinetic temperature, surface spectral emissivity, and visible and near infrared (VNIR) and shortwave infrared (SWIR) surface spectral radiance were obtained from NASA. The methodology for converting ASTER infrared measurements to

Table 7.1 Meteorological data used in estimation of heat fluxes.

Variable	6 February 2006	5 April 2004	16 June 2001	3 October 2000
	1200ET	1200ET	1200ET	1200ET
Shortwave radiation (W/m^2)	467	738	895	649
Wind speed (m/s)	4.6	2.6	2.1	2.6
Atmospheric temperature (K)	266.15	272.15	290.35	288.15
Air pressure (hPa)	1016.9	1022.7	1020.2	1014.9
Relative humidity (%)	92	63	86	93

Source: Weng et al. (2014). Reproduced with permission of IEEE.

LSTs has been reported by Gillespie et al. (1998). We used the ASTER data products of the VNIR and SWIR spectral radiance to calculate the spectral reflectance. Surface spectral reflectance and spectral emissivity data were converted into surface albedo and broadband emissivity according to Liang (2000) and Ogawa et al. (2003), respectively.

All ASTER images were resampled to 90-m resolution to correspond to the spatial resolution of their TIR bands. By resampling up from 15 to 90 m, the radiometry of original reflective bands will be preserved, and the process will not introduce autocorrelation in the resultant data. Normalized difference vegetation index (NDVI) was calculated using atmospherically corrected at-surface reflectance of NIR band and red band of Thematic Mapper (TM) data. Linear Spectral Mixture Analysis (LSMA) was applied to derive fractional images. Details about the selection of end-members and the estimation of the vegetation fraction (denoted as f_v in this study) were discussed in Weng et al. (2009). The non-vegetation fraction (denoted as f_{nv}) was calculated as $1 - f_v$.

With ASTER VNIR and SWIR bands, six land-use and land-cover (LULC) types were identified by employing an unsupervised classification algorithm, i.e. Iterative Self-Organizing Data Analysis Weng et al. (2008). The overall classification accuracy of 92% (5 April 2004, image), 88.33% (16 June 2001, image), 87% (3 October 2002, image), and 87.33% (6 February 2006, image) was achieved for each classified LULC map (Weng et al. 2014). A comparison among the classified maps indicates that the magnitude and spatial pattern of each class corresponded well to each other but also reflected the seasonal and temporal differences.

Impervious surface area (ISA) was estimated on 16 June 2001 using the method developed by Weng et al. (2009). Through this method, impervious surface was calculated based on the relationship between the reflectance of two end-members (high and low albedo) and the reflectance of the impervious areas. By examining the relationships between impervious surfaces and four end-members, it is found that impervious surfaces were located on or near the line connecting the low- and high-albedo end-members in the feature space. An estimation procedure was thus developed based on this relationship by adding the fractions of high- and low-albedo end-members.

7.2.3 Estimating Surface Heat Fluxes

7.2.3.1 Net Radiation

The net radiation was calculated as follows:

$$R_n = (1 - \alpha) R_{short} + \varepsilon \varepsilon_a \sigma T_a^4 - \varepsilon \sigma T_s^4 \tag{7.1}$$

where R_n is the net radiation, σ ($= 5.67 \times 10^{-8}$ W/m^{2}K^4) is the Stefan–Boltzmann constant, ε_a is atmospheric emissivity, α is broadband albedo, R_{short} is shortwave radiation, ε is surface broadband emissivity (that is assumed to be equivalent to the absorptance of the downward longwave radiation), T_a is atmospheric temperature, and T_s is surface temperature (Timmermans et al. 2007).

Atmospheric emissivity (ε_a) was calculated as

$$\varepsilon_a = 1.24 \left(\frac{e_a}{T_a} \right)^{\frac{1}{7}} \tag{7.2}$$

where e_a is atmospheric water vapor pressure, which was estimated based on saturation water vapor pressure e^* and relative humidity (Brutsaert 1982).

Surface reflectance can be calculated using the following equation:

$$\rho = \frac{\pi * \left(L_{sat} - L_{haze} \right) * d^2}{Esun_\lambda \left[\left(\cos\theta_s \right)^2 \right]} \tag{7.3}$$

where ρ is surface reflectance, L_{sat} is at-sensor radiance, d is the earth–sun distance, $Esun_\lambda$ is a band-dependent constant, θ_s is the solar zenith angle, and L_{haze} is estimated upwelling scattered path radiance due to atmospheric haze, aerosols, etc. (Milder 2008). After the surface reflectance was calculated, the broadband albedo α can be calculated using the method developed by Liang (2000). Broadband emissivity (ε) was calculated based on the spectral emissivities of all bands (Weng 2009).

7.2.3.2 Ground Heat Flux (G)

Estimation of ground heat flux, G, is often difficult, since it requires knowledge of the temperature gradient in the soil (Schmugge et al. 1998). In urban areas, the heat conductivity of land surface materials and the vertical temperature gradient inside the walls, roofs, and floors for buildings must also be known (Kato and Yamaguchi 2005). In practice, the information is hardly obtainable. Therefore, the ground heat flux is largely estimated based on the net radiation as

$$G = c_g * R_n \tag{7.4}$$

The coefficient, c_g, is subject to the influence of surface cover material and seasonality (Anandakumar 1999; Silberstein et al. 2001), as well as diurnal change (Santanello and Friedl 2003). Given the heterogeneity and complexity of urban component surfaces, Grimmond et al. (1991) proposed the objective hysteresis model and tested it in Vancouver; this calculated G as a function of continuously measured net radiation and the surface properties of the site. Since remote sensing measurements are instantaneous in time, we had to use a look-up table for c_g, which was compiled by Kato and Yamaguchi (2005) and was mainly based on the findings of Anandakumar (1999). Table 7.2 shows assigned c_g values based on LULC type.

Table 7.2 Parameters used for estimation of heat fluxes.

Land-use and land-cover (LULC) type	C_g	d_0 (m)	Z_{0m} (m)	Z_{0h} (m)
Water	0.35	0.05	0.00003	0.000088
Bare soils	0.3	0.05	0.001	0.00002
Grass	0.3	0.1	0.1	0.001
Forest	0.15	1.5	0.3	0.0003
Urban	0.4	1.95	0.33	0.0033
Agriculture	0.3	0.1	0.1	0.001

Source: Weng et al. (2014). Reproduced with permission of IEEE.
Note: C_g, coefficient for estimating ground heat flux (Kato and Yamaguchi 2005); Z_{0m} and Z_{0h}, roughness lengths for momentum and heat transport Kato and Yamaguchi (2005); and d_0, displacement height for different land-cover types (Kustas et al. 2003). The aerodynamic parameters (d_0, Z_{0m}, and Z_{0h}) for the urban category were determined based on a research result by Burian et al. (2003) who computed the displacement height for the city of Houston, Texas, for a project on urban building databases funded by US Environmental Protection Agency.

7.2.3.3 Sensible Heat Flux (*H*)

The sensible heat flux was calculated as

$$H = f_{nv} H_{non\text{-}veg} + f_v H_{veg} \tag{7.5}$$

where $H_{non\text{-}veg}$ and H_{veg} are sensible heat flux for non-vegetated and vegetated areas, respectively. $H_{non\text{-}veg}$ and H_{veg} were calculated using the following equations:

$$H_{non\text{-}veg} = \rho_a c_p \frac{T_s - T_a}{R_{AH} + R_s} \tag{7.6}$$

$$H_{veg} = \rho_a c_p \frac{T_c - T_a}{R_{AH}} \tag{7.7}$$

where ρ_a is the air density in kg/m^3, c_p is the specific heat of air at constant pressure in J/(kg·K), T_s and T_c are surface temperatures for non-vegetated and vegetated areas, respectively (both were obtained from the ASTER kinetic surface temperature product), T_a is atmospheric temperature, R_{AH} is the aerodynamic resistance in s/m, and R_s is the resistance to heat flow in the boundary layer immediately above the soil surface (Kustas and Norman 1999). R_{AH} was calculated as

$$R_{AH} = \frac{\left[\ln\left(\frac{z_u - d_0}{z_{0M}}\right) - \Psi_M\right]\left[\ln\left(\frac{z_t - d_0}{z_{0H}}\right) - \Psi_H\right]}{k^2 u} \tag{7.8}$$

where z_u (=10 m) and z_t (=2 m) are the respective heights at which the wind speed u and atmospheric temperature are measured, d_0 is the displacement height, and z_{0M} and z_{0H} are the roughness lengths for momentum and heat transport, respectively. All measurements are in meters. In addition, Ψ_M and Ψ_H are stability correction functions for momentum and heat, and k (=0.4) is von Karman's constant (Brutsaert 1982). Equation (7.8) may be simplified by removing Ψ_M and Ψ_H (Liu et al. 2007). Table 7.2 shows that the values of d_0, z_{0m}, and z_{0h} were fixed according to each LULC type and were estimated based on the existing literatures.

R_s can be computed as per the following equation:

$$R_s = \frac{1}{a + bu_s} \tag{7.9}$$

where a (=0.004 m/s) is the free convective velocity, b (=0.012) is a coefficient to represent the typical soil surface roughness, u_s is the wind speed over soil surface at the height of 0.05–0.2 m (Kustas and Norman 1999).

7.2.3.4 Latent Heat Flux (*LE*)

Latent heat flux was calculated as follows (Humes et al. 2004):

$$LE = f_{nv}LE_{non\text{-}veg} + f_v LE_{veg} \tag{7.10}$$

$LE_{non\text{-}veg}$ and LE_{veg} are latent heat flux for the non-vegetated and the vegetated areas, respectively. They were computed as follows:

$$LE_{non\text{-}veg} = R_{N,S} - G - H_{non\text{-}veg} \tag{7.11}$$

$$LE_{veg} = \alpha_{PT} f_G \frac{\Delta}{\Delta + \gamma} R_{N,C} \tag{7.12}$$

where α_{PT} (=1.26) is Priestley–Taylor parameter, γ is psychrometric constant, Δ is the slope of saturation vapor pressure–temperature curve, and f_G (equals unity if not available) is the fraction of the leaf area index (LAI) that is green. $R_{N,S}$ and $R_{N,C}$ are net radiation for soil and vegetation, respectively (Kustas and Norman 1999). If a negative value of $LE_{non\text{-}veg}$ was obtained, it would be set equal to zero. $H_{non\text{-}veg}$ was then recomputed as the residual of Eq. (7.11) when $LE_{non\text{-}veg}$ was set to 0, and $R_{N,S}$ and G are from previous calculations (Humes et al. 2004).

7.2.3.5 Anthropogenic Heat Discharge (A)

In this study, anthropogenic heat discharge was estimated based on a surface energy balance modeling method (Oke 1988; Kato and Yamaguchi 2005; Kato et al. 2008; Xu et al. 2008). The surface energy balance due to surface properties and anthropogenic heating in the near-surface in an urban area can be expressed as follows:

$$A = H + LE + G - R_n \qquad (7.13)$$

where A is the anthropogenic heat discharge (W/m^2), H is the sensible heat (W/m^2), LE is the latent heat flux (W/m^2), G is the ground heat flux (W/m^2), and R_n is net radiation (W/m^2). Based on the surface energy balance model, anthropogenic heat discharge A can be estimated as the residual term.

7.3 Heat Fluxes in Four Seasons

Table 7.3 shows mean values and standard deviations of heat fluxes in each imaged date. Figure 7.1 shows heat fluxes estimated for 16 June 2001. The mean value of net radiation was 432 in April, 613 W/m^2 in June, 403 W/m^2 in October, and 319 W/m^2 in February, respectively. The result indicated that the highest net radiation was received in summer, followed by spring, fall, and winter, a typical seasonal pattern in the mid-latitude region in the northern hemisphere. This was due primarily to different amounts of solar (shortwave) radiation received by the Earth surface. Moreover, the amount and vigor of vegetation varied significantly in different seasons in Indiana, and different amounts of vegetation can significantly affect net radiation. The ground heat flux showed a similar seasonal pattern because G was estimated from R_n. The standard deviation of net radiation was the highest in summer and the lowest in winter, which was, again, affected by the vegetation pattern. The vegetation was the most heterogeneous during the summer as a result of various species, life forms, structures, and habitats and was the least during the winter since most vegetation were dead. The standard deviation of R_n was higher in the

Table 7.3 Mean and standard deviation values (in parenthesis) of heat fluxes in four dates.

Heat flux (W/m^2)	5 April 2004	16 June 2001	3 October 2000	6 February 2006
Net radiation (R_n)	432 (41)	613 (48)	403 (40)	319 (18)
Sensible heat flux (H)	165 (71)	128 (56)	117 (47)	93 (38)
Latent heat flux (LE)	104 (89)	195 (94)	101 (66)	96 (62)
Ground heat flux (G)	124 (39)	176 (56)	114 (37)	97 (30)

Source: Weng et al. (2014). Reproduced with permission of IEEE.

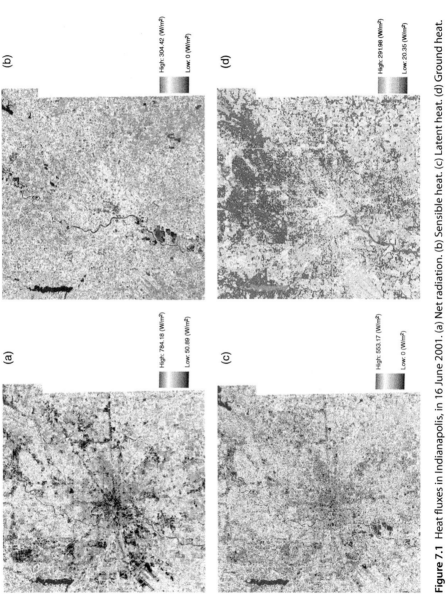

Figure 7.1 Heat fluxes in Indianapolis, in 16 June 2001. (a) Net radiation. (b) Sensible heat. (c) Latent heat. (d) Ground heat.

(a) High: 784.18 (W/m²) Low: 50.89 (W/m²)

(b) High: 304.42 (W/m²) Low: 0 (W/m²)

(c) High: 553.17 (W/m²) Low: 0 (W/m²)

(d) High: 291.98 (W/m²) Low: 20.35 (W/m²)

spring than in the fall. In the mid-latitude region, phenological changes were more obvious in the spring than in any other seasons.

In terms of sensible heat flux, values of 165, 128, 117, and 93 W/m^2 were obtained for each imaging date in April, June, October, and February, respectively. Sensible heat flux is driven by temperature differences between the atmosphere and the surface and as such is highly variable in space and in time. Its seasonal variability was obvious based on our result. It showed the highest H in the spring. The extreme low atmospheric temperature (−1 °C) and larger wind speed at the time when the April image was acquired were two main factors contributing to this anomaly. LULC change associated with urban sprawl between 2000 and 2004 may be another factor. The urbanization process favored higher sensible heat flux as evapotranspirative surfaces were reduced. Moreover, increased anthropogenic heat discharges also contributed to the sensible heat flux. The spatial variation in H was high in the summer months but was generally low in the winter months. From April to October in Indiana (the summer months), there were a larger amount of vegetation and more growth-related activities and also a bigger contrast among the urban, suburban, and rural areas in radiative, thermal, moisture, and aerodynamic properties. In comparison with net radiation, sensible heat flux showed a stronger spatial variability (thus higher standard deviation) in all imaging dates. Although there were a variety of surface materials and seasonal changes in vegetation amount and vigor, which affected surface albedo and emissivity, the differences in the net shortwave and longwave radiation were constrained by the albedo-temperature feedback mechanism (Schmid et al. 1991; Liang 2000).

The latent heat flux showed an obvious seasonal variability too. LE value yielded 104 W/m^2 in April, 195 W/m^2 in June, 101 W/m^2 in October, and 96 W/m^2 in February. The decrease of LE value was attributed to the change in vegetation transpiration and water evaporation, reflecting the vegetation growth and decline over a year cycle and the accompanying moisture condition by season. The highest LE value corresponded to the existence of the largest amount of vegetation and its vigor in the summer. LE reached the lowest point in February. The standard deviation value of LE, again, was the lowest in the winter due to a less contrasting surface condition in terms of evapotranspiration. It should be noted that LE had a stronger variability in the spring time (standard deviation: 89) than in the fall (standard deviation: 66), as various species of vegetation and crops started to grow following different calendars. This variability contributed to a higher standard deviation value in spring and was closely related to the spatial pattern of LE.

7.4 Heat Fluxes by LULC Type

In all dates, water exhibited the highest R_n, followed by forest. The lowest R_n values were detected in urban in the summer months, while in bare soil in the winter months. Agriculture and grass land possessed the third and fourth places. The variations of R_n were only controlled by surface albedo and surface temperature. Water bodies generally had low albedo and surface temperature resulting in a high R_n. Forest yielded a high R_n value as a consequence of its ability of absorbing more solar energy, lowering albedo and temperature. Owing to the higher surface temperature, the longwave upward radiation from the urban surfaces was higher, and thus urban areas often discovered the lower R_n than the nearby rural areas. However, the lowest R_n position can be overtaken by bare soil in the winter months (in February and April) because of its high albedo. Bare grounds may be covered by snow and were frequently seen in dry and plowed conditions in the early spring before planting. Agriculture and grassland emulated closely in both mean and standard deviation values of R_n, followed by their similar cover conditions in albedo and surface temperature. The variability of R_n was relatively high in bare soil, water, and urban, especially in the summer. High variability of net radiation over water (especially the two reservoirs in the study areas) may be associated with the growth of algae. But the fluctuation became much weaker in the winter (February image). The fall and spring images shared similar moderate variability in R_n. Figure 7.2 shows the mean value of each flux by LULC type in 16 June 2001.

The estimation of sensible heat flux was found to be linked with surface temperature. Urban possessed the highest H values due to higher surface temperature and surface roughness. Various types of impervious surfaces generally produced higher thermal emittance into the atmosphere as sensible heat. Moreover, anthropogenic heat discharge in the urban areas was much higher than in the rural, which contributed substantially to the sensible heat flux, especially in the winter months. In contrast, water yielded the lowest H. The contrast of H between urban and water was most pronounced in the spring and

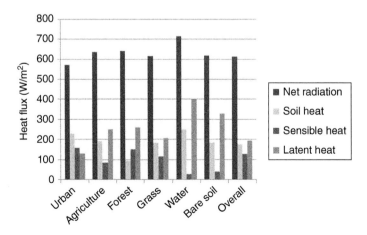

Figure 7.2 The mean value of each flux by land-use and land-cover (LULC) type in 16 June 2001.

summer, but became less obvious in the fall, reaching the lowest in the winter. It is abnormal to learn that both forest and grass displayed a high H, regardless of the seasonality. We believe that the fact that forest and grass were mingled with urban in the study area had caused image misclassification. Forest was mostly confused with residential land (especially in the low-density and medium-density areas), which in our classification scheme belonged to urban. Grass was largely found in the residential areas, confusing itself with urban too. Another interesting finding was that bare soil always exhibited a low H, preceded only by water. Geographically, bare soil was largely located along rivers and other water bodies with small amount scattered within agricultural land. The spatial adjacency between soil and water/agricultural land may have caused misclassification and have thus resulted in computation error in the estimation of H for bare soil. The standard deviation values were high in forest, grass, and urban, but all decreased to the lowest in the winter.

Because latent heat flux was associated with evapotranspiration, high LE corresponded to water, agriculture, and forest. Grass also displayed fairly high LE in the summer, but declined to considerably low levels in other months. At all times, urban possessed the lowest LE, but a large standard deviation value. This strong intra-class variability can be largely attributed to the image classification scheme, where residential lands were combined with commercial, industrial, and transportation, etc. to form the urban class. The increase of the standard deviation in LE was related to the change of vegetation abundance and vigor within the urban class, especially in the residential areas. It is worthy to note that bare soil detected the second highest LE in all four dates. This overestimation was mainly caused by image misclassification between bare soil and water and its confusion with agriculture, forest, and grass. Overall, since the fluctuation of LE had much to do with precipitation and vegetation activities, the standard deviation values for all LULC types appeared to be high from April to October. By comparing standard deviation values among all the heat fluxes for all LULC types, it is apparent that both LE and H generated a strong variability in the four imaging dates, while R_n and G observed a relatively weak variability.

7.5 Extreme Values of Heat Fluxes

Analysis of extreme values of heat fluxes can help to assess the validity of modeling procedures. It is found that a few very low values (0.5–1% of the pixels depending on season) existed in the R_n images for all dates. Most of the extreme values of R_n occurred in the urban areas, in particular, in commercial, industrial areas, and the airport area (Figure 7.3). These areas exhibited both very high surface temperature and albedo, which led to relatively low values in R_n. The reason for the extreme values of net radiation may be the heterogeneity of

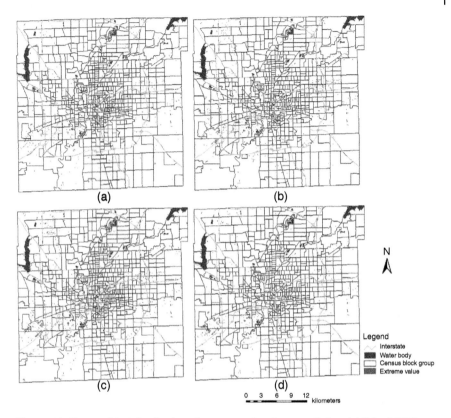

Figure 7.3 Geographical distribution of extreme values in net radiation. (a) 5 April 2004. (b) 16 June 2001. (c) 3 October 2000. (d) 6 February 2006.

downward solar radiation and longwave radiation, which is out of the scope of this study. Extreme high values (0.002–0.1% of the pixels) were also observed in the estimation of sensible heat flux (Figure 7.4). Since our model assumed a constant wind speed, extreme values of H may arise because the observed surface temperature may be mismatched with the appropriate wind speed. Some of the extreme H values were associated with image classification errors. Vegetated pixels, for instance, when misclassified as urban, which was associated with higher surface temperatures, would result in extreme high values in H. Moreover, the extreme high values in H were discovered in industrial areas. Within these industrial areas, a huge amount of energy consumption was transferred into atmosphere as sensible heat. In contrast, extreme high values (0.13–1.9% of the pixels) in latent heat flux were detected mostly from water bodies (Figure 7.5). Water bodies generally had high values of LE due to their low temperatures as a result of evaporation.

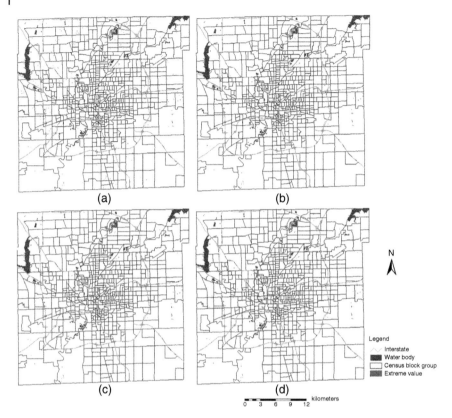

Figure 7.4 Geographical distribution of extreme values in sensible heat flux. (a) 5 April 2004. (b) 16 June 2001. (c) 3 October 2000. (d) 6 February 2006.

7.6 Anthropogenic Heat Discharge

The anthropogenic heat discharge on 16 June 2001 was estimated using the energy balance method (Zhou et al. 2012). To reduce the uncertainties in the evaluation of anthropogenic heat discharge, especially in less developed areas, ISA data was used to exclude pixels with the ISA less than 25%. The result of anthropogenic heat discharge estimation is shown in Figure 7.6. The result shows the spatial pattern and magnitude of the anthropogenic heat discharge in the city of Indianapolis in a typical summer time. The anthropogenic heat discharge ranges from 0 to $400\,\mathrm{W/m^2}$, with a mean value of $32\,\mathrm{W/m^2}$. The highest anthropogenic heat discharge occurs in the dense residential and commercial areas such as the center of the city and the residential areas in the northwest corner of the city. The township of Center has the highest anthropogenic heat discharge of $78\,\mathrm{W/m^2}$ as it has high density of commercial buildings

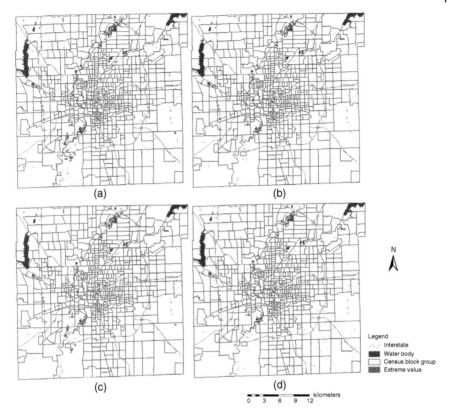

Figure 7.5 Geographical distribution of extreme values in latent heat flux. (a) 5 April 2004. (b) 16 June 2001. (c) 3 October 2000. (d) 6 February 2006.

while the Franklin Township with large rural areas observes the lowest value. There are negative values in the anthropogenic heat discharge estimation, although it theoretically should be great than $0\,W/m^2$. These negative values are regarded as estimation errors and were set $0\,W/m^2$ in the final result.

7.7 Summary

Knowledge of urban surface energy balance is fundamental to understanding of UHIs and urban thermal behavior (Oke 1982, 1988). The main contribution of this study lies in improving the knowledge of the intra-urban variability of surface heat fluxes. Satellite remote sensing imagery and routine meteorological data were integrated for use to provide "snapshots" of spatially distributed energy fluxes at 90-m resolution. By computing the energy fluxes by LULC

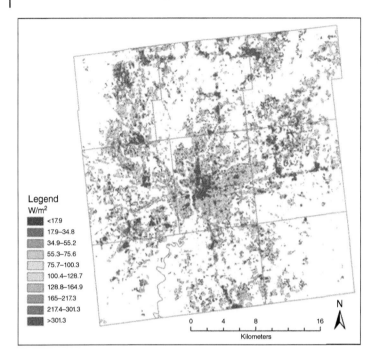

Figure 7.6 Geographical distribution of anthropogenic heat discharge on 16 June 2001. *Source:* Zhou et al. (2012). Reproduced with permission of Elsevier.

type, we can improve understating of the geographical patterns of energy fluxes. Further, the within-class variability was examined with respect to seasonality and was related to the changes in biophysical variables in addition to the heterogeneity of wind speed and air temperature.

This chapter has applied a method to study urban surface energy balance using a TSEB approach. This method maximized the use of satellite data from a multispectral remote sensor (i.e. ASTER) in conjunction with a minimum amount of routine meteorological data from a permanent weather station. ASTER images acquired in four distinct seasons were used to derive LULC, LST, vegetation fraction, broadband albedo, broadband emissivity, NDVI, and LAI. Although aerodynamic parameters such as displacement height and roughness length were empirically estimated, they were also computed based on remote sensing data. By applying the method to Indianapolis, Indiana, USA, this paper has examined the spatial variability of surface heat fluxes in four dates of distinct seasons. The methodology developed in this study by the integration of remotely sensed and routinely measured meteorological data has the advantage of being quite "data-driven" and no need to be calibrated for a particular study site. This methodology would be very helpful for analyzing and modeling UHIs as a result of urbanization-associated LULC and global climate changes.

There is a challenge to obtaining accurate estimation of each component of heat flux in the urban energy balance such as ground heat flux. Accurate estimation of anthropogenic heat discharge will help to better understand the urban energy balance and separate the contribution of anthropogenic heat discharge from that of ground heat flux (Zhou et al. 2012). By better estimation of energy use from sources such as buildings, transportation, industries, and metabolism, the estimation of anthropogenic heat discharge may be improved, and therefore, help to reduce the uncertainties in modeling the urban energy balance.

References

Anandakumar, K. (1999). A study on the partition of net radiation into heat fluxes on a dry asphalt surface. *Atmospheric Environment* 33: 3911–3918.

Arnfield, J.A. (2003). Two decades of urban climate research: a review of turbulence, exchanges of energy and water, and the urban heat island. *International Journal of Climatology* 23: 1–26.

Bastiaanssen, W.G.M., Menenti, M., Feddes, R.A., and Holtslag, A.A.M. (1998a). A remote sensing surface energy balance algorithm for land (SEBAL), part 1: formulation. *Journal of Hydrology* 213: 198–212.

Bastiaanssen, W.G.M., Pelgrum, H., Wang, J. et al. (1998b). A remote sensing surface energy balance algorithm for land (SEBAL), part 2: validation. *Journal of Hydrology* 213: 213–229.

Belan, B., Pelymskii, O., and Uzhegova, N. (2009). Study of the anthropogenic component of urban heat balance. *Atmospheric and Oceanic Optics* 22 (4): 441–445.

Brutsaert, W.H. (1982). *Evaporation into the Atmosphere: Theory, History and Applications*, 299. London, England: D Reidel Publishing Company.

Burian, S.J., Han, W.S., and Brown, M.J. (2003). Morphological analyses using 3D building databases: Houston, Texas LA-UR-03-8633. Los Alamos National Laboratory. http://www.nudapt.org/pdf/Houston_MorphologicalAnalysis_FinalReport.pdf (accessed 20 January 2012).

Carlson, T.N., Dodd, J.K., Benjamin, S.G., and Cooper, J.N. (1981). Satellite estimation of the surface energy balance, moisture availability and thermal inertia. *Journal of Applied Meteorology* 20: 67–87.

Chrysoulakis, N. (2003). Estimation of the all-wave urban surface radiation balance by use of ASTER multispectral imagery and in situ spatial data. *Journal of Geophysical Research* 108: 4582.

Czajkowski, K.P., Goward, S., Stadler, S.J., and Walz, A. (2000). Thermal remote sensing of near surface environmental variables: application over the Oklahoma mesonet. *The Professional Geographer* 52: 345–357.

French, A.N., Hunsaker, D.J., Clarke, T.R. et al. (2010). Combining remotely sensed data and ground-based radiometers to estimate crop cover and surface temperatures at daily time steps. *Journal of Irrigation and Drainage Engineering* 136: 232–239.

French, A.N., Jacob, F., Anderson, M.C. et al. (2005). Surface energy fluxes with the Advanced Spaceborne Thermal Emission and Reflection radiometer (ASTER) at the Iowa 2002 SMACEX site (USA). *Remote Sensing of Environment* 99 (1–2): 55–65.

Frey, C.M., Parlow, E., Vogt, R. et al. (2011). Flux measurements in Cairo. Part 1: in situ measurements and their applicability for comparison with satellite data. *International Journal of Climatology* 31: 218–231.

Gillespie, A.R., Rokugawa, S., Matsunaga, T. et al. (1998). A temperature and emissivity separation algorithm for advanced spaceborne thermal emission and reflection radiometer (ASTER) images. *IEEE Transactions on Geoscience and Remote Sensing* 36: 1113–1126.

Grimmond, C.S.B. (2006). Progress in measuring and observing the urban atmosphere. *Theoretical and Applied Climatology* 84: 3–22.

Grimmond, C.S.B., Cleugh, H.A., and Oke, T.R. (1991). An objective urban heat storage model and its comparison with other schemes. *Atmospheric Environment* 25: 311–326.

Hafner, J. and Kidder, S.Q. (1999). Urban heat island modeling in conjunction with satellite-derived surface/soil parameters. *Journal of Applied Meteorology* 38: 448–465.

Heiple, S. and Sailor, D.J. (2008). Using building energy simulation and geospatial modeling techniques to determine high resolution building sector energy consumption profiles. *Energy and Buildings* 40 (8): 1426–1436.

Humes, K., Hardy, R., and Kustas, W.P. (2000). Spatial patterns in surface energy balance components derived from remotely sensed data. *The Professional Geographer* 52: 272–288.

Humes, K., Hardy, R., Kustas, W.P. et al. (2004). High spatial resolution mapping of surface energy balance components with remotely sensed data. In: *Thermal Remote Sensing in Land Surface Processes* (eds. D.A. Quattrochi and J.C. Luvall), 110–132. Boca Raton, FL: Taylor and Francis Group, LLC, CRC Press.

Ichinose, T., Shimodozono, K., and Hanaki, K. (1999). Impact of anthropogenic heat on urban climate in Tokyo. *Atmospheric Environment* 33 (24–25): 3897–3909.

Jensen, J.R. and Cowen, D.C. (1999). Remote sensing of urban/suburban infrastructure and socioeconomic attributes. *Photogrammetric Engineering & Remote Sensing* 65: 611–622.

Kato, S. and Yamaguchi, Y. (2005). Analysis of urban heat-island effect using ASTER and ETM+ data: separation of anthropogenic heat discharge and natural heat radiation from sensible heat flux. *Remote Sensing of Environment* 99: 44–54.

Kato, S. and Yamaguchi, Y. (2007). Estimation of storage heat flux in an urban area using ASTER data. *Remote Sensing of Environment* 110: 1–17.

Kato, S., Yamaguchi, Y., Liu, C.-C., and Sun, C.-Y. (2008). Surface heat balance analysis of Tainan City on March 6, 2001 using ASTER and Formosat-2 data. *Sensors* 8 (9): 6026–6044.

Klysik, K. (1996). Spatial and seasonal distribution of anthropogenic heat emissions in Lodz, Poland. *Atmospheric Environment* 30 (20): 3397–3404.

Kustas, W.P., French, A.N., Hatfield, J.L. et al. (2003). Remote sensing research in hydrometeorology. *Photogrammetric Engineering & Remote Sensing* 69: 631–646.

Kustas, W.P. and Norman, J.M. (1999). Evaluation of soil and vegetation heat flux predictions using a simple two-source model with radiometric temperatures for partial canopy cover. *Agricultural and Forest Meteorology* 94: 13–29.

Kustas, W.P., Norman, J.M., Schmugge, T.J., and Anderson, M.C. (2004). Mapping surface energy fluxes with radiometric temperature. In: *Thermal Remote Sensing in Land Surface Processes* (eds. D.A. Quattrochi and J.C. Luvall), 205–253. Boca Raton, FL: Taylor and Francis Group, LLC, CRC Press.

van der Kwast, J., Timmermans, W., Gieske, A. et al. (2009). Evaluation of the Surface Energy Balance System (SEBS) applied to ASTER imagery with flux-measurements at the SPARC 2004 site (Barrax, Spain). *Hydrology and Earth System Sciences Discussion* 6 (1): 1165–1196.

Lee, S.H., Song, C.K., Baik, J.J., and Park, S.U. (2009). Estimation of anthropogenic heat emission in the Gyeong-In region of Korea. *Theoretical and Applied Climatology* 96 (3): 291–303.

Liang, S. (2000). Narrowband to broadband conversions of land surface albedo I: algorithms. *Remote Sensing of Environment* 76: 213–238.

Liu, S.M., Lu, L., Mao, D., and Jia, L. (2007). Evaluating parameterizations of aerodynamic resistance to heat transfer using field measurements. *Hydrology and Earth System Sciences* 11: 769–783.

Martilli, A. (2007). Current research and future challenges in urban mesoscale modeling. *International Journal of Climatology* 27: 1909–1918.

Masson, V., Grimmond, C.S.B., and Oke, T.R. (2002). Evaluation of the Town Energy Balance (TEB) scheme with direct measurements from dry districts in two cities. *Journal of Applied Meteorology* 41 (10): 1011–1026.

Milder, J.C. (2008). ASTER processing method. Department of Natural Resource, Cornell University. https://sharepoint.ngdc.wvu.edu/sites/digital_soils/Remote%20Sensing/Models/ASTER_processing_method_FINAL_7-25-08.doc (accessed 20 January 2012).

Ogawa, K., Schmugge, T., Jacob, F., and French, A. (2003). Estimation of land surface window (8–12 μm) emissivity from multispectral thermal infrared remote sensing-a case study in a part of Sahara Desert. *Geophysical Research Letters* 30: 1067.

Oke, T.R. (1982). The energetic basis of the urban heat island. *Quarterly Journal of the Royal Meteorological Society* 108: 1–24.

Oke, T.R. (1988). The urban energy balance. *Progress in Physical Geography* 12 (4): 471–508.

Oke, T.R., Spronken-Smith, R.A., Jauregui, E., and Grimmond, C.S.B. (1999). The energy balance of central Mexico City during the dry season. *Atmospheric Environment* 33 (24–25): 3919–3930.

Pigeon, G., Legain, D., Durand, P., and Masson, V. (2007). Anthropogenic heat release in an old European agglomeration (Toulouse, France). *International Journal of Climatology* 27 (14): 1969–1981.

Pigeon, G., Moscicki, M.A., Voogt, J.A., and Masson, V. (2008). Simulation of fall and winter surface energy balance over a dense urban area using the TEB scheme. *Meteorology and Atmospheric Physics* 102 (3): 159–171.

Quattrochi, D.A. and Luvall, J.C. (1999). Thermal infrared remote sensing for analysis of landscape ecological processes: methods and applications. *Landscape Ecology* 14: 577–598.

Rigo, G. and Parlow, E. (2007). Modelling the ground heat flux of an urban area using remote sensing data. *Theoretical and Applied Climatology* 90: 185–199.

Sailor, D.J. (2011). A review of methods for estimating anthropogenic heat and moisture emissions in the urban environment. *International Journal of Climatology* 31 (Special Issue): 189–199.

Sailor, D.J. and Lu, L. (2004). A top-down methodology for developing diurnal and seasonal anthropogenic heating profiles for urban areas. *Atmospheric Environment* 38 (17): 2737–2748.

Santanello, J.A. Jr. and Friedl, M.A. (2003). Diurnal covariation in soil heat flux and net radiation. *Journal of Applied Meteorology* 42: 851–862.

Schmid, H.P., Cleugh, H.A., Grimmond, C.S.B., and Oke, T.R. (1991). Spatial variability of energy fluxes in suburban terrain. *Boundary-Layer Meteorology* 54: 249–276.

Schmugge, T.J., Kustas, W.P., and Humes, K.S. (1998). Monitoring land surface fluxes using ASTER observations. *IEEE Transactions on Geoscience and Remote Sensing* 36 (5): 1421–1430.

Silberstein, R., Held, A., Hatton, T. et al. (2001). Energy balance of a natural jarrah (*Eucalyptus marginata*) forest in Western Australia: measurements during the spring and summer. *Agricultural and Forest Meteorology* 109: 79–104.

Taha, H. (1997). Urban climates and heat islands: albedo, evapotranspiration, and anthropogenic heat. *Energy and Buildings* 25 (2): 99–103.

Timmermans, W.J., Kustas, W.P., Anderson, M.C., and French, A.N. (2007). An intercomparison of the Surface Energy Balance Algorithm for Land (SEBAL) and the Two-Source Energy Balance (TSEB) modeling schemes. *Remote Sensing of Environment* 108: 369–384.

Wang, K. and Liang, S. (2009). Evaluation of ASTER and MODIS land surface temperature and emissivity products using long-term surface longwave radiation observations at SURFRAD sites. *Remote Sensing of Environment* 113: 1556–1565.

Weng, Q. (2009). Thermal infrared remote sensing for urban climate and environmental studies: methods, applications, and trends. *ISPRS Journal of Photogrammetry and Remote Sensing* 64: 335–344.

Weng, Q. (2012). Remote sensing of impervious surfaces in the urban areas: requirements, methods, and trends. *Remote Sensing of Environment* 117: 34–49.

Weng, Q., Hu, X., and Liu, H. (2009). Estimating impervious surfaces using linear spectral mixture analysis with multitemporal ASTER images. *International Journal of Remote Sensing* 30 (18): 4807–4830.

Weng, Q., Hu, X., Quattrochi, D.A., and Liu, H. (2014). Assessing intra-urban surface energy fluxes using remotely sensed ASTER imagery and routine meteorological data: a case study in Indianapolis, USA. *IEEE Journal of Selected Topics in Applied Earth Observations & Remote Sensing* 99: 1–12.

Weng, Q., Liu, H., Liang, B., and Lu, D. (2008). The spatial variations of urban land surface temperatures: pertinent factors, zoning effect, and seasonal variability. *IEEE Journal of Selected Topics in Applied Earth Observations & Remote Sensing* 1: 154–166.

Xu, W., Wooster, M.J., and Grimmond, C.S.B. (2008). Modelling of urban sensible heat flux at multiple spatial scales: a demonstration using airborne hyperspectral imagery of Shanghai and a temperature-emissivity separation approach. *Remote Sensing of Environment* 112 (9): 3493–3510.

Zhang, X., Aono, Y., and Monji, N. (1998). Spatial variability of urban surface het fluxes estimated from Landsat TM data under summer and winter conditions. *Journal of Agricultural Meteorology* 54: 1–11.

Zhou, Y. and Gurney, K. (2010). A new methodology for quantifying on-site residential and commercial fossil fuel CO_2 emissions at the building spatial scale and hourly time scale. *Carbon Management* 1 (1): 45–56.

Zhou, Y.Y., Weng, Q., Gurney, K.R. et al. (2012). Estimation of the relationship between remotely sensed anthropogenic heat discharge and building energy use. *ISPRS Journal of Photogrammetry and Remote Sensing* 67 (1): 65–72.

8

Cities at Night

8.1 Introduction

Nighttime light (NTL) data from The Defense Meteorological Satellite Program/Operational Linescan System (DMSP-OLS) has a number of unique features that meet the needs of wide-scale surveys of urban environment with high temporal coverage. These features include long-term data archive from 1992 to 2013 and show strong correlation to human settlements and activities such as population distribution, economic activities, and energy consumption (Elvidge et al. 1997, 2001; Xie and Weng 2016a). A few researches have reported the capacity of DMSP-OLS NTL data for mapping regional and global urban extents, proposing the methods such as iterative unsupervised classification (Zhang and Seto 2011), SVM-based classification (Cao et al. 2009; Pandey et al. 2013), and thresholding techniques (Imhoff et al. 1997; Henderson et al. 2003; Zhou et al. 2014). In particular, the thresholding method has been proved to be efficient because of its simplicity and reasonable accuracy (Imhoff et al. 1997; Henderson et al. 2003; Zhou et al. 2014, 2015). The method assumes that pixels with NTL intensity larger than a specific value (i.e. the optimal threshold) belong to urban areas. However, most of the previous thresholding-based researches focused on mapping of a single year and paid less attention to the detection of urban changes. This is partly attributed to the challenge of determining appropriate thresholds for delineating urban area from DMSP/OLS NTL data (Imhoff et al. 1997; Henderson et al. 2003; Zhou et al. 2014). More importantly, the inconsistency and relatively low radiometric resolution of DMSP/OLS NTL data and the variation of thresholds over time due to the socioeconomic development have hampered the use of thresholding for urban change detection.

Conventionally, the optimal threshold is determined as the NTL value that would result in the contiguous urban pixels to match with the reference urban extent within a predefined boundary. However, the extracted urban areas by using the optimal threshold method could be largely overestimated and

Techniques and Methods in Urban Remote Sensing, First Edition. Qihao Weng.
© 2020 by The Institute of Electrical and Electronics Engineers, Inc.
Published 2020 by John Wiley & Sons, Inc.

underestimated due to the low radiometric resolution of DMSP/OLS NTL data. Meanwhile, the optimal threshold in a specific year cannot be temporally extrapolated to other years because of different socioeconomic development status. For example, peri-urban areas can be lighted due to the intensification of human activities over the time. To compensate for the low radiometric resolution of DMSP/OLS NTL data, vegetation variables have been utilized with the assumption that vegetation abundance and urban intensity are inversely correlated (Zhang et al. 2013, 2015; Liu et al. 2015; Zhou et al. 2015). Nevertheless, no previous effort has been made to generate temporally consistent vegetation-adjusted NTL for the detection of urban changes. In addition, for the purpose of detecting urban changes from time-series DMSP/OLS NTL data, methods for intercalibration of NTL data have been proposed (Elvidge et al. 2009; Liu et al. 2012). However, previous intercalibration methods are not sufficient to make NTL digital number (DN) temporally compatible, and the assumption of stable thresholds over time may have a negative impact on the accuracy of urban area detection (Wei et al. 2014).

Energy consumption, including electricity consumption (EC), relates to CO_2 emission and pollution and is an important driving force in global climate change (Meng et al. 2014). Thus, the acquisition of EC information is crucial for understanding both the impacts of EC and its interactions with economic activities and the environment. The primary sources of EC data include statistics at temporal resolution of year and spatial resolution of administrative units (e.g. country, province, or county). However, the statistics suffers from a number of problems, one of which is lacking detailed spatial information (Cao et al. 2014). NTL imagery provides a globally consistent and repeatable way to estimate EC information. The potential use of DMSP-OLS NTL images for the inventory of EC patterns was firstly noted in the 1980s by Welch (1980). Since then, DMSP/OLS NTL imagery has been extensively used to estimate EC at multiple scales such as county, state/provincial, and global scale. For example, Elvidge et al. (1997, 2001) found the log–log relationship between lit area and electric power consumption for 21 countries; Lo (2002) established the logarithmic relationship between Equilibrium Phosphorus Concentration (EPC) and lit area for 35 Chinese cities; Amaral et al. (2005) modeled EPC in Brazilian Amazon from extracted lighted area; Letu et al. (2010) estimated EPC of 10 Asian countries using saturation-corrected DMSP NTL intensity. However, these studies were single year based, ignoring the temporal dynamics of EC. More recently, the availability of DMSP-OLS time-series NTL images makes it possible for the examination of spatiotemporal dynamics of EC. For instance, He et al. (2012) examined EC pattern at county level in the Mainland China using DMSP-OLS NTL data during 1995–2008; He et al. (2014) modeled the county-level spatiotemporal dynamics of EC from 2000 to 2008 in China by using saturation-eliminated DMSP-OLS data. However, these researches were at county level, so within-county spatiotemporal patterns were unclear.

To capture the spatiotemporal pattern of EC at subcounty level, Cao et al. (2014) proposed a top-down method based on the indicators of NTL and population density. Although the method proposed by Cao et al. captured the broad spatial and temporal pattern of EC for China, it should be noted that several limitations should be rectified before its wide application. First, the proposed Gross Domestic Product (GDP)-based method for saturation correction is problematic. One reason is that the proposed method does not actually increase variation of EC in urban cores, but just added all the bias between non-radiance-calibrated (NRC) and radiation-corrected (RC) DN value for each city to saturated pixels. Thus, this may result in the improper assignment of EC, providing inaccurate spatial and temporal pattern of EC though the GDP-based method could, to some degree, improve county-level estimation. Meanwhile, it actually simultaneously adjusted the saturated pixels to the same value. This may not lose much detailed EC information in urban cores for Chinese cities, because very few pixels are with value of 63 in China. For developed countries such as United States, GDP-based method does not work because they possess large number of saturated pixels, missing the variation information of the estimated EC in urban cores. Additionally, the application of the linear relationship between per area GDP and per area total RC DN values in 2006 to other years for saturation correction is also problematic. This is because of the incomparability of DMSP-OLS NTL imageries from different year, or even two imageries from the same year but from different sensors. Second, the availability of statistical data such as city-level GDP and population will limit the wide application of the model. In rural regions, these data sets are usually unavailable.

In this chapter, a new method is proposed for large-scale urbanization analysis by spatiotemporally adjusting NTL data and associating the thresholds of different images. This method aims at increasing the consistency over the time and the details of NTL data. Secondly, this chapter will analyze the spatiotemporal pattern of EC for United States and China from 2000 to 2012 by using the NTL imagery. This analysis offers a spatially explicit method to characterize the spatial and temporal pattern of energy consumption and provide the potential of creating EC data sets at pixel, local, regional, and global scale.

8.2 Detecting Urban Extent Changes

8.2.1 Study Area and Data Sets

The contiguous United States was selected as the study area because of its higher urbanization level and slower urbanization process during the past decades. In particular, 20 metropolitan areas across the study area with a relatively rapid urbanization pace were chosen to assess the proposed method (Figure 8.1).

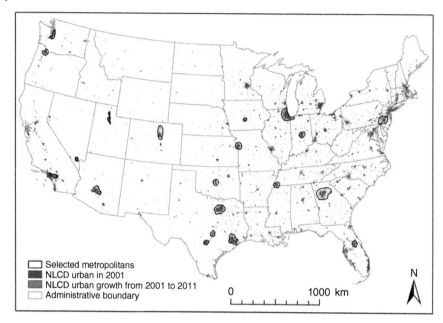

Figure 8.1 The study area of contiguous United States. Twenty cities investigated are Seattle, WA, Portland, OR, Chicago, IL, Des Moines, IA, Salt Lake City, UT, Denver, CO, Kansas City, MO, Indianapolis, IN, Philadelphia, PA, Las Vegas, NV, Oklahoma City, OK, Memphis, TN, San Bernardino, CA, Phoenix, AZ, Atlanta, GA, Dallas, TX, Austin, TX, San Antonio, TX, Houston, TX, and Orlando, FL (from north to south, west to east).

The main data sets used in this study consisted of DMSP/OLS stable NTL images, the Moderate Resolution Imaging Spectroradiometer (MODIS) and Advanced Very High Resolution Radiometer (AVHRR) Normalized Difference Vegetation Index (NDVI) data, the national land-cover data set (NLCD), and the MODIS water mask product. In specific, the stable NTL data from 1992 to 2013 were used (https://www.ngdc.noaa.gov/eog/dmsp.html). The DMSP/OLS data sets were produced using all available archived data for each calendar year (Baugh et al. 2010). The images are in the 30 arc-second grids and their DN ranged from 0 to 63 (Baugh et al. 2010). The mean composite of NTL data was generated for the year when two images were available. The AVHRR 14-day NDVI composite data for 1995 and the MODIS 16-day NDVI data product for 2001, 2006, and 2011, both at 1 km spatial resolution, were utilized (earthexplorer.usgs.gov). Yearly average NDVI composite was computed for each year to ameliorate its sensitivity to seasonal and interannual fluctuations (Zhang et al. 2013; Ma et al. 2014). The NLCD in 2001, 2006, and 2011 were generated from Landsat Thematic Mapper (TM)/Enhanced Thematic Mapper (ETM+) imagery, with the spatial resolution of 30 m (Xian et al. 2011; Homer et al. 2015). Urban masks in each year were firstly retrieved with the impervious threshold of larger than 20%, which were then aggregated to 1-km resolution grids, with each grid retaining percent urbanized area (Fry et al. 2011; Zhou et al. 2014). The 1-km resolution urban pixels were identified as impervious-ness larger than 20% (Fry et al. 2011). Additionally, a water mask in 2000 was derived from the MODIS water mask product (MOD44W) (earthexplorer. usgs.gov) (Salomon et al. 2004).

All image-based data were projected to the Albers Equal Area projection and processed to the spatial resolution of 1 km. Then, the NTL values of water were set to 0 to reduce the blooming effect in water bodies.

8.2.2 Methodology

The proposed method included four main steps: (i) generating consistent and saturation- and blooming-reduced time-series DMSP/OLS NTL data; (ii) calculating the optimal threshold for each region in the reference year; (iii) obtaining the optimal threshold for each region in a target year; and (iv) assessing changes in urban extents by applying estimated thresholds (Figure 8.2). Details of each step are described below.

The inconsistency of time-series NTL data was first reduced by using the method proposed by Elvidge et al. (2009) through assuming Mauritius, Puerto Rico, and Okinawa (in Japan) as the invariant regions of the NTL. To reduce

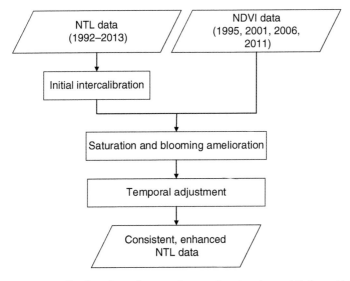

Figure 8.2 The flowchart of generating consistent, enhanced Defense Meteorological Satellite Program/Operational Linescan System (DMSP/OLS) nighttime light (NTL) data.

the saturation effect of NTL in urban cores and the blooming effect in peri-urban regions, NTL data was further adjusted by NDVI data. Specifically, NDVI data of 1995, 2001, 2006, and 2011 were used to adjust NTL for the periods of 1992–1997, 1998–2003, 2004–2008, and 2009–2013, respectively. Assuming the inverse relationship between vegetation abundance and impervious surface, NDVI-adjusted NTL (NTL_{ndvi}) was calculated as (Zhuo et al. 2015)

$$NTL_{ndvi} = \left(\frac{2}{1-k} - 1 \right) \cdot NTL_{int} \tag{8.1}$$

where NTL_{int} was the intercalibrated $NTL\ DN$ value, and k was derived as

$$k = NTL_{norm} - NDVI \tag{8.2}$$

where NTL_{norm} was the normalized NTL_{int}.

Further, inter-year correction was conducted to eliminate the impact of the inconsistent NDVI data and the drawback of the initial intercalibration. A logistic model was used for temporal adjustment after examining the changes of NTL DN values over the time (Figure 8.3) (Song et al. 2016).

$$NTL_{tAdj} = \frac{m}{1 + e^{r \cdot t + c}} + d \tag{8.3}$$

where NTL_{Adj} in year t was a function over time, with m describing the magnitude of NTL_{ndvi} change, r indicating the rate of change, c referring to the timing of change, and d representing either the pre- or post-change NTL_{ndvi}. Similar

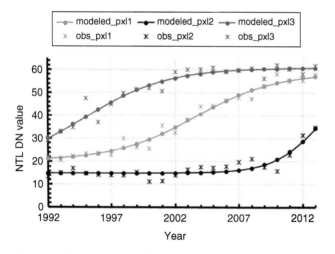

Figure 8.3 Demonstration of observed and modeled NTL values across time.

models have been proposed to describe the temporal dynamics of either vegetation phenology or impervious surface (Song et al. 2016; Xie and Weng 2016b).

An iterative method was employed to determine the optimal threshold for each metropolitan in the reference year, which was 2001 in this study. The NTL value that would result in contiguous urban pixels to match with the reference urban extent within a predefined boundary was selected as the optimal threshold.

Even if the time-series NTL data was perfectly calibrated, the optimal threshold determined in the reference year can hardly be directly used in a target year owing to the impacts of socioeconomic development and the variation of atmospheric condition. Thus, a linear relationship was built to relate the optimal threshold of the reference year t_1 ($optTH_{i,t_1}$) and that of a target year t_2 ($optTH_{i,t_2}$) (Wei et al. 2014):

$$optTH_{i,t_2} = \left\lfloor a_i + b_i \cdot optTh_{i,t_1} + 0.5 \right\rfloor \tag{8.4}$$

where i refers to the ith region, and a and b are the coefficients determined through linear regression:

$$NTL_{i,t_2} = a_i + b_i \cdot NTL_{i,t_1} \tag{8.5}$$

The method utilized the concept of pseudo-invariant features (PIFs) that assumed the stability or proportional change of NTL intensity over the time for certain pixels (Wei et al. 2014). Urban pixels in the reference year were regarded as the PIFs with the assumption of irreversible process of urbanization. To obtain robust regressions, outliers were iteratively removed based on the 95% confidence level. The process continued until a robust regression was obtained.

Urban areas were extracted for each metropolitan by using the estimated optimal thresholds. Meanwhile, the irreversibility of urban extent was assumed, that is, once a pixel changed into urban, it would not convert back to a nonurban one during the study period. Finally, the accuracy of the detected urban changes was assessed.

8.2.3 Results

8.2.3.1 Corrected Time-Series NTL Data

Figure 8.4 shows the comparisons between the original and adjusted DMSP/OLS NTL data in 1992 and 2013 for Atlanta, GA, Indianapolis, IN, and Orlando, FL, respectively. It is demonstrated that the proposed method effectively increased the details of NTL in the urban areas and reduced the inconsistency of the data set. It seems that urban centers tended to have the largest NTL intensity, followed by residential areas and peri-urban regions.

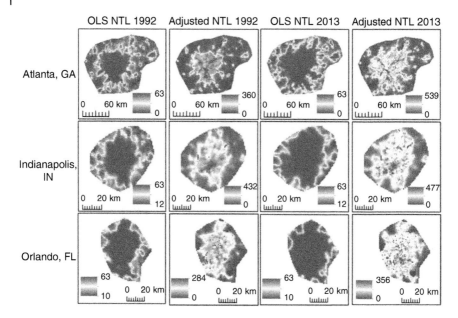

Figure 8.4 The comparisons between adjusted and unadjusted NTL data.

8.2.3.2 Optimal Thresholds in Reference and Target Year

After the adjustment, the reference optimal thresholds in 2001 and 2011 were highly correlated with R^2 of 0.98 and the slope of linear regression of 0.95 (Figure 8.5a). Despite the high correlation, the utilization of threshold of 2001 for 2011 tended to overestimate urban areas. This was especially true for the cities with the optimal thresholds between 100 and 250. The degree of overestimation for those cities was greatly relieved after adjusting thresholds, as demonstrated in Figure 8.5b ($R^2 = 0.99$, regression slope = 1.05). It is further revealed in Figure 8.5 that the threshold value varied with city, ranging from 20 in San Bernardino, CA, to 400 in Dallas, TX.

8.2.3.3 Detecting Changes in Urban Extents

Urban areas in 2001 and 2011 of the selected metropolitans were delineated from adjusted NTL data by using the reference and estimated optimal thresholds, respectively. Urban growth during the period was then estimated by subtracting urban area in 2011 from that in 2001. The average value of Kappa index for the estimated urban areas in 2011 was about 0.7 and producer's accuracy (PA), user's accuracy (UA), and overall accuracy (OA) over 0.85 (Figure 8.6b), indicating that the estimated thresholds can map urban extents with reasonably high accuracy. Figure 8.6a shows that the method can estimate the area of urban growth with high accuracy (regression slope = 0.99, $R^2 = 0.90$). However, the location accuracy of urban changes was less accurate, with mean

Figure 8.5 The comparisons between the reference threshold in 2001 and 2011 (a) and the estimated and reference optimal threshold in 2011 (b).

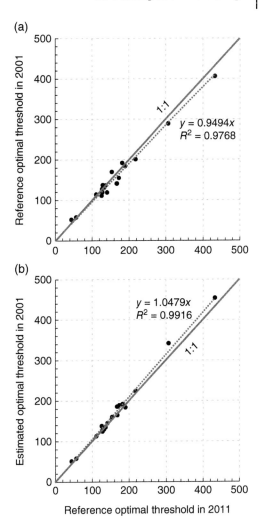

Kappa, PA, and UA value of less than 0.35. The different definition of urban may be partly contributed to its lower accuracy as the reference urban area was determined by impervious surfaces while the NTL-derived urban was identified through NTL.

8.2.4 Discussion and Conclusions

This study addressed the issue of mapping large-scale urban dynamics from uncalibrated time-series DMSP/OLS NTL data by using the threshold technique. An experiment was conducted in the contiguous USA, which

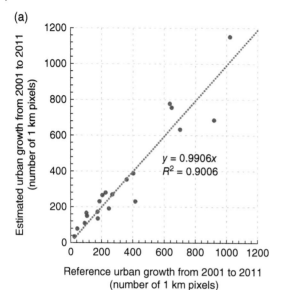

(a)

Figure 8.6 The accuracy assessment of urban mapping, with PA: producer's accuracy, UA: user's accuracy, and OA: overall accuracy. (a) Validation of estimated urban growth; (b) accuracy indexes.

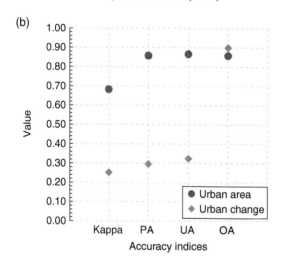

(b)

showed that the proposed method can successfully detect urban growth, especially for the increase of urban area. Thus, the method can be applied to map large-scale urban changes in a cost-effective manner.

In this study, urban changes were detected based on predefined metropolitan boundary. However, to detect urban changes across larger areas with more heterogeneous urbanization processes, the definition and selection of predefined boundaries (e.g. city or urban boundary) can be challenging. Further, the utilization of predefined boundaries may limit the ability to catch the variations of NTL thresholds within them. Thus, an urban object-based method for the determination of the unit of change analysis can be more promising and should be employed to automatically identify the unit for urban change detection (Zhou et al. 2014; Xie and Weng 2016b).

Although this study showed that the method can detect urban area increase, however, the location accuracy of urban growth was less accurate. This result implies that DMSP/OLS NTL data was not sufficient to detect binary, imperviousness-based urban changes. Further research on the relationship between physical and NTL-based urbanization can help interpret NTL-derived urban changes and fuse optical imagery and NTL data for urbanization researches. Additionally, the generated temporally consistent vegetation-adjusted NTL data provided the possibility to detect the temporal change of imperviousness per pixel, thus offering potentials to obtain the magnitude, rate, and timing of urban changes.

8.3 Spatiotemporal Pattern of Energy Consumption in United States and China

8.3.1 Study Area and Data Sets

In this study, the contiguous United States and Mainland China were chosen as the experimental areas. They are the top two countries consume the most energy. Additionally, these two study areas have different urbanization patterns. Particularly, development levels in China vary greatly across space, resulting in heterogeneous EC pattern whereas development in the United States is somewhat more uniform. Thus, the different urbanization densities and patterns in the United States and China provide ideal experimental regions for evaluating the method.

The data sets used in this study consist of DMSP/OLS NTL imageries, MODIS 16-day Enhanced Vegetation Index (EVI) composites (MOD13A2), Gridded Population Density data (GPWv3), and state/provincial and county-level statistical EC data. NTL data, EVI imageries, and statistical data are with the time period from 2000 to 2012, while population density data is at the time point of 2000, 2005, and 2010 because of its availability. The publicly available DMSP/OLS NTL imageries (from 1992 to present) were processed and published by the National Geophysical Data Center (NGDC). The annual cloud-free composite contains persistent lights from cities and settlements and ephemeral events such as wildfires have been removed. The DN of pixels are from 0 to 63 with spatial resolution of nearly 1 km (at the equator), spanning −180° to +180° in longitude and −65° to +75° in latitude. GPWv3 is created by SEDAC (NASA Socioeconomic Data and Application Center), providing globally consistent and spatially explicit human population information and data (Zhang et al. 2013). The grid cell resolution is approximately 5 km at the equator. MODIS Vegetation Index data (MOD13A2) is provided every 16 days at 1-km spatial resolution. The included vegetation indices – NDVI and EVI – are capable to monitor vegetation conditions and land cover and land-cover changes. Also, these data could be used as input for modeling global biogeochemical and hydrologic processes, global and regional climate, and land surface biophysical properties and processes.

Besides image-based data sets, the state/provincial and county-level EC was collected in this study. The EC estimates for United States was from US EIA's (Energy Information Administration) SEDS (State Energy Data System), and the statistic EC data for China was obtained from China City Statistical Yearbook and China Energy Statistical Yearbook.

8.3.2 Methodology

8.3.2.1 Data Preprocessing
This study consists of three parts: (i) data preprocessing, (ii) model calibration, (iii) validation and application, and (iv) spatiotemporal pattern analysis (Figure 8.7).

At the preprocessing step, image-based data sets were processed to the same projection (e.g. Albers Equal Area) and spatial resolution (e.g. 1 km). Specifically, yearly maximum EVI composite was computed for each year. Meanwhile, NTL imageries and yearly maximum EVI composites were combined to obtain EVI-adjusted NTL. The state-/provincial-level administrative borders were then applied to extract dependent and independent variables for model calibration.

8.3.2.2 Estimation of Pixel-Based EC
To further characterize the spatiotemporal pattern of EC for urban cores, saturation and blooming effects in DMSP/OLS NTL must be corrected. Except for the idea but cost-intensive method – utilizing dynamic satellite gain settings – numerous auxiliary data have been applied to increase NTL variation in some saturated urban cores, including population density, GDP, and NDVI (Zhang et al. 2013; Cao et al. 2014; Meng et al. 2014). Specifically, NDVI could effectively reduce the NTL saturation for core urban areas with simplicity under the assumption that urban surfaces and vegetation is inversely correlated (Zhang et al. 2013). However, like NTL, NDVI tends to saturate at high densities of vegetation cover, posing a limitation for peri-urban areas where there may be a presence of both high vegetation and high NTL values (Zhang et al. 2013). To address this problem, EVI data was used. The EVI-adjusted NTL is defined as

$$NTL_{adj} = \left(1 - EVI_{max}\right) \cdot NTL$$
$$EVI_{max} = max\left(EVI_1, EVI_2, \ldots, EVI_{23}\right)$$

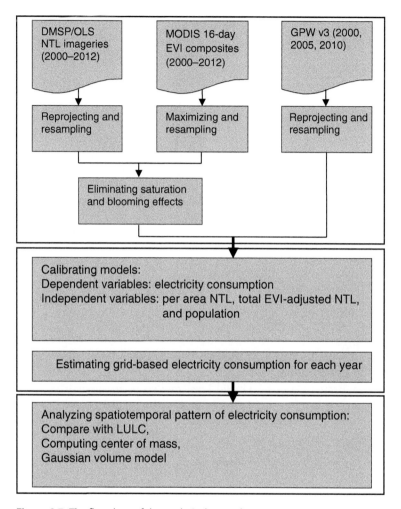

Figure 8.7 The flowchart of the analytical procedures.

where EVI_{max} is the yearly maximum EVI of total twenty-three 16-Day composites for each year.

The IPAT identity ($I = $ PAT) has been extensively used as a framework for analyzing the effects of human activities on the environment (e.g. greenhouse gas emission, energy consumption, etc.) (York et al. 2003). It specifies the driving forces of environmental change as three aspects: population (P), affluence (A), and technology (T). However, the key limitations of IPAT are that it does not permit hypothesis testing and its proportional assumption in the functional relationship between factors (York et al. 2003). To overcome these limitations, Dietz and Rosa refined IPAT identity into a stochastic model – stochastic impacts by regression on population, affluence and technology (STIRPAT) (York et al. 2003). The formula of STIRPAT is defined as

$$I = k \cdot P^{k_1} A^{k_2} T^{k_3} \varepsilon$$

where k is the scale factor, k_1, k_2, and k_3 are the components of P, A, and T, respectively, and ε is the error term. In this study, I refers to EC, and A is measured by per capita GDP. To incorporate less statistical information, which could be unavailable especially for rural regions, we used per capita EVI-adjusted NTL for the input of A. Since there is no single operational measure of T that is free of controversy, we added T to error term. Additionally, additional factors which may impact the consumption of electricity were incorporated, including NTL, spatial concentration of human activities measured by variation coefficient of NTL (i.e. CV), and latitude (Wu et al. 2013). Thus, the final formula of the model is

$$EC = k \cdot P^{c_1} A^{c_2} e^{c_3 \cdot CV} e^{c_4 \cdot latitude} NTL_{adj}^{c_5}$$

The logarithmic form of the model is

$$\ln EC = \ln k + c_1 \cdot \ln P + c_2 \cdot \ln A + c_3 \cdot CV + c_4 \cdot latitude + c_5 \cdot \ln NTL_{adj}$$

where c_1, c_2, c_3, c_4, and c_5 are the coefficient for each variable, respectively. Models were calibrated yearly to absorb any problem with intercalibration of the nightlights. After assuming that grid-level EC, EC_p, follows the same relationship, the calibrated state-/provincial-level models were applied to pixel-based estimation. Using state-/provincial-level statistical EC data as constraints, the adjusted EC_p was estimated by the formula:

$$EC'_p = s \cdot EC_p$$

$$s = \frac{EC_{ref}}{\Sigma EC_p}$$

where EC'_p is the estimated pixel-based EC, EC_{ref} is the state-/provincial-level EC reference data, and ΣEC_p is the aggregated state-/provincial-level EC before adjustment.

8.3.3 Results

8.3.3.1 Regression Results for State-/Provincial-Level Models

Table 8.1 shows the yearly regression result for US state-level model. It is shown that all the models perform quite well with R^2 larger than 0.93 and p-value of 0. Population is positively related to EC, which means that larger size of people tends to consume more electricity. Meanwhile, the brighter the area lightened by NTL, the more EC tends to be consumed. It is interesting to find that higher

Table 8.1 Yearly regression models at the state level in the USA.

Year	R^2	con	lnPop	lnNTL	Variation coefficient (CV)*	Latitude	p-Value
2000	0.9364	5.1073	0.7191	0.2082	−0.0232	−0.0240	0.0000
2001	0.9429	5.3103	0.7263	0.1955	−0.0299	−0.0272	0.0000
2002	0.9402	5.5424	0.7026	0.2092	−0.0316	−0.0278	0.0000
2003	0.9419	5.6158	0.6756	0.2347	−0.0388	−0.0258	0.0000
2004	0.9404	5.5641	0.7152	0.1918	−0.0200	−0.0267	0.0000
2005	0.9393	5.5217	0.7129	0.1992	−0.0234	−0.0265	0.0000
2006	0.9383	5.4605	0.6947	0.2189	−0.0071	−0.0250	0.0000
2007	0.9368	5.7150	0.6772	0.2328	−0.0187	−0.0279	0.0000
2008	0.9344	5.7233	0.6492	0.2589	−0.0263	−0.0272	0.0000
2009	0.9310	5.6223	0.7009	0.1992	0.0072	−0.0279	0.0000
2010	0.9315	5.7279	0.6833	0.2236	−0.0232	−0.0305	0.0000
2011	0.9318	5.5567	0.6775	0.2359	−0.0125	−0.0278	0.0000
2012	0.9378	5.4711	0.6348	0.2770	−0.0137	−0.0225	0.0000

* Significance >0.05.

latitude tends to reduce EC. One possible reason could be that the difference between cooling degree day surpasses the gap between heating degree day in lower and higher latitude regions. Another possible reason may be that the majority of economic activities occur in Midwestern and Eastern United States, which are relatively closer to equator than Western and Northwestern regions. The coefficients of CV are not significant at 5% level. Thus, to apply the models, the variable of CV was deleted and keeping all other coefficients the same.

8.3.3.2 Spatialization of EC for United States and China

Using calibrated state-level models in Table 8.1, we calculated pixel-based EC distribution for Contiguous United States from 2000 to 2012, followed by state-level EC regulation. Figure 8.8 presents the estimated result in 2000 and 2012. Visually, the overall pattern of EC in 2000 and 2011 are almost the same. There are some regions with remarkably higher EC than other regions, such as New York, NY, Los Angeles, CA, Chicago, IL, and Huston, TX. It is also shown in Figure 8.8 that urban cores tend to consume more electricity than their suburban regions because more intensive economic activities (e.g. service industry) occur there. Totally, eastern states consumed more electricity than western regions during the past decade and the country-level pattern of EC keeps relatively stable.

(a)

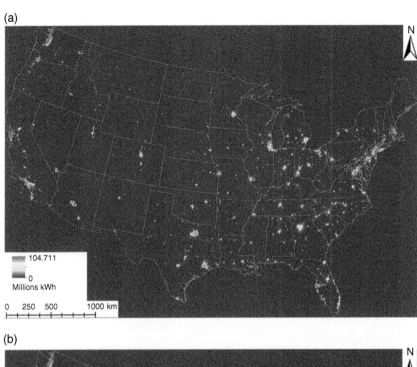

(b)

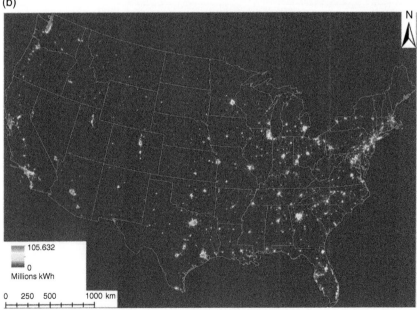

Figure 8.8 Estimated pixel-based energy consumption in the USA in 2000 (a) and 2012 (b).

Follow the same procedure, pixel-based EC maps from 2000 to 2012 for China were also estimated. Figure 8.9 shows the estimated results of 2000 and 2012 for China. Visually, coastal regions consume the most electricity in China, both in 2000 and 2012. These regions include Beijing, Tianjin, Shanghai, and Pearl Delta Region. Meanwhile, high EC regions could also be capital city of each province and some industrial cities such as Chengdu and Chongqing. In both Figures 8.8 and 8.9, rural regions tend to consume less electricity and non-EC regions could be forests, deserts, and water.

Since per-pixel evaluation of the estimated results in United States is impossible, we validated the results by using county-level reference. County-level EC of California from 2006 to 2012 was applied to conduct accuracy evaluation because of data availability. County-level EC of Californian was obtained from Energy Consumption Data Management System of California. The amount of electricity consumed by the counties in California ranges from very small value (e.g. approximately 13 million kWh for Alpine County in 2012) to large value (e.g. approximately 70 000 million kWh for Los Angeles County in 2012), so it is reasonable to validate the estimated results by using county-level EC of California to represent the overall accuracy for United States. Figure 8.10 shows the validation results for 58 counties in California. The figure shows good consistency between the reference and the aggregated estimated EC, with the most points are evenly distributed along the 1 : 1 line. It should also be mentioned that the R^2 of the linear relationships between estimated and reference EC are all above 0.97. Overall, the models slightly underestimated EC with the regression coefficients larger than 1.0, except for 2011. For China, county-level validation is usually unavailable. The city- or county-level EC is usually recorded as the urban consumption in China, excluding rural regions. Thus, we visually compared the EC maps with land use and land cover to validate the accuracy.

8.3.3.3 Spatiotemporal Pattern of EC for United States and China

To compare the spatiotemporal pattern of EC for United States and China, the difference map between 2000 and 2012 was calculated. Figures 8.11a and 8.12a show the increase of EC from 2000 to 2012 for United States and China, respectively. Meanwhile, Figures 8.11b and 8.12b respectively map the increase rate of EC during the same period for United States and China. It is shown in Figure 8.5a that metropolitan regions and their surrounding areas have experienced the most increase of EC volume. However, the cores of big cities, actually, have experienced EC reduction, such as the core areas of New York, Los Angeles, Chicago, and Atlanta. Some Midwestern areas have broadly decreased EC, such as the State of Ohio. The EC in Ohio reduced from 165 194 in 2000 to 150 307 million kWh, decreased by 9%. While for China, most areas have increased EC, especially for coastal regions such as Beijing–Tianjin, Jiangsu–Shanghai–Zhejiang, Pearl Delta Region, and the capital cities of each

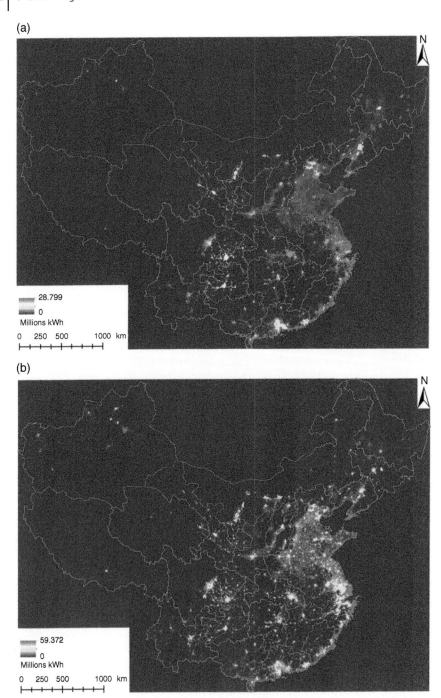

Figure 8.9 Estimated pixel-based energy consumption in China in 2000 (a) and 2012 (b).

(a)

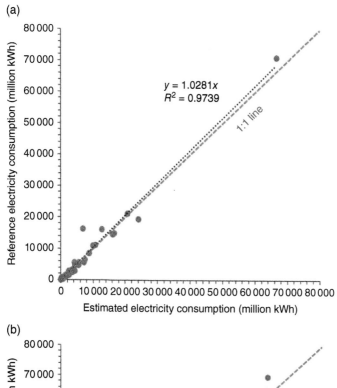

(b)

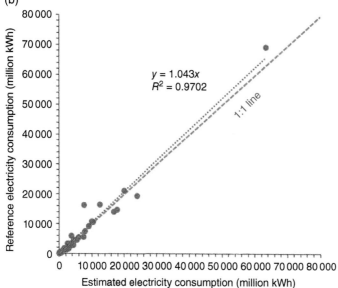

Figure 8.10 Validation results for 58 counties in California in 2006 and 2012.

province (Figure 8.6a). Meanwhile, provinces such as Jiangsu, Shandong, Henan, and Hebei have increased EC across the whole area during the past decade.

The pattern of increase rate of EC for United States (Figure 8.12b) is quite different from that of the increased EC volume. The increase rate for Northeastern, Southern, and some Midwestern areas is relatively smaller than Western regions. It is worthy to mention that some regions have experienced remarkable increase rate of EC, such as Northwestern North Dakota and Southern. This may be due to the activities of fuel exploitation. For example, the Williston, Tioga, Stanley, and Minot–Burlington communities are experiencing rapid growth caused by an oil boom in Northwestern North Dakota. For China, regions with remarkable increase rate of EC are Xinjiang Uygur Autonomous Region (northwest), the Inner Mongolia Autonomous Region (middle north), and Northern Shanxi and Shaanxi, but the Tibet Autonomous Region (southwest) has barely increase rate. However, the increase rate of coastal regions is not so remarkable when compared with their patterns of increased EC.

The comparison between Figures 8.11 and 8.12 shows that the intensity of increased EC in United States is extensively higher than that in China. Per-pixel increased EC for United States could be as high as 700 million kWh, while the maximum value for China is 34 million kWh. Urbanization industrialization in China during the past decade has intensively increased EC, especially the urban areas in Eastern China. However, the phenomenon of urban decentralization in United States has reduced its EC in metropolitan urban core areas while remarkably increased EC in suburban regions. Meanwhile, the increase rate in United States is more homogeneous than that in China, especially for the eastern part of each country. One possible reason for this is that urban development is more concentrated in some specific regions (e.g. Eastern China).

8.3.4 Discussion and Conclusions

In this study, we adjusted the STIRPAT identity to conduct pixel-based EC estimation. Yearly maximum EVI was applied to eliminate the saturation effect of DMSP/OLS NTL image in urban core areas. Variation coefficient of NTL (i.e. CV) and latitude were also considered as additional factors. The experiments in United States and China showed that the model could provide relatively accurate spatiotemporal pattern of EC. The spatiotemporal pattern of EC in United States and China is significantly different: EC decreased in metropolitan urban cores and increased in suburban regions in United States during the past decade, while, both urban and suburban regions have experienced fast increase of EC for China, with urban cores increased the most. Thus, the method provides the potential for analyzing the impact of

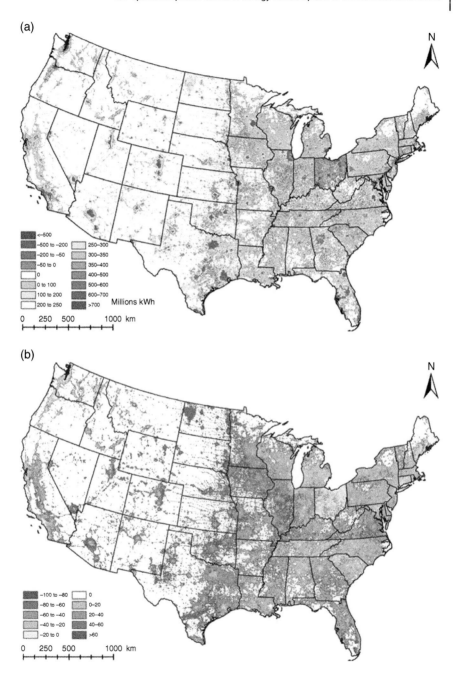

Figure 8.11 Per-pixel electricity consumption (EC) change from 2000 to 2012 for United States, (a) EC increase from 2000 to 2012, and (b) EC increase rate (%) from 2000 to 2012.

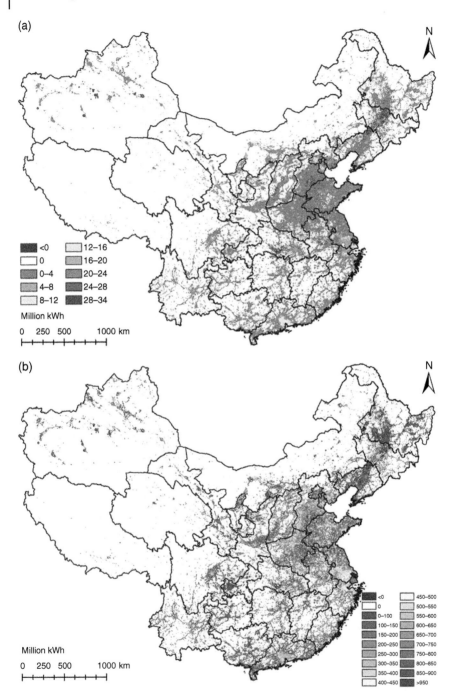

(a)

(b)

Figure 8.12 Per-pixel EC change from 2000 to 2012 for China, (a) EC increase from 2000 to 2012, and (b) EC increase rate (%) from 2000 to 2012.

urbanization and urban decentralization on EC. The advantage of the proposed method is its easy implementation and few data inputs. Also, it could provide detailed spatiotemporal pattern of EC in urban cores.

The EVI-based method provides region-independent way to increase NTL variations in urban cores. Thus, it does not need the input of statistical data such as GDP, which strongly impact the wide application of the model proposed by Cao et al. However, EVI-based method, similar to VANUI, is not good at increasing interurban variation for the cities which have experienced remarkable growth over a relatively short time period, such as most Chinese cities. Publicly available population data was applied to improve the accuracy of rural areas where no NTL signal was recorded by DMSP/OLS. Without the assumption that population is evenly distributed in rural areas (Cao et al. 2014), the estimated EC for each Chinese province is no longer homogeneous, as was reported by Cao et al. However, the relatively coarse resolution of GPWv3 data would definitely introduce errors in per-pixel EC estimation. However, to widely use the method, data availability is a critical consideration. One limitation of the proposed method, like many administrative unit-based methods, is that adjacent pixels near state/provincial borders tend to have significantly different values when they belong to different state/province. This is because of the use of state-/province level to adjust pixel-based estimations. To obtain more accurate and reasonable spatiotemporal pattern of EC, the relationship between neighbor pixels should be considered.

References

Amaral, S., Câmara, G., Monteiro, A.M.V. et al. (2005). Estimating population and energy consumption in Brazilian Amazonia using DMSP night-time satellite data. *Computers, Environment and Urban Systems* 29 (2): 179–195.

Baugh, K., Elvidge, C.D., Ghosh, T., and Ziskin, D. (2010). Development of a 2009 stable lights product using DMSP-OLS data. *Proceedings of the Asia-Pacific Advanced Network* 30: 114–130.

Cao, X., Chen, J., Imura, H., and Higashi, O. (2009). A SVM-based method to extract urban areas from DMSP-OLS and SPOT VGT data. *Remote Sensing of Environment* 113: 2205–2209.

Cao, X., Wang, J., Chen, J., and Shi, F. (2014). Spatialization of electricity consumption of China using saturation-corrected DMSP-OLS data. *International Journal of Applied Earth Observation and Geoinformation* 28: 193–200.

Elvidge, C.D., Baugh, K.E., Kihn, E.A. et al. (1997). Relation between satellite observed visible-near infrared emissions, population, economic activity and electric power consumption. *International Journal of Remote Sensing* 18 (6): 1373–1379.

Elvidge, C.D., Imhoff, M.L., Baugh, K.E. et al. (2001). Night-time lights of the world: 1994–1995. *ISPRS Journal of Photogrammetry and Remote Sensing* 56 (2): 81–99.

Elvidge, C.D., Ziskin, D., Baugh, K.E. et al. (2009). A fifteen year record of global natural gas flaring derived from satellite data. *Energies* 2: 595–622.

Fry, J.A., Xian, G.S., Jin, S. et al. (2011). Completion of the 2006 National Land Cover Database for the conterminous United States. *Photogrammetric Engineering & Remote Sensing* 77: 858–864.

He, C., Ma, Q., Li, T. et al. (2012). Spatiotemporal dynamics of electric power consumption in Chinese Mainland from 1995 to 2008 modeled using DMSP/OLS stable nighttime lights data. *Journal of Geographical Sciences* 22 (1): 125–136.

He, C., Ma, Q., Liu, Z., and Zhang, Q. (2014). Modeling the spatiotemporal dynamics of electric power consumption in Mainland China using saturation-corrected DMSP/OLS nighttime stable light data. *International Journal of Digital Earth* 7 (12): 993–1014.

Henderson, M., Yeh, E.T., Gong, P. et al. (2003). Validation of urban boundaries derived from global night-time satellite imagery. *International Journal of Remote Sensing* 24: 595–609.

Homer, C., Dewitz, J., Yang, L. et al. (2015). Completion of the 2011 National Land Cover Database for the Conterminous United States – representing a decade of land cover change information. *Photogrammetric Engineering & Remote Sensing* 81: 346–354.

Imhoff, M.L., Lawrence, W.T., Stutzer, D.C., and Elvidge, C.D. (1997). A technique for using composite DMSP/OLS "city lights" satellite data to map urban area. *Remote Sensing of Environment* 61: 361–370.

Letu, H., Hara, M., Yagi, H. et al. (2010). Estimating energy consumption from night-time DMPS/OLS imagery after correcting for saturation effects. *International Journal of Remote Sensing* 31 (16): 4443–4458.

Liu, X., Hu, G., Ai, B. et al. (2015). A Normalized Urban Areas Composite Index (NUACI) based on combination of DMSP-OLS and MODIS for mapping impervious surface area. *Remote Sensing* 7: 17168–17189.

Liu, Z., He, C., Zhang, Q. et al. (2012). Extracting the dynamics of urban expansion in China using DMSP-OLS nighttime light data from 1992 to 2008. *Landscape and Urban Planning* 106: 62–72.

Lo, C. (2002). Urban indicators of China from radiance-calibrated digital DMSP-OLS nighttime images. *Annals of the Association of American Geographers* 92 (2): 225–240.

Ma, Q., He, C., Wu, J. et al. (2014). Quantifying spatiotemporal patterns of urban impervious surfaces in China: an improved assessment using nighttime light data. *Landscape and Urban Planning* 130: 36–49.

Meng, L., Graus, W., Worrell, E., and Huang, B. (2014). Estimating CO_2 (carbon dioxide) emissions at urban scales by DMSP/OLS (Defense Meteorological

Satellite Program's Operational Linescan System) nighttime light imagery: methodological challenges and a case study for China. *Energy* 71: 468–478.

Pandey, B., Joshi, P., and Seto, K.C. (2013). Monitoring urbanization dynamics in India using DMSP/OLS night time lights and SPOT-VGT data. *International Journal of Applied Earth Observation and Geoinformation* 23: 49–61.

Salomon, J., Hodges, J.C.F., Friedl, M. et al. (2004). Global land-water mask derived from MODIS Nadir BRDF-adjusted reflectances (NBAR) and the MODIS land cover algorithm. *Geoscience and Remote Sensing Symposium, 2004. IGARSS '04. Proceedings. 2004 IEEE International*, Anchorage, AK, USA (20–24 September 2004).

Song, X.-P., Sexton, J.O., Huang, C. et al. (2016). Characterizing the magnitude, timing and duration of urban growth from time series of Landsat-based estimates of impervious cover. *Remote Sensing of Environment* 175: 1–13.

Wei, Y., Liu, H., Song, W. et al. (2014). Normalization of time series DMSP-OLS nighttime light images for urban growth analysis with pseudo invariant features. *Landscape and Urban Planning* 128: 1–13.

Welch, R. (1980). Monitoring urban population and energy utilization patterns from satellite data. *Remote Sensing of Environment* 9 (1): 1–9.

Wu, J., Wang, Z., Li, W., and Peng, J. (2013). Exploring factors affecting the relationship between light consumption and GDP based on DMSP/OLS nighttime satellite imagery. *Remote Sensing of Environment* 134: 111–119.

Xian, G., Homer, C., Dewitz, J. et al. (2011). Change of impervious surface area between 2001 and 2006 in the conterminous United States. *Photogrammetric Engineering & Remote Sensing* 77: 758–762.

Xie, Y. and Weng, Q. (2016a). Detecting urban-scale dynamics of electricity consumption at Chinese cities using time-series DMSP-OLS (Defense Meteorological Satellite Program-Operational Linescan System) nighttime light imageries. *Energy* 100: 177–189.

Xie, Y. and Weng, Q. (2016b). Updating urban extents with nighttime light imagery by using an object-based thresholding method. *Remote Sensing of Environment* 187 (12): 1–13.

York, R., Rosa, E.A., and Dietz, T. (2003). STIRPAT, IPAT and ImPACT: analytic tools for unpacking the driving forces of environmental impacts. *Ecological Economics* 46 (3): 351–365.

Zhang, Q., Li, B., Thau, D., and Moore, R. (2015). Building a better urban picture: combining day and night remote sensing imagery. *Remote Sensing* 7: 11887–11913.

Zhang, Q., Schaaf, C., and Seto, K.C. (2013). The vegetation adjusted NTL urban index: a new approach to reduce saturation and increase variation in nighttime luminosity. *Remote Sensing of Environment* 129: 32–41.

Zhang, Q. and Seto, K.C. (2011). Mapping urbanization dynamics at regional and global scales using multi-temporal DMSP/OLS nighttime light data. *Remote Sensing of Environment* 115: 2320–2329.

Zhou, Y., Smith, S.J., Elvidge, C.D. et al. (2014). A cluster-based method to map urban area from DMSP/OLS nightlights. *Remote Sensing of Environment* 147: 173–185.

Zhou, Y., Smith, S.J., Zhao, K. et al. (2015). A global map of urban extent from nightlights. *Environmental Research Letters* 10: 054011.

Zhuo, L., Zheng, J., Zhang, X. et al. (2015). An improved method of night-time light saturation reduction based on EVI. *International Journal of Remote Sensing* 36: 4114–4130.

9

Urban Runoff Modeling and Prediction

9.1 Introduction

9.1.1 Land-Use and Land-Cover Change and Urban Runoff

Land-use and land-cover (LULC) changes may have four major direct impacts on the hydrological cycle and water quality: they can cause floods, droughts, and changes in river and ground water regimes, and they can affect water quality (Rogers 1994). In addition to these direct impacts, there are also indirect impacts on climate and the altered climate's subsequent impact on the waters. Urbanization, the conversion of other types of land to uses associated with the growth of population and economy is a main type of LULC change especially in recent human history. The process of urbanization has a considerable hydrological impact in terms of influencing the nature of runoff and other hydrological characteristics, delivering pollutants to rivers, and controlling rates of erosion (Goudie 1990).

At different stages of urban growth, various impacts can be observed (Kibler 1982). In the early stage of urbanization, removal of trees and vegetation may decrease evapotranspiration and interception and increase stream sedimentation. Later, when construction of houses, streets, and culverts initiates, the impacts may include decreased infiltration, lowered ground water table, increased storm flows, and decreased base flows during dry periods. After the development of residential and commercial buildings has been completed, increased imperviousness will reduce time of runoff concentration so that peak discharges are higher and occur sooner after rainfall starts in basins. The volume of runoff and flood damage potential will greatly increase. Moreover, the installation of sewers and storm drains accelerate runoff (Goudie 1990). As a result, the rainfall-runoff process in an urban area tends to be quite different from that in natural conditions depicted in classical hydrological cycles. This effect of urbanization, however, varies according to the size of a flood. As the

Techniques and Methods in Urban Remote Sensing, First Edition. Qihao Weng.

size of the flood becomes larger and its recurrence interval increases, the effect of urbanization decreases (Hollis 1975).

The integration of remote sensing (RS) and geographic information systems (GIS) has been widely applied and been recognized as a powerful and effective tool in detecting urban growth (Weng 2001b). RS collects multispectral, multiresolution, and multi-temporal data and turns them into information valuable for understanding and monitoring urban land processes and for building urban land-cover data sets. GIS technology provides a flexible environment for entering, analyzing, and displaying digital data from various sources necessary for urban feature identification, change detection, and database development. In hydrological and watershed modeling, remotely sensed data are found valuable for providing cost-effective data input and for estimating model parameters (Engman and Gurney 1991; Drayton et al. 1992; Mattikalli et al. 1996). The introduction of GIS to the field makes it possible for computer systems to handle the spatial nature of hydrological parameters. The hydrological community now increasingly adopts GIS-based distributed modeling approaches (Berry and Sailor 1987; Drayton et al. 1992; Mattikalli and Richards 1996). However, no attempt has been made to relate urban growth studies to distributed hydrological modeling, although both studies share LULC data.

This chapter attempts to develop an integrated approach of RS and GIS for examining the effects of urban growth on surface runoff at local level by using the Zhujiang Delta of South China as a case.

9.1.2   Spatially Distributed Surface Runoff Models

The model used for estimating surface runoff in this study was developed by the United States Soil Conservation Service (SCS). It has been widely applied to estimate storm runoff depth for every patch within a watershed based on runoff curve numbers (CN) (USDA, SCS 1972). The SCS equation for storm runoff depth is mathematically expressed as

$$Q = \frac{(P-0.2S)^2}{(P+0.8S)} \tag{9.1}$$

where Q is storm runoff, P is rainfall, S is potential maximum storage, and $S = (1000/CN) - 10$, and CN is runoff CN of hydrologic soil group–land cover complex.

To solve this equation, two input values are necessary: P and CN. Precipitation data are often available from meteorological observations. A runoff CN is a quantitative description of land cover and soil conditions that affect the runoff process. The CN values are normally estimated using field survey data with reference to United States Department of Agriculture (USDA)'s SCS tables (Table 9.1). From Table 9.1, it is apparent that CN values approaching 100 are

Table 9.1 Runoff curve numbers (CNs) for hydrologic soil–cover complexes (USDA, SCS 1972) (Antecedent moisture condition II and Ia = 0.2S).

	Cover		Runoff CN for hydrologic soil group			
Land use	Treatment or practice	Hydrologic condition	A	B	C	D
Fallow	Bare soil	—	77	86	91	94
Row crops	Straight row	Poor	72	81	88	91
		Good	67	78	85	89
	Contoured	Poor	70	79	84	88
		Good	65	75	82	86
	Contoured and terraced	Poor	66	74	80	82
		Good	62	71	78	81
Small grain	Straight row	Poor	65	76	84	88
		Good	63	75	83	87
	Contoured	Poor	63	74	82	85
		Good	61	73	81	84
	Contoured and terraced	Poor	61	72	79	82
		Good	59	70	78	81
Close-seeded legumes[a] or rotation meadow	Straight row	Poor	66	77	85	89
		Good	58	72	81	85
	Contoured	Poor	64	75	83	85
		Good	55	69	78	83
	Contoured and terraced	Poor	63	73	80	83
		Good	51	67	76	80
Pasture or range		Poor	68	79	86	89
		Fair	49	69	79	84
		Good	39	61	74	80
	Contoured	Poor	47	67	81	88
		Fair	25	59	75	83
		Good	6	35	70	79
Meadow		Good	30	58	71	78
Woods		Poor	45	66	77	83
		Fair	36	60	73	79
		Good	25	55	70	77

[a] Close drilled or broadcast.

associated with high runoff from cultivated agricultural land, whereas low-to-moderate *CN* values indicate the reduced runoff from heavily vegetated areas (Slack and Welch 1980). A Hydrological Soil Group code A, B, C, or D was set up by the SCS for over 4000 soils in the United States based on permeability and infiltration characteristics. Group A soils are coarse, sandy, and well-drained soils, with the highest rate of infiltration and the lowest potential for runoff. Group D soils, on the other hand, are heavy, clayey, and poorly drained soils, with the lowest rate of infiltration and highest potential for runoff. In between, Groups A and D are Groups B and C soils.

Since the availability of Landsat data in the 1970s, several attempts have been made to use these satellite data to determine CNs because of the cost-effective nature of these data. Mintzer and Askari (1980) employed Landsat Multispectral Scanner (MSS) data, in combination with color infrared photography, to derive the runoff coefficients for the watersheds of Mill Creek in Ohio by using poly-nomial regression modeling. Ragan and Jackson (1980) compared the estimation of CNs from Landsat MSS with those from field surveys, and from high-level aerial photography. It was suggested that there was no significant difference among the three sources but a modified CN classification system compatible with Landsat data needs to be established. Slack and Welch (1980) conducted a similar study for the Little River watershed near Tifton, Georgia. They generated four hydrologically important land classes – agricultural vegetation, forest, wetland, and bare ground from Landsat MSS data – and found that CNs for six sub-watersheds and for the entire watershed were estimated within two CN units. Rango et al. (1983) claimed only a five percent error in land-cover estimation by Landsat data at the basin level, but a much greater error at the cell level.

Among existing runoff simulations models, the SCS-CN is the most enduring one for estimating the volume of direct surface runoff in ungauged rural catchments (Boughton 1989). It is also noted that several complex models, such as Soil and Water Assessment Tool (SWAT), Hydrologic Modeling System (HEC-HMS), Erosion Productivity Impact Calculator (EPIC), and Agricultural Nonpoint Source Pollution Model (AGNPS) have been developed based on SCS-CN method (Kousari et al. 2010). However, due to the complexity of land surface features in the urban environments, the CN in Table 9.1 cannot be applied to all surface types. Therefore, previous studies have attempted to improve CN estimation based on Table 9.1. Many researchers developed CN estimation methods by incorporating land-cover information into the CN table. For example, Hong and Adler (2008) developed a global SCS-CN runoff map using land cover, soils, and antecedent moisture conditions. Canters et al. (2006) estimated CN at the catchment level by combining impervious surface, vegetation, bare soil, and water/shade information. Reistetter and Russell (2011) proposed that CN was estimated by integrating the percentages of impervious surface, tree canopy density and pervious surface. Furthermore, CN may be calculated via the combination of water, dense forest, and sand

(Dutta et al. 2006). Ludlow (2009) discussed a composite CN calculation method by integrating vegetation, impervious surface, and soil. Apparently, most previous studies showed that impervious surface played an important role in computing the CN, because it affected the infiltration of surface water. In addition, the number of land-cover types described in Table 9.1 is too many to be classified from medium spatial resolution RS imagery with proper accuracy (Cronshey et al. 1985).

9.2 Estimating Composite CN and Simulating Urban Surface Runoff

Because of the complexity in urban landscape composition and the limitation in spatial resolution in medium and coarse spatial resolution imagery, mixed pixels are common in urban landscapes, affecting the effective use of per-pixel classifiers (Lu and Weng 2004). Therefore, the "soft" approach of image classifications has been developed, in which each pixel is assigned a class membership of each land-cover type rather than a single label (Wang 1990). Nevertheless, both Ridd (1995) and Mather (1999) maintained that characterization and quantification, rather than classification, should be applied in order to provide a better understanding of the composition of heterogeneous landscapes such as urban areas. Ridd (1995) proposed an interesting conceptual model for RS analysis of urban landscapes, i.e. the vegetation–impervious surface–soil (V-I-S) model. It assumes that land cover in urban environments is a linear combination of the three components, namely, vegetation, impervious surface, and soil. The V-I-S model was developed for Salt Lake City, Utah, USA, but has been tested in other cities (Weng and Lu 2009). Weng and Lu (2009) suggested that the reconciliation between the V-I-S model and LSMA provided a continuum field model, which offered an alternative, effective approach for characterizing and quantifying the spatial and temporal changes of the urban landscape compositions.

This section summarizes a method to estimate composite CN which was developed with the V-I-S model, normalized difference vegetation index (NDVI), and soil types developed by Fan et al. (2013). First, soil types and NDVI are classified into several classes, respectively. Second, each class of soil type and NDVI are given an initial CN_s (initial CN of soil) and CN_v (initial CN of vegetation), respectively. Lastly, the CN is calculated by computing the percentages of impervious surface, NDVI class, and soil class.

9.2.1 Study Area and Data Sets

Guangzhou, the capital of the Guangdong province, is the political, economic, and cultural center of the Guangdong province and a transportation hub in Southern China. Tianhe district (Figure 9.1) is a new central business district

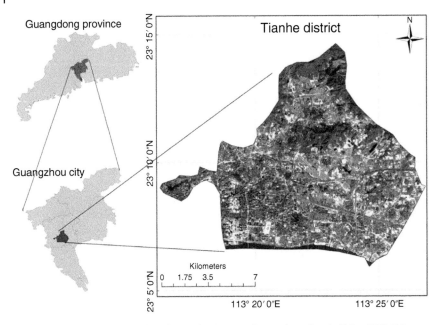

Figure 9.1 Study area located in the Tianhe District, Guangzhou, South China. RGB Color composite of Landsat TM4/TM3/TM2 is displaced (Fan et al. 2013).

(CBD) of Guangzhou which covers 139 km^2 with the population of 0.62 million in 2005. It is located in the north bank of Pearl River, east of Guangzhou city. It has a subtropical climate, which is strongly influenced by the Asian monsoon, with an average annual temperature between 21 and 23 °C and an average precipitation from 1600 to 2600 mm.

A Landsat-5 Thematic Mapper (TM) image (Path 122/Row 44, acquired on 11 February 2009) is used for extracting the fraction images of V-I-S. The TM image is rectified to the Universal Transverse Mercator projection system. A geometric correction is performed to register the TM image to high-resolution SPOT image of 2010 obtained from Google Earth, and the root-mean-square error (RMSE) for geometric correction is 0.2928. In addition, radiometric correction is performed using the calibration parameter file released by the Earth Resources Observation Systems Data Center, United States Geological Survey. Soil types are derived from the soil image of Guangdong Province with the scale of 1 : 1 000 000 (Wu et al. 2001). The precipitation data of 11 February 2009 (including preceding five days data) are obtained from the Meteorological Bureau of Tianhe District. The high-resolution SPOT images are used as ground reference to assess the accuracy of linear unmixing result and the classification of NDVI (Fan et al. 2013).

9.2.2 Estimation of Composite CN

A new procedure is designed to calculate composite CN (Figure 9.2), which includes four steps: (i) obtaining NDVI values and classifying vegetation types: the NDVI values are grouped into four categories and each is assigned an initial CN value in reference to Table 9.1; (ii) extracting V-I-S fraction images using the LSMA model: vegetation, impervious surface, and soil fractions are extracted from the satellite image; (iii) soil classification: each type of soil is given an initial CN value in reference to Table 9.1 based on its characteristics; and (iv) calculating composite CN: the composite CN is calculated as the weighted average of the initial *CN* values of vegetation, impervious surface, and soil fractions within a pixel. With this composite CN estimation method, surface runoff is then simulated under the 57- and 81-mm precipitation for the study area.

In order to calculate composite CN of Guangzhou, we assume that any 30 × 30 m pixel is considered as an independent drainage area and comprises impervious surface, vegetation, and soil only. The formula of the composite CN calculation is as follows:

$$CN_C = ISA \times CN_{ISA} + V \times CN_V + S \times CN_S \tag{9.2}$$

where *ISA*, *V*, and *S* are the percentage of impervious surface, vegetation, and soil, respectively; CN_C, CN_{ISA}, CN_V, and CN_S denote composite CN, initial imperviousness CN, vegetation CN, and soil CN, respectively.

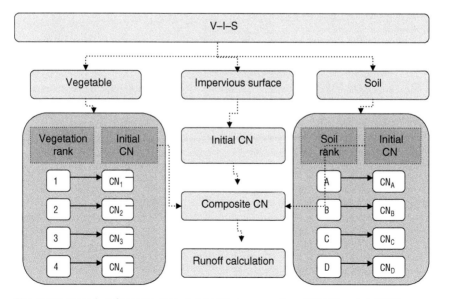

Figure 9.2 Procedure for computing composite curve numbers (CNs) (Fan et al. 2013).

Table 9.2 The value of CN and soil class (Fan et al. 2013).

Soil type	CN	Class
Paddy soil	94	D
Deposited soil	91	C
Red soil	86	B
Aquic soil	91	C

In this study, all soils, vegetation, and impervious surfaces are assumed under the condition of AMC-II in which adjustments to the corresponding CN for dry soils (AMC-I) and wet soils (AMC-III) must be conducted (Silveira et al. 2000). CN_{ISA} is assigned a value of 98 due to its impermeable characteristic (Reistetter and Russell 2011). CN_S depends largely on soil type and antecedent moisture condition. We assume that the soil condition is between dry and wet. To estimate CNs of different soil types, soils are classified into four different categories, ranked from A to D based on infiltration. Class A mostly consist of well-drained sands and gravels with low runoff potential and high infiltration rate. Class B has moderate-to-coarse textures and moderate infiltration rate when completely wet. Class C yields moderately fine to fine textures with low infiltration rate. Class D is primarily clay soils or soils with clay pans that possessed low infiltration rate when wet. Four types of soil can be found in Guangzhou, including paddy soil, deposited soil, red soil, and aquic soil according to the Soil Taxonomy of Guangdong Province. The initial CN for the soils is shown in Table 9.2.

The values of NDVI are classified into four categories: (i) larger than 0.65 representing forest; (ii) between 0.57 and 0.65 grass land, (iii) between 0.4 and 0.57 farmland; and (iv) less than 0.4 non-vegetation surfaces. Each category is further classified into several subclasses based on vegetation abundance derived from the vegetation fraction image. When vegetation fraction is larger than 0.75, vegetation is considered in healthy condition, while the fraction between 0.5 and 0.75 is considered in the fair condition and below 0.5 the poor condition. There are only one poor condition and good condition for farmland in the look-up table (Table 9.1). Hence, for farmland, we set the fraction value of 0.5 as the threshold to separate the good condition from the poor condition to determine CN_V. The initial CN of each vegetation class is shown in Table 9.3.

9.2.3 Simulation of Surface Runoff

Composite CN image is shown in Figure 9.3. The composite CN values are divided into 10 categories, and the values larger than 75 are further examined because they affected the runoff significantly. Figure 9.4 shows that high values

Table 9.3 Initial values of initial CN of vegetation (CNv) (Fan et al. 2013).

Vegetation	Normalized difference vegetation index (NDVI)	Vegetation vigor	CN			
			A	B	C	D
Forest	$NDVI > 0.65$	Poor: $V < 50\%$	45	66	77	83
		Fair: $50\% < V < 75\%$	36	60	73	79
		Good: $V > 75\%$	30	55	70	77
Grass and bush	$0.57 < NDVI < 0.65$	Poor: $V < 50\%$	57	73	82	86
		Fair: $50\% < V < 75\%$	43	65	76	82
		Good: $V > 75\%$	32	58	72	79
Farmland	$0.4 < NDVI < 0.57$	Poor: $V < 50\%$	72	81	88	91
		Good: $V > 50\%$	67	78	85	89
Non-vegetated	$NDVI < 0.4$		59	74	82	86

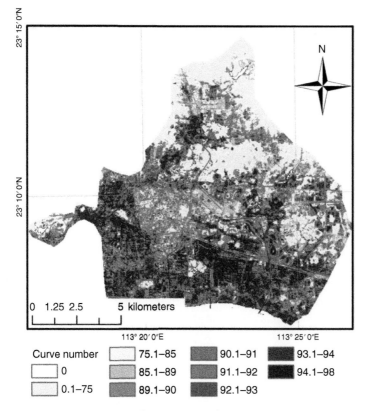

Figure 9.3 The composite CN map (Fan et al. 2013).

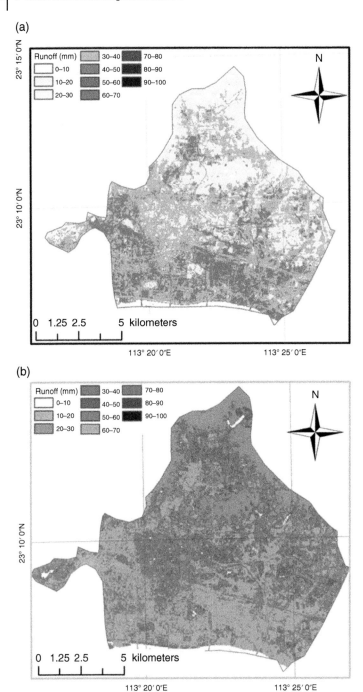

Figure 9.4 Simulated runoff under 57 and 81 mm precipitation (Fan et al. 2013). (a) 57 mm precipitation. (b) 81 mm precipitation.

of NDVI are largely found in the northern part while lower values are in the southern part. In contrast, higher composite CN values tended to cluster in the south and lower values in the north. High percentage of impervious and low percentage of vegetation led to a high composite CN value, which is as anticipated based on background knowledge.

In Guangzhou, precipitation mainly occurs in the period from April to September and is strongly impacted by typhoons and summer rainstorms. Daily precipitation up to 100 mm is common. In this study, we chose 57 and 81 mm of precipitation, which occur once in 5 years and once in 10 years precipitation, respectively, to simulate surface runoff using the improved SCS-CN method. Because most of the rainfall in the urban city is transported through sewers rather than flowing onto the open terrain, the impacts of terrain roughness and elevation difference were not taken into account during the simulations.

Figure 9.4 shows the simulation results. The simulation with 57-mm precipitation is illustrated in Figure 9.4a. The highest runoff volume is found to be 51 mm, but most of the study areas observe runoff below 40 mm. It is found that the runoff volume is less than 31 mm in the areas that possessed NDVI value of greater than 0.57. However, under the scenario of 81-mm precipitation, the majority of study area is estimated to have a runoff volume of greater than 32 mm. This finding implies that when precipitation reached 81 mm, most of the study area would be threatened by surface flooding, especially in the southern part. Due to the lack of data, a real flood event is not included as a case study here.

9.2.4 Discussion

The SCS-CN model utilized in this study is to simulate surface runoff volume in the urban areas. Each pixel in the image is considered as a hydrologic response unit and is decomposed into fractions of V-I-S. The composite CN of a pixel means the average CN of the whole pixel. This sub-pixel method of CN modeling is distinct from previous studies where each pixel is treated as a whole, i.e. per-pixel method (for example, Weng 2001a). Some previous researches calculate the CN based on land-cover types and vegetation condition, while soil types are neglected (Kumar et al. 1991). Hawkins (1998) and Canters et al. (2006) suggest that CN tables should only be used as a guideline, and actual CN and empirical relationships should be determined based on local and regional conditions. In this study, the composite CN values are determined by the percentage of vegetation, impervious surface, and soil types. This procedure is straightforward. The main difficulty in its application lies in estimating the percentage of impervious surface, soil, and vegetation via LSMA and determining the initial CN in reference to the look-up table of TR-55. While the urban landscape is examined under three conditions in TR-55, there are more than 10 classes under each condition. The selection of initial CN requires having a good knowledge of the study area. Different soil types possess different physical characteristics.

These characteristics, like grain size and physical components, will affect water infiltration capability. As a result, biophysical characteristics of soil are important parameters for assigning initial CN. Likewise, soil moisture may be a critical parameter for assigning initial CN. Since the soil moisture content varies over time, it is difficult to be incorporated into the model. In this study, we rely solely on soil types for assigning initial *CN* values. Impervious surface features include concrete, asphalt, metal, and other artificial features, and water can hardly infiltrate. Thus, it is appropriate to assign a value of 98 for all impervious surface areas. NDVI is a simple vegetation index that can be used to analyze vegetation vigor from RS images. The vegetation types may be classified according to the value of NDVI (Wu 2009). In this study, forest, grass, bush, farmland, and non-vegetation area was classified via NDVI. This classification corresponded well to the vegetation types in the CN look-up table of the TR-55. It should be noted that NDVI can vary with seasons. Although most of the vegetation in Guangzhou is evergreen, the seasonal impact on computing the NDVI and the relationship between NDVI and vegetation cover should not be ignored.

9.3 Surface Water Quality and Urban Land-Cover Changes

9.3.1 Introduction

Previous studies have demonstrated that certain types of human activities on land generate particular types of pollutants (Bolstad and Swank 1997; Carpenter et al. 1998; Gan et al. 2008). Other researchers have singled out one land use and assess its effect on surface water quality (Crabtree et al. 2008; Gan et al. 2008, Thompson et al. 1997). Highways have been found to produce more soluble pollutants than insoluble ones in the short-term and high amounts of insoluble pollutants in the long run (Crabtree et al. 2008). Other studies have explored a couple of nonpoint source pollutants and establish their relationship to land use and watershed characteristics (Coats et al. 2008). Moreover, some scholars have probed into the pattern of LULC change and its effects on water quality. He (2003) concluded that the extent of land-use change and its proximity to water bodies determine the magnitude of water-quality impairment that will occur. The more connected impervious surfaces are close to a water body, the greater will be the effect on water quality than impervious surfaces that are fragmented (Wang et al. 2001). These studies also concluded that the closer an urban development is to aquatic environment, the greater the extent of water-quality impairment that will be exerted on water bodies. Some researchers have revealed that runoff volume, peak rates, and sediment load in runoff increase as urban areas expand (Harbor 1994; Osborne and Wiley 1998).

Land-cover change has two broad dimensions, land-cover conversion and land-cover modification (Jenerette and Wu 2001; Lambin et al. 2001; Lambin and Geist 2006). Land-cover conversion is the replacement of one land-cover type with another, for example, the shift from agricultural to residential or a change from open space to transportation. Land-cover modification on the other hand reflects certain changes that affect the character of the land cover without a complete change in the land cover itself (Lambin and Geist 2006). When this occurs, land fragmentation develops which in turn changes the structural complexity of the landscape (Jenerette and Wu 2001; Lopez et al. 2001). The latter is mostly evident in urban landscapes where development of land parcel takes place more frequently compared to land cover outside cities. Land-cover modification which is the main type of land change process evident in urban areas is either ignored or proved difficult to incorporate into modeling efforts. Furthermore, out of the few studies that attempt incorporating the effects of land-cover modification on surface water quality, most are centered on Hortonian uniform overland flow paradigm which assumes that all areas within a watershed contribute toward runoff (Horton 1933). An area that contributes toward runoff in a watershed varies with time, the level of precipitation and other meteorological characteristics (Betson 1964; Van De Griend and Engman 1985). Few studies exploring runoff have integrated contributing and noncontributing areas in modeling surface water flow from land surface (Boughton 1990; Lyon et al. 2004; Steenhuis et al. 1995). This was actualized with painstaking fieldwork in demarcating active and non-active area's contributions to surface runoff. The use of geospatial technology can facilitate a more accurate delineation of contributing areas to surface runoff which in turn can improve model prediction of surface water quality.

Among studies that have predicted the types of nonpoint source pollutants generated by various land uses, they mostly utilize existing water quality simulation models which are often structured and only allow modest modification to the model. Moreover, land-use classes and nonpoint source pollutants predicted rarely exceed five or six. Furthermore, almost all of these studies examine a one-time step or a couple of years in predicting water quality resulting from LULC change. As a result, the impacts of LULC change over a more extensive time within a partial area hydrology framework and land modification scenario are partially understood. This study explores the ramifications of LULC change within a partial area hydrology and land-cover modification framework over a period of nine years. The main goal of this research is to unearth the extent of water quality effects that will result from LULC change taking into consideration areas that contribute toward runoff in an urban watershed, land-cover modification, and a relatively long time period.

9.3.2 Study Area and Data Sets

This study was conducted in the Lake Calumet watershed, Greater Chicago Area. The size of the study area is 54 518 acres (85.2 square miles). Lake Calumet is located 15 miles south of downtown Chicago (Figure 9.5). This lake

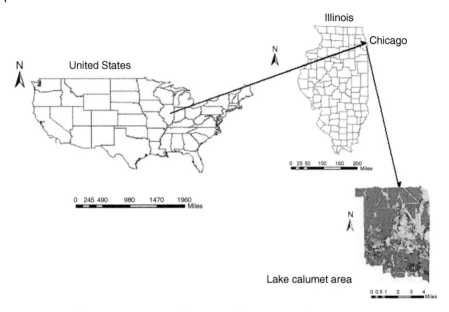

Figure 9.5 Lake Calumet Area, South Chicago, Illinois, USA (Wilson and Weng 2010). *Source:* US Census Bureau TIGER/Line Shapefiles and National Land Cover Database. Reproduced with permission of Springer Science.

was formed more than 13 500 years ago as a result of retreating glaciers (Ross et al. 1989). As early as 1869, the area was promoted as a suitable site for the establishment of manufacturing industries. A number of factors accounted for the area's suitability for industries; these include easy accessibility to Lake Michigan for shipping of raw materials and finished produce, water used by industries within Calumet for manufacturing, and Chicago's railroad hub that facilitates inland transportation of industrial materials and finished product. From this period onwards, a number of industries developed in Calumet reducing the lake's surface area to coastal docks and eventually landfill. Over this long period of heavy industrial activities around Lake Calumet, a number of industrial contaminants entered the lake. With the establishment of stringent environmental legislation over the past decade, there has been some improvement in the level of pollutants entering the lake from industrial activities. Notwithstanding, a number of anthropogenic metals and polynuclear aromatic hydrocarbons are higher in the waters of lake Calumet than water samples from other water bodies in the area (Illinois Department of Natural Resources 2002). Similarly, the lake has higher concentrations of metals when compared to other lakes in Illinois (Ross et al. 1989). Water pollution in Lake Calumet does not only stem from industrial land use but also from other urban land uses in the region. For instance, metal contamination of sediments has been found in association with municipal wastewater operations, coal-fired power plants, landfill leachate, urban runoff, highway runoff, mining and metal-working operations, airborne particulates, and industrial wastewaters (Ross et al. 1989).

Two National Land-Cover Data sets for the study area were used. These include 1992 and 2001 classified Landsat TM 30-m resolution images. Precipitation data was obtained from Illinois State Climatologist Office. Precipitation data originated from Chicago's Midway International Airport which lies approximately three miles west of the study area. Illinois State Climatologist Office converts precipitation data from hourly to daily rain and snowfall. Soil data was solicited from the United States Department of Agriculture Natural Resource Conservation Services (NRCS) in the form of Soil Survey Geographic (SSURGO) shapefile. Ten-meter Digital Elevation Model (DEM) was downloaded from the United States Geological Survey DEM data base, while population of study area was obtained from the United States Census Bureau. The United States Environmental Protection Agency's STORET (storage and retrieval) historical and modernized water quality data sets for 1992 and 2001 were used in model calibration and validation of simulated results.

9.3.3 Methods

National Land-Cover Data sets (NLCD) for the two dates of images were adjusted by recoding and combining some classes in each date of image to produce similar classes for both images. Recoding was also done to produce LULC classes that can be accommodated by the modified long-term hydrologic impact assessment and nonpoint source pollution (L-THIA-NPS) model. The various land classes were compared with Chicago Metropolitan Area Planning land-cover map for verification (Chicago Metropolitan Agency for Planning 2008). The study area was then subset by using United States Census Bureau's census tracts as guide. Figure 9.6 illustrate land uses around Lake Calumet in 1992 and 2001.

L-THIA-NPS model use the former SCSs CN approach in estimating runoff depth and volume (SCS 1972). Runoff depth is predicted by Eq. (9.1). Runoff volume is calculated by Eq. (9.3):

$$Q_v = Q * A \tag{9.3}$$

where Q_v is runoff volume, Q is runoff depth, and A is land area over which water flows. In modifying the L-THIA-NPS model, two additional land-cover classes (barren land and woody vegetation) were included because they are present in the study area and the default model framework does not account for them. Figure 9.7 demonstrate how the model estimates nonpoint source pollution concentration.

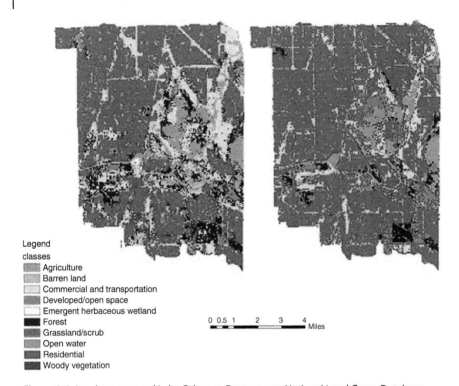

Figure 9.6 Land uses around Lake Calumet. Data source: National Land Cover Database. Wilson and Weng (2010). Reproduced with permission of Springer Science.

Precipitation data was converted from average daily to annual average rainfall. The snowfall component of the precipitation data was excluded as a result of the inability of the model to estimate the contribution of snowmelt to runoff. DEMs covering the study area were mosaicked and subset according to the study area's perimeter. ArcHydro GIS extension was used to fill sinks on the DEM, calculate flow direction of runoff, and storm water routing and delineate the contributing areas (active) of Lake Calumet watershed for 1992 and 2001, respectively. Soil data was processed in order to be usable in ArcGIS 9.3. Processing took the form of linking the tabular portion of the data to the spatial component in order to obtain the hydrologic soil groups and other important information contained in the SSURGO data.

The United States Geological Survey Estimated Mean Concentration of nonpoint source pollutants generated by specific land uses were used as a guide in arriving at pre-calibrated estimates for the 12 nonpoint source pollutants predicted in the study. The United States Environmental Protection Agency's STORET water quality data was used in model calibration and validation of simulated results.

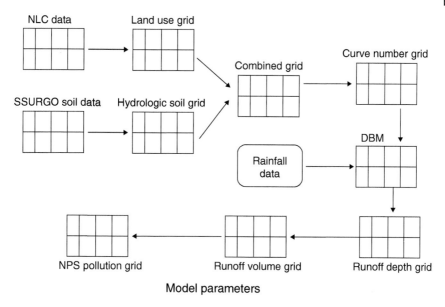

Figure 9.7 Modified L-THIA-NPS Model parameters. *Note:* NLC data, National Land-Cover Landsat image; SSURGO, Soil Survey Geographic shapefile of the US Department of Agriculture Natural Resource Conservation Services; L-THIA-NPS, Long-Term Hydrologic Impact Assessment and Nonpoint Source Pollution Model; DEM, Digital Elevation Model. *Source:* Wilson and Weng (2010). Reproduced with permission of Springer Science.

In order to enhance model calibration, the L-THIA-NPS model was spatially distributed into two parts within the study area. One part of the study area had the U.S. Environmental Protection Agency (EPA) STORET gauge monitoring site while the other part was ungauged. The distribution was created taking into cognizance elevation of landscape, flow direction, and flow accumulation of surface water. Calibrated model parameters obtained from the gauged section was then applied to the entire study area during the final model run.

Sensitivity analysis was performed on the model to ascertain the parameters that experience the largest change in the model with changes in model input. Four sets of rainfall data that varies from low, medium, moderately high, and high rainfall values were used to run the model in an uncalibrated framework while the sensitivity of the model parameters to changes in rainfall data was measured. L-THIA-NPS input parameters include rainfall, land use, and soil; process parameters encompass runoff depth, runoff volume, potential infiltration, CN, and estimated mean concentration coefficient. Output parameter includes nonpoint source pollutant concentration. Input and process parameters were included in sensitivity analysis. Changes in model output resulting from changes in model input and process parameters were assessed.

Model calibration was conducted manually by fitting measured data obtained from the United States Environmental Protection Agency STORET data set to the simulated output of L-THIA-NPS model. This was done using the split-sample technique (McCuen 2003). The second set of water quality data was used in validating the model.

L-THIA-NPS model was employed in the study as a result of its accessibility, simplicity of use, relative ease in customizing major parameters, and availability of model inputs and its easy applicability to urban areas compared to complex models. L-THIA-NPS model combined land use and soil data to assign CN to the output grid generated using the normal antecedent soil moisture condition 2. The CN grid was then added to the precipitation data producing runoff depth. Runoff volume was then estimated based on the runoff depth and spatial extent of each LULC. In the final stage of the model operation, nonpoint source pollution concentration loads for 12 pollutants was estimated. The model was run for 1992 and 2001, respectively.

9.3.4 Changes in Water Quantity

LULC change detection reveals that four land covers increased over the study period while six experience reduction in their spatial extent. Barren land increased exponentially between 1992 and 2001 (506.7%). The land-cover map for 2001 shows that most of this area lies within parts of the former steel mill site around Lake Calumet (Figure 9.6). Grassland/scrub increased by 57.1%, while residential land use increased by 37.3%. A closer look at residential land use in the original NLCD image illustrates that most of the increase was low density development. Developed/open space which encompasses parks and parking lots also increased over the study period (Table 9.4).

Land uses that reduced in spatial extent between 1992 and 2001 include agriculture (89.1%), woody vegetation (87.8%), forest (69.9%), open water (26.2%), and the combined land used for commercial, industrial, and transportation (38.3%). Using Chicago Metropolitan Agency for Planning land-use map as reference data to unravel the three grouped land covers in the National Land-Cover Data classification, most of this decline was observed within land used for industrial activities. Commercial land use demonstrated an increase over the study period while there was minimal changes to land used for transportation.

Differences in average annual rainfall between 1992 and 2001 resulted in changes in runoff depth over the study period. Runoff depth for each land use was observed to be higher in 2001 compared to 1992. The model predicted that emergent herbaceous wetlands had the highest runoff depth compared to all other land-use categories over the study period. Land used for commercial, industrial, and transportation activities generated the second highest runoff depth. Average runoff depth for the combined land use increased by 17% over the study period. Barren land recorded the third highest runoff depth between

Table 9.4 Land-use/land-cover change in Lake Calumet Area, South Chicago, Illinois, USA.

Land use	1992	2001	% Change
Open water	4182.57	3085.3	−26.2
Residential	27375.9	37592.5	37.31
Comm/Indst/Trans	11556.1	7133.5	−38.3
Barren land	2.67	16.2	506.7
Forest	5858.8	1762	−69.9
Grassland/scrub	449.15	705.5	57.1
Agriculture	888.7	96.5	−89.1
Developed/open space	3395.5	3977.3	17.1
Woody vegetation	778.6	94.7	−87.8
Emergent herbaceous wetland	429.7	213.1	−50.4

Source: Wilson and Weng (2010). Reproduced with permission of Springer Science.

1992 and 2001. These lands are mostly composed of deserted pavements that were initially used for other activities. The model predicted a 20% increase in runoff depth for this land use.

Residential land use generated runoff depth of less than 1 in. over the study period, when the two time periods are compared, an increase of 28% was observed. Runoff depth recorded within land used for agriculture, grassland/scrub, developed/open space, and woody vegetation is less than that of residential land use. Lowest runoff depth predicted over the entire period of study is within woody vegetation (0.08 and 0.1 in. in 1992 and 2001, respectively). Other land-use categories of comparatively low runoff depth include forest and grassland.

Runoff volume is a factor of size of the land use and its relationship to hydrologically active areas of a watershed. In 1992, the combined land used for commercial, industrial, and transportation activities produced the highest runoff volume (Figures 9.8 and 9.9). However, simulation results for 2001 demonstrated a reduction in the average runoff volume for the aforementioned combined land use (−21%). In 1992, residential land use produced the second highest runoff volume; while in 2001 it generated the highest runoff volume (Figure 9.9). This can be attributed to an increase in the spatial extent of residential land use over the study period and also an increase in the amount of residential land cover within areas that contributed toward runoff in 2001. The above scenarios resulted in an increase of 102% in runoff volume within residential land cover. Runoff volume recorded for emergent herbaceous wetlands, agriculture, developed/open space, and forest are relatively lower than residential and the combined transportation, industrial, and commercial land use over

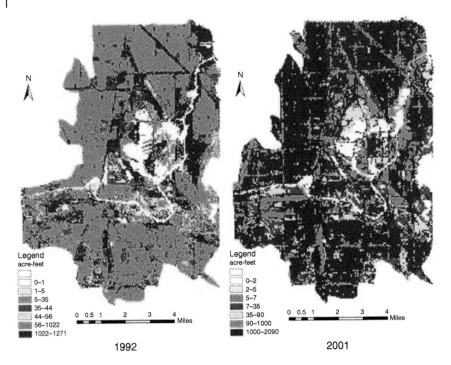

Figure 9.8 Runoff volume map calculated from runoff depth grid. *Source:* Wilson and Weng (2010). Reproduced with permission of Springer Science.

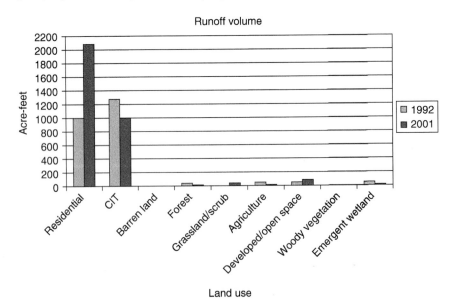

Figure 9.9 Runoff volume per land-use/land-cover type. *Source:* Wilson and Weng (2010). Reproduced with permission of Springer Science.

the study period. Woody vegetation, grassland/scrub, and barren land generated runoff volume of less than 5 acre-feet. Highest change in runoff volume between 1992 and 2001 was produced by grassland/scrub (over 10-fold), while the combined land used for commercial, transportation, and industrial activities produced the lowest (21%).

9.3.5 Changes in Water Quality

LULC produce some type and concentration of nonpoint source pollutants (Loehr 1974; Baird and Jennings 1996). The implication of these pollutants for surface water quality depends on the magnitude of the pollutants produced, spatial extent of the land use, runoff depth, runoff volume, and the topography of the landscape. The modified L-THIA-NPS simulation model was used to estimate 12 nonpoint source pollutant concentration loads from the various land uses (Table 9.5). In 1992, total nitrogen concentration was highest within agricultural land use (4.51 mg/l). Other land uses of relatively high total nitrogen concentrations include barren land, the combined commercial, industrial, and transportation land uses, developed/open space and woody vegetation. Total nitrogen concentration was insignificant within forest, grass/scrub, and emergent herbaceous wetlands (less than 1 mg/l). Total nitrogen concentration for 2001 mirrors that of 1992. In 2001, agriculture had the highest concentration (4.14 mg/l), while forest, grassland/scrub, and emergent herbaceous wetland produced lower concentrations (less than 1 mg/l).

Other nitrogen-related pollutants that were estimated include total kjeldahl nitrogen and nitrate nitrite. Highest total kjeldahl nitrogen concentration in 1992 was observed within residential land use (1.9 mg/l) while forest, grassland/scrub, and emergent herbaceous wetlands produced relatively small concentrations (0.2 mg/l and lower). The pattern of total kjeldahl nitrogen generated by the model vis-à-vis land uses in 2001 mirrors that of 1992 (Table 9.4). Agricultural land use produced the highest nitrate nitrite concentration for both study periods, while all other land uses had concentrations of less than 1 mg/l as N.

Total phosphorus concentration was highest within agricultural land use over the study period while emergent herbaceous wetlands, forest, grassland/scrub, and barren land recorded very low concentrations of less than 0.2 mg/l.

Dissolved phosphorus concentration for individual land uses was significantly low (<1 mg/l). Residential land use generated the highest dissolved phosphorus concentration in 1992 (0.48 mg/l) and 2001 (0.57 mg/l). Woody vegetation and barren land produced the lowest dissolved phosphorus concentration while some land use did not generate any of this pollutant (Table 9.5).

Suspended solids recorded the highest concentration compared to all other nonpoint source pollutants estimated in the study (Figure 9.10). In 1992, highest suspended solids concentration was predicted within agricultural land use (90 mg/l). In 2001, barren land generated the highest concentration of suspended solids (78 mg/l). Other land uses that produced high concentration of suspended

Table 9.5 Nonpoint source pollution estimation for 1992 (first row) and 2001 (second row).

Land use	TN	TKN	NN	TP	DP	SS	Pb	Cd	Ni	BOD	COD	O&G
Residential	1.82	1.90	0.23	0.72	0.48	41	4	0.75	0.65	18.5	24.6	1.7
	1.96	2.1	0.67	0.83	0.57	52	9	0.73	0.69	25.5	35.5	2.1
Commercial	1.49	1.25	0.37	0.51	0.14	63.17	9	1.25	8.03	11.4	67.5	4.13
	1.41	1.2	0.24	0.27	0.09	56.27	14.5	1.23	4.03	18.47	53.5	4.59
Barren land	1.50	0.96	0.54	0.12	0.03	70	1.52	0	0	0	31	0
	1.9	1.16	0.83	0.26	0.14	78	1.7	0	0	0	40	0
Forest	0.7	0.06	0.40	0.01	0	1	1.5	0.3	0	0.5	0	0
	0.5	0.4	0.32	0.01	0	0.8	2.2	0.18	0	0.46	0	0
Grassland/scrub	0.7	0.10	0.40	0.01	0	1	0.9	0.5	0	0.5	0	0
	0.9	0.2	0.8	0.11	0	1.4	5	0.9	0	0.53	0	0
Agriculture	4.51	1.65	1.60	1.69	0	90	1.1	1	0	4	0	0
	4.14	1.23	1.48	1.3	0	75	0.93	0.8	0	3.2	0	0
Developed/open space	1.23	0.90	0.34	0.35	0.23	41	1.6	1.05	3.8	6.4	59.3	1.2
	1.57	1.12	0.39	0.38	0.29	57.9	2.37	1.15	1.2	17.2	20.5	1.9
Woody vegetation	1.2	0.82	0.44	0.09	0.03	70	0.8	0.3	0	2	30	0
	0.8	0.35	0.29	0.03	0.01	58.2	0.62	0.21	0	1.42	40	0
Emergent wetland	0.7	0.2	0.40	0.01	0	1	0	0.8	0	0.5	0	0
	0.61	0.18	0.35	0.01	0	0.8	0	0.62	0	0.39	0	0

Source: Wilson and Weng (2010). Reproduced with permission of Springer Science.
Note: Pb, Cd, and Ni are in µg/l (microgram per liter) and all other pollutants are in mg/l (milligram per liter). TN, total nitrogen; TKN, total kjeldahl nitrogen; NN, nitrate nitrite; TP, total phosphorus; DP, dissolved phosphorus; SS, suspended solids; Pb, lead; Cd, cadmium; Ni, nickel; BOD, biochemical oxygen demand; COD, chemical oxygen demand; O&G, oil and grease.

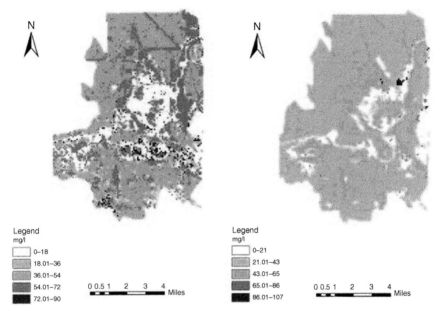

Figure 9.10 Nonpoint source pollution estimate for suspended solids. *Source:* Wilson and Weng (2010). Reproduced with permission of Springer Science.

solids over the nine-year period include the combined commercial, industrial, transportation, woody vegetation, and developed/open space land uses (Table 9.5).

Lead, cadmium, and nickel concentrations were predicted in trace quantities compared to other nonpoint source pollutants estimated in the study. Highest lead concentration was estimated within the combined commercial, industrial, and transportation land for both study periods. In 1992, lead concentration within this land was 9 μg/l and increased to 14.5 μg/l in 2001. Woody vegetation, forest, and grassland/scrub generated small quantities of lead (Table 9.5). Cadmium and nickel concentrations were highest within the combined land used for commercial, industrial, and transportation. In 2001, predicted values were lower for the two pollutants compared to 1992 though not proportionally. Over this period, nickel concentration declined by half while cadmium experienced an infinitesimal drop of 0.02 μg/l.

Biochemical oxygen demand (BOD) and chemical oxygen demand (COD) properties of water demonstrate different relationships to runoff depth over the study period. BOD concentration was highest on residential land use over the study period while COD concentration was greatest within the combined commercial, industrial, and transportation land use (Figure 9.11). Lowest BOD concentration was found on forest, grassland/scrub, and emergent herbaceous wetlands over the study period. COD concentration was higher than BOD over the nine-year period although COD was nonexistent within six land uses (Table 9.5).

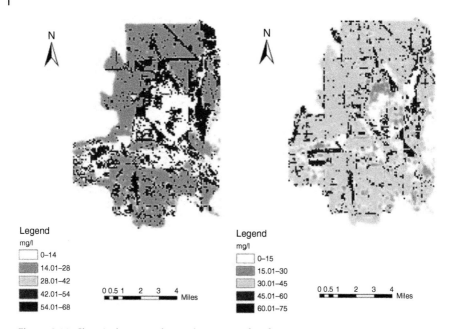

Figure 9.11 Chemical oxygen demand property of surface water. *Source:* Wilson and Weng (2010). Reproduced with permission of Springer Science.

A very important nonpoint source pollutant mainly emanating from roads and highways in urban areas is oil and grease. Simulation result suggested a slight increase in oil and grease concentration between 1992 and 2001 (Table 9.5). The combined land used for commercial, industrial, and transportation activities generated the highest concentration of this pollutant over the study period. Less than half of this quantity was found within residential and developed/open space (Figure 9.12). There was no trace of oil and grease found in the other land uses.

9.3.6 Discussion

Modeling the effects of urban LULC change on surface water quality within a partial area hydrology framework is pivotal in monitoring urban water resources. The study demonstrates that surface water quality depends on the extent of LULC change over time and also the spatial extent of hydrologically active areas within the watershed. The model reveals that an increase in runoff volume will contribute to differential increases in concentration among most pollutants. Conversely, BOD and COD properties of surface water demonstrated a contrary pattern to the aforementioned one.

This study has demonstrated the importance of urban watershed delineation in estimating the concentration of major nonpoint source pollutants evident in

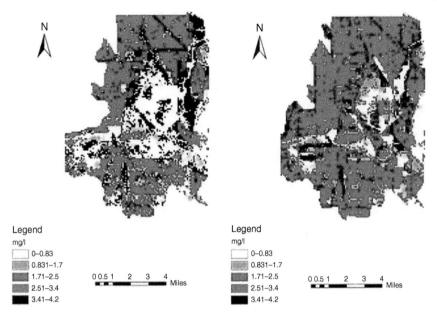

Figure 9.12 Oil and grease estimate. *Source:* Wilson and Weng (2010). Reproduced with permission of Springer Science.

urban environments. Impervious land-cover types like residential, commercial, industrial, transportation, and developed/open space that experience higher volume of human activities tend to have greater concentration of BOD, COD, and oil and grease compared to those with lesser human activities. The more land area that exists within active areas of a watershed over time, the greater the concentration levels of most pollutants, while that of biological oxygen demand property of water demonstrates an inverse relationship. The modeling framework employed in this study can be applied to other urban areas with moderate size watershed and a gauged water quality monitoring system. The latter can facilitate calibration of model parameters to fit in with observed data. The study has attempted an investigation into the dynamics of partial area hydrology and its effects on nonpoint source pollution concentration amid LULC change.

Runoff depth demonstrated a direct relationship to average annual rainfall over the study period, while runoff volume displays a nonlinear relationship to increase in rainfall over the same period. Runoff volume is mostly related to changes in the spatial extent of each land cover over the study period and more importantly the spatiotemporal variations in hydrologically active area for 1992 and 2001. Runoff volume within residential land use increased far above the increase in spatial extent of this land-use type and also the increase in rainfall over the study period. Runoff volume within residential land use increased by 108%, while the spatial extent only increased by 37.3%, whereas increase in rainfall was less than 30%. Differences between the increase in runoff volume and spatial extent of land can be attributed to the shifting nature of hydrologically active areas within Lake Calumet's watershed over the study period. In a similar vein, land-cover change detection demonstrated a 38.3% decline in size for the combined commercial, industrial, and transportation land use. Even though runoff depth for this land use increased over the study period, runoff volume declined over the same period (21%). This again suggests that comparatively smaller area of the combined land use falls within active areas of the watershed in 2001.

Nonpoint source pollutant concentration depends on a wide range of anthropogenic and natural factors. However, runoff volume, knowledge of the spatial extent of land uses, and a picture of hydrologically active areas within a watershed can shed insight into the level of concentration of these pollutants. Concentration for 11 of the 12 pollutants estimated within residential land use increased over the study period. Some pollutants increased more than others. While nitrate nitrite concentration within residential land use increased close to 200%, nickel concentration only increased by 6%. Cadmium concentration declined by 2% which might be attributed to lower production of this pollutant from the industrial site within Lake Calumet's vicinity. While the increase in concentration for 11 of the 12 nonpoint source pollutants within residential land use might be attributed to additional anthropogenic production of these pollutants, or an increase in atmospheric concentration of these pollutants, a causal relationship cannot be directly made without a longitudinal onsite monitoring of these pollutants within the study area. Notwithstanding, the study has demonstrated that the concentration of most nonpoint source pollutants within residential land uses increases with size of land use and runoff volume. Another factor that might be responsible for increase in concentration of nonpoint source pollutants within residential land cover besides the increase in the spatial extent is an increase in population. Census tract population figures covering residential land demonstrated an increase in population between 1992 and 2001. The generation of additional household waste and increased use of automobile might be responsible for the rise of 11 of the 12 nonpoint source pollutants simulated within residential land in the study.

Ten of the 12 nonpoint source pollutant concentration estimated within the combined commercial, industrial, and transportation land use reduced while two increased over the study period. The decline of these nonpoint source pollutants ranges between 1.6% for cadmium and 47% for total phosphorus, while concentration for lead and oil and grease increased by 61 and 11.1%, respectively. It is likely that the decline in concentration of some of the industrial pollutants arises from downsizing of the industrial plant within Lake Calumet's vicinity. Increase in oil and grease concentration may have partly arisen from increased vehicular traffic resulting from the growth of residential area and commercial establishments.

High concentration of suspended solids within the combined commercial, industrial, and transportation land use in 1992 may have resulted from comparatively large industrial activities. While the transition of part of industrial lands to barren land cover in 2001 partly explains the high suspended solid concentration within barren land. Concentration of BOD and COD properties of water demonstrates an inverse relationship over the study period. While BOD concentration reduced, COD increased even though runoff depth increased between 1992 and 2001. A number of factors including increase in the proportion of low-density residential vis-à-vis high density and increase in the size of barren lands which comprise a large proportion of brownfield as at 2001 might be responsible for the relationship displayed by BOD and COD.

References

Baird, C. and Jennings, M. (1996). Characterization of nonpoint sources and loadings to the Corpus Christi Bay National Estuary Program study area. Texas National Resource Conservation Commission.

Berry, J.K. and Sailor, J.K. (1987). Use of a Geographic Information System for storm runoff prediction from small urban watersheds. *Environmental Management* 11: 21–27.

Betson, R.P. (1964). What is watershed runoff? *Journal of Geophysical Research* 68: 1541–1552.

Bolstad, P.V. and Swank, W.T. (1997). Cumulative impacts of land use on water quality in a Southern Appalachian watershed. *Journal of the American Water Resources Association* 33 (3): 519–533.

Boughton, W. (1989). A review of the USDA SCS curve number method. *Soil Research* 27: 511–523.

Boughton, W.C. (1990). Systematic procedure for evaluating partial areas of watershed runoff. *Journal of Irrigation and Drainage Engineering* 116 (1): 83–98.

Canters, F., Chormanski, J., Van de Voorde, T., and Batelaan, O. (2006). Effects of different methods for estimating impervious surface cover on runoff estimation at catchment level. *Proceedings of the 7th International Symposium on Spatial Accuracy Assessment in Natural Resources and Environmental Sciences*, Lisbon, Portugal (5–7 July 2006), pp. 557–566.

Carpenter, S.R., Caraco, N.F., Correll, D.L. et al. (1998). Nonpoint pollution of surface water with phosphorus and nitrogen. *Ecological Applications* 8 (3): 559–568.

Chicago Metropolitan Agency for Planning (2008). Land use inventory shapefiles.

Coats, R., Larsen, M., Heyvaert, A. et al. (2008). Nutrient and sediment production, watershed characteristics, and land use in the Tahoe basin, California-Nevada. *Journal of the American Water Resources Association* 44 (3): 754–770.

Cronshey, R.G., Roberts, R.T., and Miller, N. (1985). Urban hydrology for small watersheds TR-55. *Proceedings of the ASCE Hydraulic Division Specialty Conference*, Orlando, FL.

Crabtree, B., Dempsey, P., Johnson, I., and Whitehead, M. (2008). The development of a risk-based approach to managing the ecological impact of pollutants in highway runoff. *Water Science and Technology* 57 (10): 1595–1600.

Drayton, R.S., Wilde, B.M., and Harris, J.H.K. (1992). Geographic information system approach to distributed modeling. *Hydrological Processes* 6: 361–368.

Dutta, S., Mishra, A., Kar, S., and Panigrahy, S. (2006). Estimating spatial curve number for hydrologic response analysis of a small watershed. *Journal of Spatial Hydrology* 6: 57–67.

Engman, E.T. and Gurney, R.J. (1991). *Remote Sensing in Hydrology*. London: Chapman & Hall.

Fan, F., Deng, Y., Hu, X., and Weng, Q. (2013). Estimating composite curve number using an improved SCS-CN method with remotely sensed variables in Guangzhou, China. *Remote Sensing* 5 (3): 1425–1438.

Gan, H., Zhuo, M., Li, D., and Zhou, Y. (2008). Quality characterization and impact assessment of highway runoff in urban and rural area of Guangzhou, China. *Environmental Monitoring and Assessment* 140 (1-3): 147–159.

Goudie, A. (1990). *The Human Impact on the Natural Environment*, 3e. Cambridge, MA: The MIT Press.

Harbor, J. (1994). A practical method for estimating the impact of land-use change on surface runoff, groundwater recharge and wetland hydrology. *Journal of the American Planning Association* 60: 95–108.

Hawkins, R.H. (1998). Local sources for runoff curve numbers. *Eleventh Annual Symposium of the Arizona Hydrological Society*, Tucson, Arizona.

He, C. (2003). Integration of geographic information systems and simulation model for watershed management. *Environmental Modeling and Software* 18: 809–813.

Hollis, G.E. (1975). The effects of urbanization on floods of different recurrence interval. *Water Resources Research* 11: 431–435.

Hong, Y. and Adler, R.F. (2008). Estimation of global SCS curve numbers using satellite remote sensing and geospatial data. *International Journal of Remote Sensing* 29: 471–477.

Horton, R.E. (1933). The role of infiltration in the hydrologic cycle. *EOS, American Geophysical Union Transactions* 14: 446–460.

Illinois Department of Natural Resources (2002). The Calumet Area: ecological management strategy, Executive summary: phase 1. City of Chicago Department of Environment.

Jenerette, G.D. and Wu, J. (2001). Analysis and simulation of land-use change in the central Arizona-Phoenix region, USA. *Landscape Ecology* 16: 611–626.

Kibler, D.F. (ed.) (1982). *Urban Stormwater Hydrology*. Washington, DC: American Geophysical Union.

Kousari, M.R., Malekinezhad, H., Ahani, H., and Zarch, M.A.A. (2010). Sensitivity analysis and impact quantification of the main factors affecting peak discharge in the SCS curve number method: an analysis of Iranian watersheds. *Quaternary International* 226: 66–74.

Kumar, P., TIWART, K., and Pal, D. (1991). Establishing SCS runoff curve number from IRS digital data base. *Journal of the Indian Society of Remote Sensing* 19: 245–252.

Lambin, E.F. and Geist, H.J. (eds.) (2006). *Land-Use and Land Cover Change: Local Processes and Global Impacts* (222 pp). New York: Springer.

Lambin, E.F., Turner, B.L., Geist, H.J. et al. (2001). The causes of land-use and land-cover change: moving beyond the myths. *Global Environmental Change* 11: 261–269.

Loehr, R.C. (1974). Characteristics and comparative magnitude of non-point sources. *Journal of Water Pollution Control Federation* 46 (8): 1849–1863.

Lopez, E., Bocco, G., Mendoza, M., and Duhau, E. (2001). Predicting land cover and land use change in the urban fringe: a case in Morelia city, Mexico. *Landscape and Urban Planning* 55 (4): 271–285.

Lu, D.S. and Weng, Q.H. (2004). Spectral mixture analysis of the urban landscape in Indianapolis with landsat ETM plus imagery. *Photogrammetric Engineering & Remote Sensing* 70: 1053–1062.

Ludlow, C. (2009). Flood modeling in a data-poor region: a satellite data-supported flood model for Accra, Ghana. *Annual Meeting of the Association for American Geographers*, Las Vegas, Nevada (22–27 March 2009).

Lyon, S.W., Walter, M.T., Gerard-Marchant, P., and Steenhuis, T.S. (2004). Using a topographic index to distribute variable source area runoff predicted with the SCS curve number equation. *Hydrological Processes* 18: 2757–2771.

Mather, P.M. (1999). Land cover classification revisited. In: *Advances in Remote Sensing and GIS* (eds. P.M. Atkinson and N.J. Tate), 7–16. New York: Wiley.

Mattikalli, N.M., Devereux, B.J., and Richards, K.S. (1996). Prediction of river discharge and surface water quality using an integrated geographic information system approach. *International Journal of Remote Sensing* 17 (4): 683–701.

Mattikalli, N.M. and Richards, K.S. (1996). Estimation of surface water quality changes in response to land use change: application of the export coefficient model using remote sensing and geographic information system. *Journal of Environmental Management* 48: 263–282.

McCuen, R.H. (2003). *Modeling Hydrologic Change: Statistical Methods*, vol. 433. Boca Raton, FL: Lewis Publishers.

Mintzer, O. and Askari, F. (1980). *A Remote Sensing Technique for Estimating Watershed Runoff* (ed. US Department of Commerce). Washington, DC.

Osborne, L. and Wiley, M. (1998). Empirical relationship between landuse/cover and stream water quality in an agricultural watershed. *Journal of Environmental Management* 26 (1): 9–27.

Ragan, R.M. and Jackson, T.J. (1980). Runoff synthesis using Landsat and SCS model. *Journal of Hydraulic Division of the American Society of Civil Engineers* 106: 667–678.

Rango, A., Feldman, A., George, T.S., and Ragan, R.M. (1983). Effective use of Landsat data in hydrological models. *Water Resources Bulletin* 19: 165–174.

Reistetter, J.A. and Russell, M. (2011). High-resolution land cover datasets, composite curve numbers, and storm water retention in the Tampa Bay, FL region. *Applied Geography* 31: 740–747.

Ridd, M.K. (1995). Exploring a V-I-S (vegetation-impervious surface-soil) model for urban ecosystem analysis through remote sensing: comparative anatomy for cities. *International Journal of Remote Sensing* 16 (12): 2165–185.

Rogers, P. (1994). Hydrology and water quality. In: *Changes in Land Use and Land Cover: A Global Perspective* (eds. W.B. Meyer and B.L. Turner II), 231–258. Cambridge: Cambridge University Press.

Ross, P.E., Burnett, L.C. and Henebry, M.S. (1989). Chemical and toxicological analysis of Lake Calumet (Cook County Illinois) sediments. Illinois State Natural History Survey.

Silveira, L., Charbonnier, F., and Genta, J.L. (2000). The antecedent soil moisture condition of the curve number procedure. *Hydrological Sciences Journal* 45: 3–12.

Slack, R.B. and Welch, R. (1980). Soil Conservation Service runoff curve number estimates from LANDSAT data. *Water Resources Bulletin* 16 (5): 887–893.

Soil Conservation Services (1972). Estimation of direct runoff from storm rainfall, chapter 10. In: *National Engineering Handbook, Soil Conservation Survey*. Washington, DC: U.S. Government Printing Office.

Steenhuis, T.S., Winchell, M., Rossing, J. et al. (1995). SCS runoff equation revisited for variable-source runoff areas. *Journal of Irrigation and Drainage Engineering* 121 (3): 234–238.

Thompson, N.R., McBeanl, E.A., Snodgrass, W., and Monstrenko, I.B. (1997). Highway storm water runoff quality: development of surrogate parameter relationship. *Water, Air and Soil Pollution* 94: 307–347.

USDA, Soil Conservation Services (1972). *National Engineering Handbook*, Section 4, Hydrology. Washington, DC: U.S. Government Printing Office.

Van De Griend, A.A. and Engman, E.T. (1985). Partial area hydrology and remote sensing. *Journal of Hydrology* 81: 211–251.

Wang, F. (1990). Fuzzy supervised classification of remote sensing images. *IEEE Transactions on Geoscience and Remote Sensing* 28: 194–201.

Wang, L., Lyons, J., and Kanehl, P. (2001). Impacts of urbanization on stream habitat and fish across multiple spatial scales. *Environmental Management* 28 (2): 255–266.

Weng, Q. (2001a). Modeling urban growth effects on surface runoff with the integration of remote sensing and GIS. *Environmental Management* 28 (6): 737–748.

Weng, Q. (2001b). A remote sensing-GIS evaluation of urban expansion and its impact on surface temperature in the Zhujiang Delta, China. *International Journal of Remote Sensing* 22 (10): 1999–2014.

Weng, Q. and Lu, D. (2009). Landscape as a continuum: an examination of the urban landscape structures and dynamics of Indianapolis city, 1991–2000. *International Journal of Remote Sensing* 30 (10): 2547–2577.

Wilson, C.O. and Weng, Q. (2010). Modeling surface water quality and its relationship with urban land cover changes in the Lake Calumet area, Greater Chicago. *Environmental Management* 45 (5): 1096–1111.

Wu, C.S. (2009). Quantifying high-resolution impervious surfaces using spectral mixture analysis. *International Journal of Remote Sensing* 30: 2915–2932.

Wu, S., Hu, Y., Dai, J., and Jiang, C. (2001). A study on creation of database for a Guangdong Province soil resource information system. *Journal of South China Agricultural University* 4: 22–25.

10

Urban Ecology of West Nile Virus

10.1 Introduction

West Nile virus (WNV) infection has attracted the attention of public health departments and related organizations in America since the WNV first appeared in New York in 1999. It has been rapidly disseminated to the west, south, and northern parts of the country in the following three to four years (Centers for Disease Control and Prevention 2002). The cases of WNV disease in the whole nation increased and peaked in 2003 (MMWR 2005). Mosquitoes, birds, blood transfusions, organ transplants, and breast milk are believed to contribute to the spread of WNV (Komar 2003; Hayes and O'Leary 2004). The spread of WNV shows unique geographic patterns in various study areas (Komar 2003; Hodge and O'Connell 2005; Pierson 2005; Rainham 2005; Ruiz et al. 2007). Environmental factors such as the presence of animal habitats, temperature, and climate play important roles in WNV dissemination in North America (Reisen et al. 2006a).

Mosquitoes from the genus Culex, in particular the species *Culex pipiens*, are the vector of WNV from birds to birds and birds to humans. Culex species appear to prefer some land-use and land-cover (LULC) types (e.g. wetlands and specific grasslands) than some others (e.g. exposed dry soils). Mosquitoes in the canopy site were believed to possess more infections than those in the subterranean and ground sites (Anderson et al. 2006). Wetlands and stormwater ponds, especially those under heavy shade, provide an ideal environment for mosquito development. Ponds with plenty of sunshine and a shortage of vegetation are believed to be a poor environment for mosquitoes growing (Gingrich et al. 2006).

Land surface temperatures (LSTs) reflect surface–atmosphere interactions and energy fluxes between the ground and the atmosphere (Wan and Dozier 1996). Solar radiation and LST are important parameters for analysis of urban thermal behavior (Aguiar et al. 2002; Liu et al. 2006). LST has been used as an important indicator to examine the urban atmosphere and to model urban climate

Techniques and Methods in Urban Remote Sensing, First Edition. Qihao Weng.
© 2020 by The Institute of Electrical and Electronics Engineers, Inc.
Published 2020 by John Wiley & Sons, Inc.

(Voogt and Oke 1997; Jacob et al. 2002; Voogt and Oke 2003). Characteristic of natural and human-involved patches have ecological implications at various spatial levels and influence the distribution of habitats and material flows (Peterjohn and Correll 1984; Turner 1990). LULC patterns are regarded as an important determinant of ecosystem function and can be considered as the representative of in situ landscape patterns (Bain and Brush 2004). LULC categories are linked to distinct behaviors of the urban thermal environment (Liu and Weng 2008; Voogt and Oke 1997; Weng et al. 2004).

To understand patterns and processes in a heterogeneous landscape, one must be able to accurately quantify the spatial pattern and temporal change (Wu et al. 2000). In recent years, a series of landscape metrics have been developed to characterize the spatial patterns and to compare ecological quality across the landscapes (O'Neill et al. 1988; McGarigal and Marks 1995; Riitters et al. 1995; Gustafson 1998). Software has been developed to calculate the metrics, such as FRAGSTATS (McGarigal et al. 2002). Remote sensing (RS) and geographic information system (GIS) technologies have been broadly applied in public health study and related issues like urban environmental analysis (Sannier et al. 1998; Hay and Lennon 1999; Wang et al. 1999; Weng et al. 2004; Herbreteau et al. 2007). These researches include diverse epidemiological issues, such as parasitic diseases and schistosomiasis, by using RS and GIS as an exclusive source of information for studying epidemics. The potential of thermal infrared (TIR) RS data has been shown to provide valuable temperature information (Luvall and Holbo 1991; Owen et al. 1998; Quattrochi and Ridd 1998; Hirano et al. 2004; Weng et al. 2004). A series of satellite sensors have been developed to collect TIR data from the earth surface, such as Heat Capacity Mapping Mission (HCMM), LandSat Thematic Mapper/Enhanced Thematic Mapper plus (TM/ETM+), Advanced Very High Resolution Radiometer (AVHRR), and Thermal Infrared Multispectral Scanner (TIMS). These TIR sensors may also be utilized to obtain LST and emissivity data of different surfaces with varied spatial resolutions and accuracy. These LST and emissivity data have been used in the analysis of the temperature–vegetation abundance relationship, drought evaluation, modeling of urban surface temperatures with surface structural information, and forest regeneration detection (Boyd et al. 1996; Voogt and Oke 1997; McVicar and Jupp 1998; Wan et al. 2004; Weng et al. 2004).

This chapter illustrates the applications of RS and GIS techniques for examining and modeling the spread of WNV in urban contexts. The first study, by a case study of the City of Chicago, aimed to improve the understanding of how landscape, LSTs, and socioeconomic conditions interacted to influence the spread of WNV in the urban context. This case study is a short version of the article by Liu and Weng (2009). The second study investigated the WNV spread in the epidemiological weeks of 18–26, 27–35, and 36–44 in the years of 2007–2009 in the Southern California and mapped the risk areas. The inclusion here shows part of a detailed study by Liu and Weng (2012) in the same study area.

10.2 Research Background

Research finds that natural environmental constraints like climatic parameter have significant effects on the transmission of WNV (Dohm and Turell 2001; Dohm et al. 2002). Higher temperatures were believed to contribute to the distribution of WNV throughout North America, and WNV dissemination was significantly related to average summer temperatures from 2002 to 2004 in the USA (Reisen et al. 2006a). WNV was not evident in most mosquitoes in cool temperatures (Dohm and Turell 2001). The infection rates of WNV were significantly related to the incubation temperatures chosen (Dohm et al. 2002). Temperature affects replication of mosquitoes during the incubation period (Whitman 1937). The vector capability of specific mosquito species increases with the increase of temperature. In the fall, some mosquito species tend to hibernate and are in developmental diapauses. They prefer plant feed instead of blood feed (Marfin et al. 2000; Komar 2003). In general, when temperature cools down in fall, some mosquito species and many other animals will not start to hibernate by this time, and birds will intend to migrate later (Marfin et al. 2000; Komar 2003). This time table is not optimal, but there is sufficient time to explore the spread of WNV caused by mosquitoes and animal hosts in urban areas. The measurements of LSTs during the fall season would provide clues about how the local temperatures influence the virus spread (Komar 2003).

Human-related activities, such as traveling, have significant effects on WNV dissemination. The oviposition patterns of mosquito species were significantly different at urban and rural sites since oviposition activity of mosquitoes reached a peak in the evening and morning in urban areas but it did not have an obvious morning peak in rural areas (Savage et al. 2006). Scientists believed that flooding created by human activities such as logging rather than rainfall caused the increasing abundance of mosquitoes in many wet locations (Balenghien et al. 2006). Inner suburbs were found to have the highest incidence of WNV illness in Chicago and the characteristics of neighborhoods were believed to be more important than the geographic locations where the illnesses were found in the same study area (Ruiz et al. 2007).

The integration of RS and GIS contributes to the study of physical and environmental factors influencing public health, such as the factors related to the transmission of the virus: mosquito habitat, energy exchange, and LST (Liang et al. 2002; Rogers et al. 2002; Weng et al. 2004). Satellite imagery has been used to monitor potential disease risks all over the world (Anyamba et al. 2006; Zou et al. 2006). The researchers (Anyamba et al. 2006) analyzed the possible effects of El Niño-/Southern Oscillation-related climate anomalies on the dissemination of epidemics all over the world, including WNV based on the study of remotely sensed data. The extreme climate conditions were believed to affect mosquito abundance and elevate the risk of WNV in that area in the near future (Anyamba et al. 2006). Zou et al. (2006) mapped potential habitats of mosquito larval, a main vector in the transmission of WNV, in the Powder River Basin of north central Wyoming by the analysis of Landsat TM and ETM+ imagery and GIS data. The results indicated that coalbed methane water caused a significant increase of mosquito larval habitats from 1999 to 2004 in the study area.

10.3 Effect of Landscape and Socioeconomic Conditions on WNV Dissemination in Chicago

Previous researchers have examined and documented the factors associated with WNV transmission all over the world (Anderson et al. 2006; Balenghien et al. 2006; Gingrich et al. 2006; Ruiz et al. 2007), but the study of the influences of landscape patterns, LSTs, and socioeconomic conditions on WNV dissemination is still undergoing. This study, by a case study of the City of Chicago, aimed to improve the understanding of how landscape, LSTs, and socioeconomic conditions together influence WNV dissemination in the urban context and to assess the importance of environmental factors in the spread of WNV. The study also determined whether LULC types, LSTs, and socioeconomic conditions have significant effects on the spread of WNV caused by mosquitoes and animal hosts in the study area, through the interpretation and analysis of landscape metrics, correlation, factor, and regression. RS and GIS techniques were used to derive information of landscape pattern and LSTs in the study area. WNV dissemination is a persistent public health concern and the results of this study can contribute to urban planning and public health management and protection in the study area and other areas with similar characters.

10.3.1 Study Area

Cook County, Illinois, has been chosen to implement the study of the effects of landscape pattern, LST, and socioeconomic conditions on WNV transmission. The city of Chicago (41°53′N, 87°38′W), the seat of Cook County, Illinois, is the nation's third largest city and the largest city in Illinois in population. It is located at the southwestern tip of Lake Michigan with the elevation of 182 m. The city proper extends over 588 km^2. It has a population of about 2.9 million according to the US Census 2000. The climate in Chicago is continental with frequently changing temperatures ranging from relatively warm in summertime to relatively cold in wintertime. Its annual average temperature is 49.8 °F, and the average temperature in January is only 23.0 and 74.7 °F in July. The average annual precipitation is 35.8 in., and about 1.9 in. in January and 3.7 in. in July. The lake area experiences heavy snowfall in the winter due to moisture from Lake Michigan.

Serving as the primary connection between the east coast cities and western part of the nation, the city is the financial and cultural center in the central east of the nation. It has advanced industry, transportation, and infrastructure diversity and is famous for its agricultural commodities. The population has kept increasing in recent decades. Urbanization has affected the habitats and thermal environment in the Greater Chicago area. The region also contains a variety of types of natural communities that include some of the best remaining remnants of Midwest wilderness, such as open oak woodland.

This study focused on the total cumulative numbers of WNV infection in mosquitoes, birds, horses, and other mosquitoes/animal hosts of WNV in the study area. The Cook County is the second largest county of the country by population with more than five million in 2004. The county has 138 municipalities (cities, villages, and incorporated towns) with an area of 1635 square miles, 42.16% of it is water (most of it in Lake Michigan), according to the statistic results of US Census Bureau. The City of Chicago is the largest municipality in the County with 24% area percentage. Any other municipalities are around the City of Chicago and spread to the north, west, and south of the county. It has the third-largest public health system in the country.

The Illinois Department of Public Health maintains a disease surveillance system to monitor insects and animals that can potentially carry WNV: mosquitoes, dead crows, robins, blue jays, and horses. Mosquitoes can either carry the virus or get it by feeding on infected birds. While many mosquito abatement districts and other agencies collect and test mosquitoes, the department asks the public for help with the collection of dead birds in their backyards. The surveillance system then requests experts in different fields, such as infectious disease physicians and infection control practitioners, to test for and report suspect or confirmed cases of various diseases that can be caused by WNV. The Cook County Public Health Department has published WNV surveillance data based on county level for years 2001–2007 (http://www.idph.state.il.us/envhealth/wnv.htm). The natural features of the county provide good environmental conditions for mosquitoes, the main vector of WNV. These features include wetlands, forest, prairies, and aquatic systems throughout the county.

10.3.2 Remote Sensing Image Processing

This study acquired information of landscape patterns and LSTs by processing four Advanced Spaceborne Thermal Emission and Reflection Radiometer (ASTER) images purchased from NASA. ASTER imagery carries 15-m resolution in visible bands and 90-m resolution in thermal bands, which provide great opportunity to identify landscape patterns and retrieve LSTs. All the images were acquired in the same day (10 September 2004) except the one covering the southwestern corner of the county (acquired on 12 October 2003) because no image could be found to cover the corner with the same acquisition date. The WNV study was based on the data collection in three months (August–October in 2004). The landscape patterns are believed to not experience significant changes during the leave on season (Liu and Weng 2008). So it is still significant to use image from 10 September 2004 to indicate the environmental factor for the whole season. By using ERDAS Imagine software, these images were geocorrected and given the projection to NAD27 north Zone 16, Clarke 1866, with the reference of Illinois Digital Raster Graphics (DRG) data.

Urban areas usually include a large population and huge consumptions of resource, which provides various channels of WNV infection, like old tires, tin cans, buckets, drums, bottles or any water-holding containers. Forests and grassy areas appear to provide relatively ideal environments for birds to live (Komar 2003; Rainham 2005). Areas around water have relatively higher moisture and contribute to the gathering of some mosquitoes (Komar 2003). In this study, six LULC types, urban, forest, grass, agriculture, water, and barren land were separated from RS imagery. Wetlands were classified into water since no detailed references were collected as reference in the image classification. One hundred and twenty clusters were produced during the unsupervised classification process and labeled to specific LULC types with the reference of 2005 Chicago orthography data. The overall accuracy of classification reached 85.67%. Figure 10.1a shows the LULC map for the study area. Known from the figure, urban and grass were the dominant LULC types in the county.

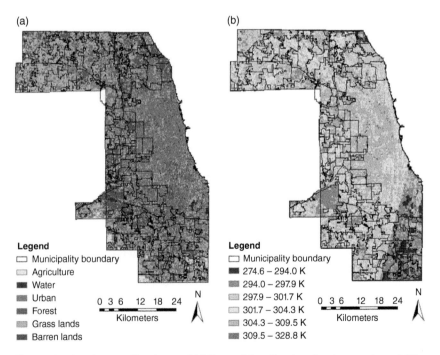

Figure 10.1 Land-use and land-cover (LULC) map (a) and land surface temperatures (LST) map (b) for Cook County in fall, 2004. A distinct different color composition was observed in the Southeast corner of the county because the Advanced Spaceborne Thermal Emission and Reflection Radiometer (ASTER) data covering that portion of the county was acquired in the different day and year from the rest images. *Source:* Liu and Weng (2009). Reproduced with permission of Springer Science.

Land surface kinetic temperature data were purchased from NASA with 90-m resolutions through a free entry. The temperature-emissivity separation algorithm was applied to compute land surface kinetic temperatures (ASTER Online Product Description 2005). According to the product description, the absolute accuracy of the kinetic temperature data is accurate within 1.5 K and relative accuracy 0.3 K. In Figure 10.1b, the whole study area was classified into six temperature zones by the use of natural break method (Smith 1986). The red colors in central and north part of the county indicate that those areas possessed relatively high temperature than their surrounding areas. Areas in blue color were the places with lowest temperatures.

10.3.3 Landscape Metrics Computation

Five landscape-level landscape metrics, Landscape Shape Index (LSI), Perimeter-area Fractal Dimension, Mean Perimeter-area Ratio, Contagion Index, and Shannon's Diversity Index (SDI) were derived from LULC map for municipalities having mosquito/animal host positive reports, with the use of ArcGIS software and FRAGSTATS, a popular software program for computing a wide variety of landscape metrics for categorical map patterns (McGarigal and Marks 1995). Since urban, forested, grassy, and water areas were the main components in the study area and are believed to show significant effects to the spread of WNV (Komar 2003; Rainham 2005), their landscape patterns in individual municipalities were examined by the derivation of five class-level landscape metrics, Percentage of Landscape, Area of Landscape, Patch Density, Perimeter-area Fractal Dimension, and Patch Cohesion Index. A total of 25 variables related to landscape pattern of the study area were produced and would be used as inputs in further statistical analysis.

10.3.4 LST and Socioeconomic Conditions

Absolute numbers of positive cases of mosquitoes and animal hosts include the cumulative totals of WNV infection in mosquitoes, birds, and horses. The absolute numbers in fall (August–October) from year 2000 to 2006 were collected from the Cook County Public Health Department as a unit of municipality. The municipalities without positive records of WNV were excluded from the study. No human case was included in the study limited by the availability of WNV data. Only the WNV data for year 2004 was used in the statistical analysis. The rest data were used to show the changes of spatial distributions of WNV in seven years (years 2000–2006). RS data, municipal boundaries, and socioeconomic data needed to be integrated since they had different formats and spatial resolutions. Six socioeconomic conditions for municipalities having mosquito/animal host positive reports were obtained from US Census 2000: total human population, area, perimeter, population above 65 years old, total house income, and income below poverty line. The mean temperatures of municipalities having mosquito/animal host (mosquitoes, birds, horses, and other mosquitoes/animal hosts) positive reports of WNV were calculated by the integration of RS data and a municipal boundary shape file with the help of ArcGIS software. The statistic features of temperature, maximum, minimum, standard deviation, and range for urban, forest, grass, and water were also computed for those municipalities. A total of 17 temperature-related variables were prepared for further statistical analysis of the spread of WNV. As a result, 48 variables were created for the study. Table 10.1 lists all 48 variables and their abbreviations.

10.3.5 Model Development

To identify the effects of landscape pattern, LST, and socioeconomic conditions on the spread of WNV, WNV record in fall (August–October), 2004, was then selected for statistical analysis since this season and year matched the acquisition date of RS data available in this study. According to the reports of the Cook County Public Health Department, 49 municipalities were reported to have mosquitoes/animal hosts positive results of WNV in fall, 2004. To determine which environmental and socioeconomic variables possessed significant relationships with the spread of WNV, these 49 municipalities were input as cases in the correlation and regression analysis based on 48 variables listed in Table 10.1. S-Plus software was used in the statistical analysis. Only variables that showed significant correlation to the absolute number of mosquitoes/animal hosts of WNV cases were used for further statistical analysis. To develop significant regression models for the study, all potential predictors selected after correlation analysis were transformed into natural logarithm and square root. Variables outside of three standard deviations were removed as outliers. The *t*-statistic was applied to identify the significance of a potential predictor.

Although significant results would be produced based on the variables selected from correlation analysis, the question, how landscape pattern, LST, and socioeconomic conditions affect WNV dissemination still needs to be answered. A stepwise regression analysis was applied to examine the relationships between the spread of WNV and landscape pattern, LST, and socioeconomic conditions. A series of regression equations were developed to model the effects of landscape pattern, LST, and socioeconomic conditions on WNV dissemination separately. The dependent variable was the absolute number of mosquito/animal host positive reports of WNV in municipalities. Independent variables were chosen which were highly correlated with mosquito/animal host positive cases. The natural logarithm and square-root forms of these variables were also used as inputs in the regression analysis for possible better result. The *t*-statistic was used to identify the relative importance of potential predictors in the regression analysis.

Table 10.1 Variables related to landscape pattern, land surface temperatures (LST), and socioeconomic aspects.

Variables	Abbreviations	Variables	Abbreviations
Landscape shape index*	LSI	Water cohesion	Wat_Coh
Mean perimeter-area ratio	Par_MN	Mean temperature	Tem_MN
Fractal dimension	FRD	Maximum urban temperature*	UT_Max
Contagion index	COI	Minimum urban temperature*	UT_Min
Shannon's diversity index	SDI	Standard deviation of urban temperature*	UT_SD
Urban area*	Urb_Are	Range of urban temperature*	UT_RG
Urban percentage	Urb_Per	Maximum forest temperature	FT_Max
Urban patch density	Urb_Den	Minimum forest temperature	FT_Min
Urban fractal dimension	Urb_Fra	Standard deviation of forest temperature	FT_SD
Urban cohesion	Urb_Coh	Range of forest temperature	FT_RG
Forest area	For_Are	Maximum grass temperature*	GT_Max
Forest percentage	For_Per	Minimum grass temperature	GT_Min
Forest patch density	For_Den	Standard grass of urban temperature*	GT_SD
Forest fractal dimension	For_Fra	Range of grass temperature*	GT_RG
Forest cohesion	For_Coh	Maximum water temperature	WT_Max
Grass area*	Gra_Are	Minimum water temperature	WT_Min
Grass percentage	Gra_Per	Standard deviation of water temperature*	WT_SD
Grass patch density	Gra_Den	Range of water temperature*	WT_RG
Grass fractal dimension	Gra_Fra	Total population*	Pop
Grass cohesion	Gra_Coh	Total area*	Are
Water area*	Wat_Are	Population above 65 years old*	Abo_65
Water percentage	Wat_Per	Perimeter	Per
Water patch density	Wat_Den	Total income*	Inc
Water fractal dimension	Wat_Fra	Family below poverty line*	Bel_Pov

Source: reproduced with permission of Elsevier.
Note: variables with * were the variables selected as potential predictors in the factor analysis and regression analysis.

10.3.6 Spatial Pattern of WNV Infections

The total insect/animal (mosquitoes, birds, horses, and other hosts) positive reports of WNV slightly decreased from 2002 to 2003 and heavily increased after 2003 for the whole county. For the City of Chicago, the number of the mosquito/animal host positive reports reached a maximum in 2002, and even the mosquitoe/animal host positive cases for the whole county kept increasing after year 2003. More mosquito/animal host positive cases in fall, 2001, were reported in north and south part of the county and the City of Chicago compared to those in year 2002. The WNV cases were spread to the central west of the county since year 2002. More cases were found in the northern part of the county in falls, years 2004 and 2005. In fall, 2006, the cases decreased in the northern part of the county, but an obviously increase of WNV infections in the southern part of the county in fall, 2006, indicates that the county apparently experienced much serious situation over that area in 2006. Since no references show how comprehensively and correctly the mosquito/animal host positive cases were reported in each year, no direct conclusion could be reached only based on these records. But the WNV maps show significant changes from year 2001 to 2006. The WNV dissemination map for fall, 2004, was recorded in the Figure 10.2 as an example.

10.3.7 Results of Statistical Analysis

All 48 variables in Table 10.1 were used to examine their relationships with the absolute number of mosquito/animal host positive reports of WNV in the study area. According to the Pearson correlation coefficients, some variables showed a stronger relationship with the absolute WNV number than some other variables, like human population, area, total house income, and areas of urban, water, and grass in municipalities. Some variables showed relatively lower correlation with the WNV number, like standard deviation and range of water temperature in municipalities (WT_SD, WT_RG). Population (Pop) showed very strong relationship with landscape aggregation (LSI) and area-related variables, such as areas of urban, grass, and water. Variables that showed poor Pearson correlation coefficients were dropped from the statistical analysis. Eighteen variables were finally selected as potential predictors in the regression modeling. They were labeled with asterisks in Table 10.1.

Stepwise regression analysis was produced to identify relationships between dependent variable (the absolute number of reported WNV cases) and 18 independent variables and their transformed forms. The results were summarized in Table 10.2. Model 1 had all 18 variables as model inputs and 4 out of 18 were remained after regression analysis (population above 65 years old, urban area, the maximum grass temperature, and standard deviation of urban surface temperature). Model 2 used natural logarithm of all the inputs in the Model 1 as potential predictors and two variables were remained after analysis (natural

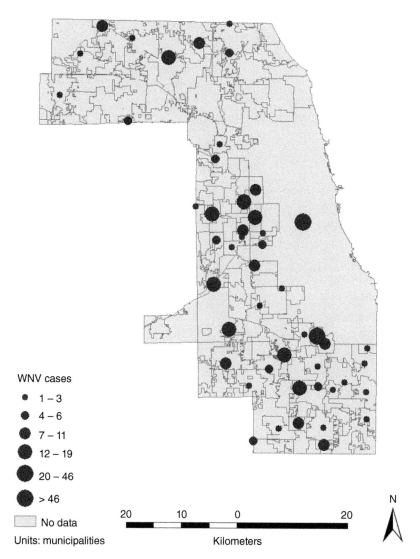

Figure 10.2 Absolute number of mosquito/animal host positive reports of West Nile Virus (WNV) in Cook County in fall, 2004. *Source:* Liu and Weng (2009). Reproduced with permission of Springer Science.

logarithms of the minimum urban surface temperature and the maximum grass temperature). Model 3 had square roots of all the inputs in the Model 1 and four predictors remained after regression analysis (square roots of population above 65 years old, urban area, urban surface temperature, and the maximum grass temperature). Model 4 used all 18 initial variables and their transformation forms, natural logarithm, and square root as inputs. There were three

Table 10.2 Potential predictors, *t*-statistics, and *R* square (R^2) values for regression models (independent variable: absolute WNV number).

| Model | Potential predictor | Unstandardized coefficient | | *t*-Statistic | R^2 |
		B	Std. error		
1	(Constant)	−25.030	7.007	−3.572	0.75
	Abo_65	0.002	0.000	6.356	
	Urb_Are	−0.012	0.002	−5.972	
	GT_Max	0.480	0.132	3.635	
	UT_SD	5.077	2.158	2.353	
2	(Constant)	−19.651	31.501	−0.624	0.50
	UTI_Log	−52.690	9.118	−5.779	
	GTM_Log	59.350	17.795	3.335	
3	(Constant)	−56.178	15.624	−3.596	0.69
	Abo65_Sqr	0.227	0.040	5.693	
	UA_Sqr	−0.453	0.109	−4.139	
	UTS_Sqr	21.148	6.756	3.130	
	GTM_Sqr	4.477	1.957	2.288	
4	(Constant)	7.905	8.168	0.968	0.75
	Abo65_Sqr	0.398	0.054	7.408	
	UA_Sqr	−0.732	0.125	−5.852	
	GT_Max	0.673	0.157	4.302	

Source: Liu and Weng (2009). Reproduced with permission of Elsevier.
Abo65_Sqr, square root of population above 65 years old; GT_Max, maximum grass temperature; UT_SD, standard deviation of urban temperature; UTI_Log, natural logarithm of the minimum urban surface temperature; GTM_Log, natural logarithm of the maximum grass temperature; UA_Sqr, square root of urban area; UTS_Sqr, square root of standard deviation of urban surface temperature; GTM_Sqr, square root of the maximum grass temperature. For the rest variables, see Table 10.1 for the explanations of abbreviations.

variables remained after analysis: square roots of population above 65 years old, urban area, and grass temperature. According to Table 10.2, transformation itself did not increase the R^2 value but decreased the value instead.

Model 1 appeared to be the optimal models in Table 10.2. In the model, the combination of four variables effectively predicted the absolute WNV number with an R^2 value of 0.75. The residual of WNV number for City of Chicago was underestimated. Negative residuals occurred more in the northern and southern parts of the county. This indicates that those municipalities were overestimated by the model.

This study found that urban and grass apparently showed a close relationship with the spread of WNV caused by insects and animals. This finding supports the observation that mosquitoes, the vector of WNV dissemination, prefer to stay in habitats with high organic content and fine moisture, such as old tires and stormwater catch basins in the residential area. This founding also indicates that the composition of different LULC types in the study area may affect the WNV dissemination more than a single factor can do. It may be related to the growing population and urbanization in recent years in the study area. For example, more discarded tires could be exposed in the back yards and more pipes would be added to existing sewer systems.

10.3.8 Discussion

This study demonstrates that ASTER imagery could be used to examine the spread of WNV by combining various RS- and GIS-related variables and thermal and socioeconomic information. However, it is still a challenging work to use RS and GIS techniques to study the spread of WNV due to some factors: the quality of RS data, its acquisition time, and methods used to process it, the complexity of WNV dissemination, and the integrity and reliability of the records of infections (mosquito/animal host positive reports). RS data with higher resolutions may provide more detailed information on landscape patterns. The accuracy of LSTs retrieved from RS data can be affected by sensor resolution and algorithm chosen. Specific acquisition time of RS data restricts the study to specific periods and decreases the possibility of combining WNV reports in different seasons, unless RS data acquired at different seasons would be available. More specific LULC categories (e.g. commercial areas and residential communities) can be identified by using different classification methods. This study found that urban and grass apparently showed a close relationship with the spread of WNV caused by Culex species and hosts. This indicates that the composition of different LULC types may affect the WNV dissemination more than a single factor. More detailed records of WNV cases, like physical addresses of cases are expected to improve the accuracy of analysis.

Based on the results identified in the last section, we concluded that

1) WNV infections caused by insect and animals were spread throughout the whole Cook County since 2001. The size of urban area, urban and grass temperatures, and population above 65 years old were the most important factors for the spread of WNV in Cook County. The landscape aggregation level and areas of three LULC types, urban, grass, and water showed high correlations with the spread of WNV caused by insects and animals, no matter how high or low their patch densities were, how complicated their spatial configurations may be, and how high or low their spatial connectivity were.

2) socioeconomic conditions, population, total area, population above 65 years old, total house income, and income below poverty line also showed strong relationships with the spread of WNV caused by insects and animals in Cook County. The combination of these variables predicted the WNV number with acceptable R^2 value.

3) thermal conditions like maximum temperatures and standard deviations of urban and grass were also related to the spread of WNV caused by insects/ animals. Thermal conditions of water (standard deviation and range) showed less but still considerable correlation to the spread of WNV compared to those of urban and grass.

This study further proves the capability of RS and GIS techniques in the study of public health. Some findings in this study could contribute to urban planning and public health management and protection in the studied area and other areas with similar characters:

1) Epidemic disease control can benefit from RS and GIS techniques. Remotely sensed imagery could be used to derive environmental factors that are believed to be closely related to any specific epidemic diseases (Liu and Weng 2012).

2) The aggregation levels and sizes of urban areas, grass, and water need to be controlled so as to decrease the possibilities of disease dissemination. In this study, the aggregations and sizes of urban, grass, and water show close relationship to the spread of WNV caused by insects and animals. In order to decrease the spread of WNV caused by insects and animals in the Cook county, possible measures include developing more green vegetation zones with small sizes in the urban area, so as to decrease the urban area and its aggregation level, decreasing the sizes of existing grasslands, ponds, and lakes, controlling the sizes of those features, and limiting their aggregation levels in the future urban planning.

3) It is necessary to pay extra attention to older people and low-income families in the Cook County. According to the conclusion, socioeconomic factors, population above 65 years old, total house income, and income below poverty line showed strong relationship with the spread of WNV caused by insects and animals in Cook County. Possible measures are to help low-income families maintain a clean living environment and to increase the medical assistants for population above 65 years old.

4) Attention is also needed in temperature control of the Cook County. High LSTs of urban, grass, and water may contribute to the WNV dissemination. Possible solutions can include decrease in energy transfer and control in pollution: decreasing the use of vehicles, avoiding urban sprawl, planting more green vegetation, and so on.

10.4 WNV-Risk Areas in Southern California, 2007–2009

10.4.1 Study Area and Data Collection

Ten counties in California were selected as the study area: San Luis Obispo, Kern, San Bernardino, Santa Barbara, Ventura, Los Angeles, Orange, Riverside, San Diego, and Imperial. All the islands (e.g. Santa Catalina Island, Santa Clemente Island, and Santa Barbara Island) belonged to these counties were excluded from the study because few incidents were reported from these islands. Figure 10.3 illustrates the geographical location of the study area. The total land area of these 10 counties is about 56 500 square miles and is inhabited by over 22 million people (US Census estimation 2006). The temperature difference between immediate coastal counties and inland counties in the

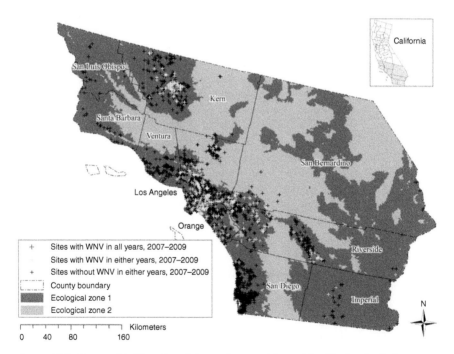

Figure 10.3 Geographical location of the study area with two identified ecological zones. Islands were not included in the analysis since limited WNV surveillance data were reported from those areas. Surveillance sites (islands excluded) are displayed at three levels, having WNV in all years of 2007–2009 (light gray), having WNV in either years of 2007–2009 (white), and having WNV infection in none years of 2007–2009 (dark gray). *Source:* Liu and Weng (2012). Reproduced with permission of Elsevier.

study area is about 4 °F (2 °C) in winter and approximately 23 °F (13 °C) in summer (National Weather Service). The southwest monsoon *increases the humidity in* arid areas of southern California in the summer in general. There is a mixture of heavily developed urban settings and undeveloped arid lands.

California has a long history of mosquito-borne disease and has become one of the most affected states for WNV epidemics in the United States, since the virus was initially detected in a mosquito pool near Imperial County in summer 2003. The moderate Mediterranean climate provides a favorable habitat for mosquitoes and thus contributes to the dissemination of WNV in the state (Reisen et al. 2004). A great number of WNV infections have been detected in birds, mosquitoes, and some other mammals in the state according to the statistics published by the Centers for Disease Control and Prevention (CDC). The severe status of the WNV epidemic in California, especially in the study area, makes this study critical. California started to develop mosquito control programs in the early 1900s to defend mosquito-borne diseases, e.g. arboviruses and malaria. Its strong surveillance program provides enriched data for this study.

The geographical dissemination of WNV is related to the occurrence of both avian reservoir host and mosquito vector, which is potentially associated with the environmental and social-economic conditions that can affect the abundances of related birds and mosquito species (Gibbs et al. 2006; Ozdenerol et al. 2008; Yang et al. 2008). We first collected necessary WNV surveillance data and next created environmental variables using RS and GIS technologies to assess the high-risk areas of WNV infection combined with statistical analysis. In order to control the error of localization of WNV incidence and to integrate environmental variables with various spatial resolutions, we constructed a hexagonal network covering all 10 counties (total 42 560 hexagons). The area of each hexagon (2-km side length) is close to the average size of census blockgroups in those counties, forming more neutral landscape units in comparison to the inconsistent shapes and uneven areas of block groups. Both WNV surveillance data and environmental variables discussed below were integrated into individual hexagonal geographic units for further analysis.

WNV surveillance data for the whole state of California were collected from the Arbovirus Bulletins published in the California Vectorborne Disease Surveillance System (CALSURV). The records include the species of tested positive mosquitoes, the number of mosquitoes in each pool, collected dates, surveillance site ID, and the method of virus detection for years 2004–2009. We collected the geographical coordinates for each surveillance site in 10 counties from the interactive surveillance map published in CALSURV. Yearly human WNV infection data were provided by CDC. As a result, every tested positive mosquito pool was linked to a specific hexagon using the geographical coordinates of surveillance sites. We focused our study of risk areas for years 2007–2009 due to the lack of data availability, although the WNV infection (in humans) has been reported from California since 2002. More specifically, we grouped the mosquito surveillance data into three periods, epidemiological weeks 18–26 (mainly in May and June), 27–35 (mainly in July and August), 36–44 (mainly in September and October) for each year of 2007–2009 in the identification of WNV-risk areas.

10.4.2 Environmental Factors of WNV Dissemination

Climate is an important driver in the reproduction and survival of mosquitoes (Whitman 1937; Epstein et al. 1998). Higher temperatures contribute to the faster distribution and wider spread of WNV throughout North America (Reisen et al. 2004; Liu and Weng 2009; Soverow et al. 2009; Ruiz et al. 2010). Because mosquitoes thrive in hot, humid environments (Dohm et al. 2002), this exacerbation in temperature demonstrates the potential role of urban environment in WNV dissemination. Precipitation shows close relationship to WNV incidence in the region although its role seems to be inconsistent among studies (Epstein and Defilippo 2001; Landesman et al. 2007; Soverow et al. 2009). We obtained the historical temperature and precipitation factors (monthly mean temperature, departure from normal temperature, total monthly precipitation, and departure from normal monthly precipitation) from National Oceanic and Atmospheric Administration (NOAA) National Climatic Data Center. The timeframe of weather records was from years 2004–2009 since WNV infections became significant starting in 2004. We calculated the mean values of temperature and precipitation in years 2004–2009 before computing their means for hexagons, the units of study. The inverse distance weighting spatial interpolation method was chosen to create four weather maps (30-m spatial resolution), each corresponding to one of the temperature and precipitation variables. In order to control the error caused by edge effects in the spatial interpolation, we selected a total of 135 weather stations within 10 counties and their surrounding areas in three states (California, Arizona, and Nevada). To identify the possible influence of LST on WNV dissemination in three studied periods (weeks 18–26, 27–35, and 36–44) in years 2007–2009, we collected Terra Moderate Resolution Imaging Spectroradiometer (MODIS) Land Surface Temperature and Emissivity 8-day Level-3 Global images (1-km spatial resolution) acquired between April and November in years 2007–2009. The mean LST was calculated for each period in assessing the WNV-risk areas, and the average LST per hexagon (unit for analysis) was computed accordingly.

Elevation and slope are found to be related to the disseminations of WNV in different regions (Ozdenerol et al. 2008; Winters et al. 2008). Winters's study was done in Colorado where there are real differences in climate due to great differences in elevation within the study area. We used US Geological Survey (USGS) Digital Elevation Model (DEM) data (30-m resolution) to create a raster data set of slope. The average elevation and slope were then derived from the DEM and slope maps for each hexagonal unit.

Landscape patterns, such as landscape complexity and aggregation, were reported to be strongly associated with the distributions of virus (Yang et al. 2008; Liu and Weng 2009). The presence of grass and other vegetation helps to retain moisture in the neighborhood which provides a favorable habitat for mosquito survival. Forest and other green spaces enhance local bird abundance and species diversity (Emlen 1974; Cavareski 1976), which in turn can contribute to the spread of WNV. Higher landscape diversity is associated with multiple land-cover types such as urban, grass, and water. The 10 studied counties, especially the areas along the coast are highly urbanized with diverse habitats which can contribute to bird abundance (Savard et al. 2000). In this study, land-cover data in remotely sensed datum format for year 2001 were collected from the Multi-Resolution Land Characteristics Consortium (MRLC). Those land-cover images were created by using the unsupervised image classification method on the Landsat TM imagery (Vogelmann et al. 1998). The initial land-cover images were encoded to only include seven different land-cover types: open water, developed open space (e.g. parks) and low-to-medium-intensity residential areas, high-intensity developed areas (e.g. commercial places), forest and scrub, grassland/pasture, wetlands, and barren lands. We used Patch Analyst 4, an extension to the ArcGIS application that facilitates the assessment of landscape pattern to derive more than 15 relevant landscape metrics from the encoded land-cover image. *Due to possible correlation among these metrics, we performed bivariate Pearson correlation analysis and Principal Component analysis (PCA), and selected two landscape metrics, SDI, and Mean Shape Index (MSI)* to represent the whole landscape pattern. The SDI measures the landscape diversity introduced by Shannon (1948):

$$SDI = \sum_{i=1}^{m} \left(P_i \ln P_i \right) \tag{10.1}$$

where P_i is the proportion of patch type (class), i is the landscape, and m is the number of patch types (classes) (McGarigal and Marks 1995).

$$MSI = \frac{\sum_{i=1}^{m} \sum_{j=1}^{n} \left(0.25 P_{ij} / \sqrt{A_{ij}} \right)}{N} \tag{10.2}$$

where P_{ij} is the perimeter of patch ij, A_{ij} is the area of patch ij, m is the number of patch types (classes), n is the number of patches per class, and N is the total number of patches in the landscape (McGarigal and Marks 1995). The mean percentage of canopy was calculated for each hexagon based on the MRLC canopy product 2001 (Huang et al. 2001).

The Normalized Difference Water Index (NDWI) is a metric measuring vegetation water content, derived from the near-infrared (NIR) and shortwave infrared (SWIR) channels of the RS satellite imagery (Gao 1996). The possible variations caused by leaf internal structure and leaf dry matter content can be eliminated by combining the NIR with the SWIR reflectance, which better measures the vegetation water content (Ceccato et al. 2001).

$$NDWI = \frac{P_{NIR} - P_{SWIR}}{P_{NIR} + P_{SWIR}} \tag{10.3}$$

where P_{NIR} is the reflectance in the NIR region and P_{SWIR} is the reflectance in the SWIR region.

Multiple studies have applied MODIS NDWI data to investigate vegetation water content (Jackson et al. 2004; Delbart et al. 2005). In order to investigate the possible influence of vegetation humidity to mosquito vector competence, we calculated NDWI based on Terra MODIS Surface Reflectance 8-Day Level-3 images (500-m spatial resolution) acquired in the same period of time as that of MODIS LST data sets. In the calculation of NDWI, Band 2 was selected for use as the NIR channel and Band 6 was chosen as the SWIR channel. As what we did for MODIS LST images, the mean NDWI was calculated for each studied period in years 2007–2009 and the average NDWI per hexagon as well.

As a result, we carefully chose a group of environmental factors as expositive variables for WNV dissemination: annual average temperature, summer (May–August) average temperature, annual average departure from normal temperature, summer (May–August) average departure from normal temperature, annual total precipitation, summer (May–August) total precipitation, annual average departure from normal precipitation, summer (May–August) average departure from normal precipitation, LST, elevation, slope, SDI, MSI, and NDWI.

10.4.3   Statistical Analysis

A PCA (with Quartimax rotation method) was first conducted to reduce the number of environmental variables. Since the 10 counties contain diverse landscape and terrain conditions, it is necessary to generate homogeneous ecological zones (Sanders 1990). A K-Means cluster analysis was performed to group the hexagons with similar environmental conditions together by using the variables selected from Principal Components. Cluster analysis is a processing of grouping the objects that possess similar characteristics into the same cluster (Aldenderfer and Blashfield 1984). Each group indicates a distinguished ecological zone. The initial surveillance data were transformed into a series of Euclidean Distance maps (30-m spatial resolution) corresponding to three periods, i.e. epidemiological weeks 18–26, 27–35, and 36–44 for individual years of 2007–2009. The correlations between the Euclidean Distance and environmental variables selected by PCA, LST, SDI, MSI, and NDWI were examined using the *bivariate Pearson correlation analysis for different ecological zones.* Factors at 0.01 significance level were used to calculate *Mahalanobis Distance* in order to assess the risk area of WNV.

Mahalanobis Distance is a distance index based on correlation between variables by which different conditions can be analyzed (Mahalanobis 1936). It has been used to examine the similarity of one environmental/habitat setting to another one (Huang et al. 2001; Liu and Weng 2012). We used Mahalanobis Distance to identify habitats associated with the dissemination of WNV in mosquitoes for epidemiological weeks 18–26, 27–35, and 36–44 in 2007–2009. We assumed the hexagons containing tested positive mosquito infection as one type of landscape. Those without incident were expected to have different characteristics of landscape. Since some hexagons contained more than one tested positive mosquito pools, we substituted a Horvitz–Thompson weighted (Horvitz and Thompson 1952) mean and covariance matrix in the calculation of Mahalanobis Distance with hexagons as study units. The formula can be defined as

$$D(x)^{\circ} = \sqrt{(x-\mu)^T S^{-1}(x-\mu)} \qquad (10.4)$$

where $D(x)$ is Mahalanobis Distance, x is the vector of environmental variables for each hexagon, μ is the vector of Horvitz–Thompson weighted mean values of environmental factors for the hexagons with tested positive mosquito infections, S^{-1} is the inverse Horvitz–Thompson weighted covariance matrix of environmental factors for the hexagons with tested positive mosquito infections, and T indicates a transposed matrix. Small Mahalanobis Distance values suggest more favorable habitats for WNV-infected mosquitoes while greater values associate with less suitable habitats. Assuming that Mahalanobis Distances follow the Chi-square (χ^2) distribution, we calculated the p-value for each Mahalanobis Distance and considered hexagons with p-value between 0.9 and 1.0 as high-risk areas and ones with p-value between 0.6 and 0.9 moderate risk.

10.4.4 Pattern of WNV Dissemination

The investigation of WNV surveillance data helped to understand the general pattern of virus dissemination. Figure 10.4 shows the total tested positive mosquito and human surveillance records in 10 counties in years 2004–2009. This figure reveals a repeating pattern for every year: the mosquito infection started to expand in May (mainly corresponding to epidemiological weeks 18–22) with a limited number reported in earlier months. The amplification reached a peak in early or middle August (e.g. week 33 in 2005 and week 32 in 2008) except that year 2004 had a crest in week 21 (late May) and year 2006 peaked in week 37 (early September). The infection generally ended in late November or December. Years 2004, 2005, and 2008 received the greatest total number of positive mosquito pools, while year 2006 received the fewest mosquito infections. The conformable increase of WNV infection in mosquitoes in late

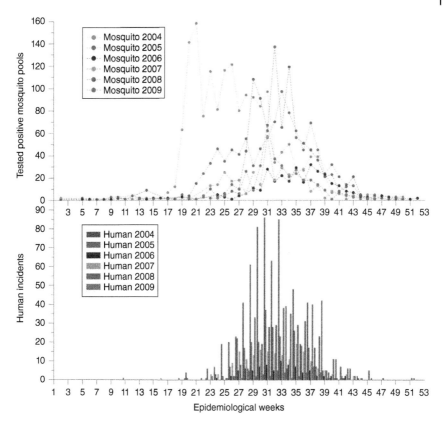

Figure 10.4 Total tested positive mosquito and human surveillance records in 10 counties in years 2004–2009. *X*-axis associates with epidemiological weeks, *Y*-axis is the raw number of tested positive mosquito pools, and *Z*-axis is the raw number of human incidence. *Source:* Liu and Weng (2012). Reproduced with permission of Elsevier.

summer across the years 2004–2009 may associate with post nesting migrations of summer and year-round resident birds (Reisen et al. 2004). This increase may contribute to the outbreak of the virus in humans in the following weeks. The summer outbreak of WNV in 2004 may be linked closely to positive departure from normal temperature (Reisen et al. 2006a). Overwintering of WNV during the winters (e.g. 2004–2005) might contribute to the subsequent amplification of virus in the following year 2006 (Reisen et al. 2006b). The continuous amplifications of WNV in years 2007–2009 may be associated with the combining effects of drought conditions in the previous year and a departure from normal temperatures in the early season.

Human and mosquito infection incidence followed a similar pattern throughout the study years; human infections peaked in August or September in years

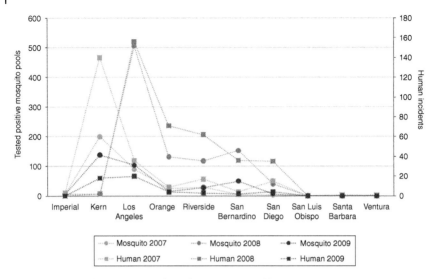

Figure 10.5 Yearly surveillance data of mosquitoes and human per county in years 2007–2009. X-axis represents individual counties and Y-axis shows the raw numbers of tested positive mosquito pools (left) and human incidents (right). *Source:* Liu and Weng (2012). Reproduced with permission of Elsevier.

2004–2009, e.g. week 31 in 2005 and week 37 in 2008 and mainly ended in November. Human incidence followed a nearly identical trend as that of mosquito incidence: the peak years were 2004, 2005, and 2008, and year 2009 recorded the smallest number of human cases. Figure 10.5 presents the yearly surveillance data of mosquitoes and humans per county in years of 2007–2009. Limited by data availability and completeness, years 2003–2006 were excluded in the figure. According to Figure 10.5, two counties, Kern and Los Angeles, received the maximum numbers of WNV infections in both mosquitoes and human, followed by Orange, Riverside, and the rest counties with San Luis Obispo, Santa Barbara, and Ventura possessing the minimum numbers (0–1) of infections.

10.4.5 Mapping of Risk Areas

Ten components were generated as the result of PCA, among which the first three representing about 71% of the total variance (Table 10.3). The PC1 included the monthly temperature and annual precipitation variables explaining 38% variance. It can therefore be defined as a weather factor. The PC2 (20% variance) possessed the departures from normal temperatures and precipitations, which seems to reflect an abnormal climate condition. The PC3 summarized 13% variance with two variables, elevation and slope relating to the terrain characteristics. Three representative variables with the highest loading

Table 10.3 Rotated component matrix as results of the principal component analysis.

	Loadings		
Variables	PC1	PC2	PC3
Average summer (May–August) mean temperature for years 2004–2009	0.906	0.279	−0.042
Average annual mean temperature for years 2004–2009	0.835	0.221	−0.267
Average annual mean monthly precipitation for years 2004–2009	−0.817	−0.101	0.272
Average annual mean departure from normal monthly precipitation for years 2004–2009	0.741	−0.308	0.018
Average summer (May–August) mean monthly precipitation for years 2004–2009	0.509	0.094	0.431
Average annual mean departure from normal temperature for years 2004–2009	0.239	0.892	−0.106
Average summer (May–August) mean departure from normal temperature for years 2004–2009	0.315	0.847	0.014
Average summer (May–August) mean departure from normal monthly precipitation for years 2004–2009	0.209	−0.627	−0.371
Elevation	−0.119	0.141	0.880
Slope	−0.273	−0.081	0.734

Source: Liu and Weng (2012). Reproduced with permission of Elsevier.

values, average summer (May–August) mean temperature for years 2004–2009, average annual mean departure from normal temperature for years 2004–2009, and elevation were selected as input of *K*-Means cluster analysis.

Two clusters were generated by using *K*-Means cluster analysis with hexagons as units, each indicating a distinguishable ecological zone. Ecological Zone 1 contained 24766 hexagons and 17794 in Zone 2. Zone 1 falls in coastal Central Coast and South Coast bioregions (total 34% area percentage), Colorado Desert (28%), central and southeastern Mojave (26%), southern San Joaquin valley (11%), and southern Sierra (2%), whereas Zone 2 comprised southern Mojave (61%), inner areas of Central Coast and South Coast (total 26%), southern Sierra (7%), eastern Colorado Desert (6%), and southern border of San Joaquin Valley (3%). Zone 1 possessed less departure from normal temperature, lower summer temperature based on the climate records in 2004–2009, and lower elevation compared to those of Zone 2. Since 95% collected surveillance sites in 10 counties fell inside Zone 1, we conducted the risk analysis mainly for Zone 1. Some environmental factors demonstrated a closer relationship to WNV propagation than others, which include summer (May–August) mean temperature, annual mean departure from normal temperature, LST, elevation, landscape complexity, landscape diversity, and vegetation water content.

Based on the correlation between selected environmental variables and Euclidean Distance to tested positive mosquito pools, all those environmental factors were consistently and significantly related to WNV propagation in mosquitoes in three studied periods in years 2007–2009. Figure 10.6 demonstrates the WNV-risk areas in mosquitoes in those time windows based on Mahalanobis Distance p-values. As observed in the figure, both high- (Mahalanobis Distance *p*-values: 0.9–1.0) and moderate-risk (Mahalanobis Distance *p*-value: 0.6–0.9) areas of WNV infection in mosquitoes constantly appeared in/around southern San Joaquin Valley in Kern County and southern Los Angeles County in almost all the studied periods and years. The coastal areas of Orange and San Diego also possessed moderate-to-high risk in some periods of years, e.g. weeks 36–44 in all years, 2007–2009. The spread pattern of identified WNV-risk areas seemed to be consistent in three years: the risk areas appear in southern San Joaquin Valley and southern Los Angeles County in weeks 18–26 (mainly May–June) and then expand to its surrounding areas in weeks 27–35 (mainly July–August) and 36–44 (mainly September–October). The last map highlighted the moderate-to-high-risk areas of WNV. As a contrast to those vulnerable coastal lands, areas in the Mojave Desert showed lower risk of WNV infection.

The risk analysis was conducted based on the environmental factors and surveillance data in years 2007–2009 without the records from 2003 to 2006. The risk map indicates the spatial distribution of possible invasion of WNV in studied counties. High-risk areas were identified in southern San Joaquin Valley in Kern County and southern Los Angeles County, along with some coastal areas of Orange and San Diego and their surrounding areas. The risk areas identified in the City of Los Angeles area seem to be compatible with the results of another WNV study conducted by the authors with higher spatial-resolution images (Liu and Weng 2012). The overall risk assessment is not contradictory to the existing findings done by Reisen et al. (2006b), in which infection rates in *Culex pipiens quinquefasciatus, one of the most frequently infected mosquito species in the area*, were reported to be higher in both Kern and Los Angeles counties than in Coachella Valley, Riverside County. In addition to their study, our assessment provides particular spatial locations of possible vulnerability.

10.4.6 Discussion

We conducted a retrospective study of WNV spread for 10 counties in the southern California based on surveillance data from 2004 to 2009. Particularly, we assessed the WNV-risk areas based on the mosquito surveillance records in years 2007–2009. Before conducting environmental analysis of WNV

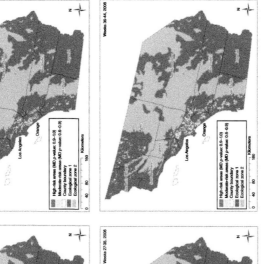

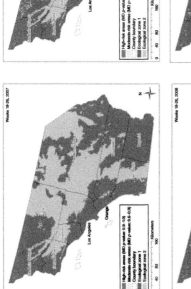

Figure 10.6 WNV-risk areas for three periods in years 2007–2009, indicated by Mahalanobis Distance *p*-value calculated based on mosquito surveillance records. *Source:* Liu and Weng (2012). Reproduced with permission of Elsevier.

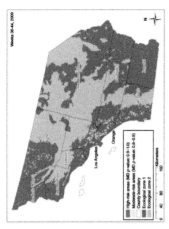

Figure 10.6 (Continued)

infections, we classified the whole study area into two zones since the area contains diverse climate conditions and elevations. The construction of hexagon network simulated more neutral landscape units than those of geographical or census units (e.g. census block groups). It is found that southern San Joaquin Valley in Kern County and the southern Los Angeles County were the most vulnerable locations for WNV outbreak. Main factors contributing to the WNV propagation included summer mean temperature, annual mean deviation from the mean temperature, LST, elevation, landscape complexity, landscape diversity, and vegetation water content. The result of the study improves understanding of WNV ecology and provides tools for detecting, tracking, and predicting the epidemic. The holistic approach developed for this multidisciplinary study, which integrated remotely sensed, GIS-based, and in situ measured environmental factors with landscape metrics, may be applied to studies of other vector-borne diseases.

Based on the results of environmental factor analysis, the positive role of temperature detected in the environmental factor analysis is consistent with the finding reported by Reisen et al. (2006a, 2006b). Temperature must be above the minimum temperature (14.3 °C) required for virus replication in mosquitoes for WNV to be disseminated throughout the year, especially in the cold months (Reisen et al. 2006b). A temperature study in southern California showed that mean temperatures in Coachella Valley in Riverside County and the hilly parts of Los Angeles County were mainly above the minimum in the coldest months of years 2003–2005, while temperatures in Long Beach, Los Angeles and Shafter, Kern County, floated around the minimum in the same period of time, and the differences increase during summer (Reisen et al. 2006b). The results also show that areas with lower elevations tended to be more susceptible to WNV invasion. This finding may be explained by the observation of mosquitos in Colorado: Mosquito species (e.g. *Culex tarsalis*) appeared to be abundant in the plain habitats with lower elevations and warmer temperatures, and the population dropped in areas with higher elevation and slightly lower temperatures (Eisen et al. 2008). Higher landscape diversity is usually associated with multiple land-cover types such as urban, grass, and water. Highly urbanized area with diverse habitats can contribute to bird abundance (Savard et al. 2000). Certain amount of vegetation moisture in the neighborhood provides a favorable habitat for mosquito survival (Dohm et al. 2002).

Future research should be directed to improve the mapping of WNV-risk areas by taking the following factors into consideration. Firstly, WNV surveillance data were collected from CA arbovirus bulletins published by California Vectorborne Disease Surveillance System; thus, it is possible that incidents existed outside the range of sampling effort, e.g. central Imperial. Secondly, the two ecological zones were defined using cluster analysis solely based on weather and terrain conditions. Although we carefully selected the variables for PCA and cluster analysis, additional factors may exist to influence the outcome of cluster analysis, e.g. the number of water impoundments (e.g. green pools). The identified risk areas in Zone 1 may be changed accordingly. Also, we used the land-cover data set for 2001 to measure the landscape complexity and diversity. Land-cover data sets for individual years of 2007–2009 may provide more up-to-date characteristics of landscape. More field and entomological investigations are needed to support these preliminary results. Other environmental factors, such as soil moisture and evaporation, can be derived from high-resolution remotely sensed imagery and serve as additional inputs for the risk area assessment.

References

Aguiar, R., Oliveira, M., and Goncalves, H. (2002). Climate change impacts on the thermal performance of Portuguese buildings. Results of the SIAM study. *Building Service Engineers Research and Technology* 23 (4): 223–231.

Aldenderfer, M.S. and Blashfield, R.K. (1984). *Cluster Analysis*. Newbury Park, CA: Sage.

Anderson, J.F., Andreadis, T.G., Main, A.J. et al. (2006). West Nile virus from female and male mosquitoes (Diptera: Culicidae) in subterranean, ground, and canopy habitats in Connecticut. *Journal of Medical Entomology* 43 (5): 1010–1019.

Anyamba, A., Chretien, J.P., Small, J. et al. (2006). Developing global climate anomalies suggest potential disease risks for 2006–2007. *International Journal of Health Geographics* 5 (1): 60.

ASTER Online Product Description (2005). Online resource. https://terra.nasa.gov/data/aster-data (accessed 5 July 2019).

Bain, D.J. and Brush, G.S. (2004). Placing the pieces: reconstructing the original property mosaic in a warrant and patent watershed. *Landscape Ecology* 19 (8): 843–856.

Balenghien, T., Fouque, F., Sabatier, P., and Bicout, D.J. (2006). Horse-, bird-, and human-seeking behavior and seasonal abundance of mosquitoes in a West Nile virus focus of Southern France. *Journal of Medical Entomology* 43 (5): 936–946.

Boyd, D.S., Foody, G.M., Curran, P.J. et al. (1996). An assessment of radiance in Landsat TM middle and thermal infrared wavebands for the detection of tropical forest regeneration. *International Journal of Remote Sensing* 17 (2): 249–261.

Cavareski, C.A. (1976). Relation of park size and vegetation to urban bird populations in Seattle, Washington. *Condor* 78: 375–382.

Ceccato, P., Flasse, S., Tarantola, S. et al. (2001). Detecting vegetation water content using reflectance in the optical domain. *Remote Sensing of Environment* 77: 22–33.

Centers for Disease Control and Prevention (2002). Provisional surveillance summary of the West Nile virus epidemic-United States, January–November 2002. *MMWR* 51: 1129–1133.

Delbart, N., Kergoat, L., Toan, T.L. et al. (2005). Determination of phenological dates in boreal regions using normalized difference water index. *Remote Sensing of Environment* 97: 26–38.

Dohm, D.J., O'Guinn, M.L., and Turell, M.J. (2002). Effect of environmental temperature on the ability of *Culex pipiens* (Diptera: Culicidae) to transmit West Nile virus. *Journal of Medical Entomology* 39 (1): 221–225.

Dohm, D.J. and Turell, M.J. (2001). Effect of incubation at overwintering temperatures on the replication of West Nile virus in New York *Culex pipiens* (Diptera: Culicidae). *Journal of Medical Entomology* 38 (3): 462–464.

Eisen, L., Bolling, B.G., Blair, C.D. et al. (2008). Mosquito species richness, composition, and abundance along habitat-climate-elevation gradients in the northern Colorado front range. *Journal of Medical Entomology* 45 (4): 800–811.

Emlen, J.T. (1974). An urban bird community in Tucson, Arizona: derivation, structure, regulation. *Condor* 76: 184–197.

Epstein, P.R. and Defilippo, C. (2001). West Nile virus and drought. *Global Change & Human Health* 2 (2): 105–107.

Epstein, P.R., Diaz, H.F., Elias, S. et al. (1998). Biological and physical signs of climate change: focus on mosquito-borne diseases. *Bulletin of the American Meteorological Society* 79 (3): 409–417.

Gao, B. (1996). NDWI – a normalized difference water index for remote sensing of vegetation liquid water from space. *Remote Sensing of Environment* 58: 257–266.

Gibbs, S.E.J., Wimberly, M.C., Madden, M. et al. (2006). Factors affecting the geographic distribution of West Nile virus in Georgia, USA: 2002–2004. *Vector-Borne and Zoonotic Diseases* 6 (1): 73–82.

Gingrich, J.B., Anderson, R.D., Williams, G.M. et al. (2006). Stormwater ponds, constructed wetlands, and other best management practices as potential breeding sites for West Nile virus vectors in Delaware during 2004. *Journal of the American Mosquito Control Association* 22 (2): 282–291.

Gustafson, E.J. (1998). Quantifying landscape spatial pattern: what is the state of the art? *Ecosystems* 1: 143–156.

Hay, S.I. and Lennon, J.J. (1999). Deriving meteorological variables across Africa for the study and control of vector-borne disease: a comparison of remote sensing and spatial interpolation of climate. *Tropical Medicine and International Health* 4: 58–71.

Hayes, E.B. and O'Leary, D.R. (2004). West Nile virus infection: a pediatric perspective. *Pediatrics* 113 (5): 1375–1381.

Herbreteau, V., Salem, G., Souris, M. et al. (2007). Thirty years of use and improvement of remote sensing, applied to epidemiology: from early promises to lasting frustration. *Health & Place* 13 (2): 400–403.

Hirano, Y., Yasuoka, Y., and Ichinose, T. (2004). Urban climate simulation by incorporating satellite-derived vegetation cover distribution into a mesoscale meteorological model. *Theoretical and Applied Climatology* 79: 175–184.

Hodge, J.G. Jr. and O'Connell, J.P. (2005). West Nile virus: legal responses that further environmental health. *Journal of Environmental Health* 68 (1): 44–47.

Horvitz, D.G. and Thompson, D.J. (1952). A generalization of sampling without replacement from a finite universe. *Journal of the American Statistical Association* 47 (260): 663–685.

Huang, C., L. Yang, B. Wylie, and C. Homer (2001). A strategy for estimating tree canopy density using Landsat 7 and ETM+ and high resolution images over large areas. *Third International Conference on Geospatial Information in Agriculture and Forestry,* Denver, Colorado (5–7 November 2001), unpaginated CD-ROM.

Jackson, T.J., Chen, D., Cosh, M. et al. (2004). Vegetation water content mapping using Landsat data derived normalized difference water index for corn and soybeans. *Remote Sensing of Environment* 92: 475–482.

Jacob, F., Olioso, A., Gu, X. et al. (2002). Mapping surface fluxes using visible, near infrared, thermal infrared remote sensing data with a spatialized surface energy balance model. *Agronomie: Agriculture and Environment* 22: 669–680.

Komar, N. (2003). West Nile virus: epidemiology and ecology in North America. *Advances in Virus Research* 61: 185–234.

Landesman, W.J., Allan, B.F., Langerhans, R.B. et al. (2007). Inter-annual associations between precipitation and human incidence of West Nile virus in the United States. *Vector-Borne and Zoonotic Diseases* 7 (3): 337–343.

Liang, S.Y., Linthicum, K.J., and Gaydos, J.C. (2002). Climate change and the monitoring of vector-borne disease. *JAMA* 287 (17): 2286.

Liu, H. and Weng, Q. (2008). Seasonal variations in the relationship between landscape pattern and land surface temperature in Indianapolis, U.S.A. *Environmental Monitoring and Assessment* 144 (1): 199–219.

Liu, H. and Weng, Q. (2009). An examination of the effect of landscape pattern, land surface temperature, and socioeconomic conditions on WNV dissemination in Chicago. *Environmental Monitoring and Assessment* 159: 143–161.

Liu, H. and Weng, Q. (2012). Enhancing temporal resolution of satellite imagery for public health studies: a case study of West Nile virus outbreak in Los Angeles in 2007. *Remote Sensing of Environment* 117 (2): 57–71.

Liu, Y., Hiyama, T., and Yamaguchi, Y. (2006). Scaling of land surface temperature using satellite data: a case examination on ASTER and MODIS products over a heterogeneous terrain area. *Remote Sensing of Environment* 105 (2): 115–128.

Luvall, J.C. and Holbo, H.R. (1991). Thermal remote sensing methods in landscape ecology. In: *Quantitative Methods in Landscape Ecology* (eds. M.G. Turner and R.H. Gardner). New York: Springer-Verlag.

Mahalanobis, P.C. (1936). On the generalised distance in statistics. *Proceedings of the National Institute of Sciences of India* 2: 49–55.

Marfin, A.A., Petersen, L.R., Campbell, G.L. et al. (2000). Widespread West Nile virus activity, Eastern United States. *Emerging Infectious Diseases* 7 (4): 730–735.

McGarigal, K., Cushman, S.A., Neel, M.C., and Ene, E. (2002). *FRAGSTATS: spatial Pattern Analysis Program for Categorical Maps*. Computer software program produced by the authors at the University of Massachusetts, Amherst. www.umass.Edu/landeco/research/fragstats/fragstats.html (accessed 5 July 2019).

McGarigal, K. and Marks, B.J. (1995). *FRAGSTATS: Spatial Pattern Analysis Program for Quantifying Landscape Structure*. General Technical Report PNW-GTR-351. Portland, OR: USDA Forest Service, Pacific Northwest Research Station.

McVicar, T.R. and Jupp, D.L.B. (1998). The current and potential operational uses of remote sensing to aid decisions on drought exceptional circumstances in Australia: a review. *Agriculture Systems* 57: 399–468.

MMWR (2005). Update: West Nile virus activity – United States 2005. *Morbidity and Mortality Weekly Report* 54 (34): 851–852.

O'Neill, R.V., Krummel, J.R., Gardner, R.H. et al. (1988). Indices of landscape pattern. *Landscape Ecology* 1: 153–162.

Owen, T.W., Carlson, T.N., and Gillies, R.R. (1998). An assessment of satellite remotely-sensed land cover parameters in quantitatively describing the climatic effect of urbanization. *International Journal of Remote Sensing* 19: 1663–1681.

Ozdenerol, E., Bialkowska-Jelinska, E., and Taff, G.N. (2008). Locating suitable habitats for West Nile virus-infected mosquitoes through association of environmental characteristics with infected mosquito locations: a case study in Shelby County, Tennessee. *International Journal of Health Geographics* 7: 12.

Peterjohn, W.T. and Correll, D.L. (1984). Nutrient dynamics in an agricultural watershed: observations on the role of a riparian forest. *Ecology* 65: 1466–1475.

Pierson, T.C. (2005). Recent findings illuminate research in West Nile virus. *Heart Disease Weekly, Atlanta*: 247.

Quattrochi, D.A. and Ridd, M.K. (1998). Analysis of vegetation within a semi-arid urban environment using high spatial resolution airborne thermal infrared remote sensing data. *Atmosphere Environment* 32 (1): 19–33.

Rainham, D.G.C. (2005). Ecological complexity and West Nile virus: perspectives on improving public health response. *Canadian Journal of Public Health* 96 (1): 34–40.

Reisen, W.K., Lothrop, H., Chiles, R. et al. (2004). West Nile virus in California. *Emerging Infectious Diseases* 10 (8): 1369–1378.

Reisen, W.K., Fang, Y., and Martinez, V.M. (2006a). Effects of temperature on the transmission of West Nile virus by *Culex tarsalis* (Diptera: Culicidae). *Journal of Medical Entomology* 43 (2): 309–317.

Reisen, W.K. et al. (2006b). Overwintering of West Nile Virus in Southern California. *Journal of Medical Entomology* 43 (2): 344–355.

Riitters, K.H., O'Neill, R.V., Hunsaker, C.T. et al. (1995). A factor analysis of landscape pattern and structure metrics. *Landscape Ecology* 10 (1): 23–39.

Rogers, D.J., Myers, M.F., Tucker, C.J. et al. (2002). Prediction the distribution of West Nile fever in North America using satellite sensor data. *Journal of the American Society for Photogrammetry and Remote Sensing* 68 (2): 112–136.

Ruiz, M.O., Walker, E.D., Foster, E.S. et al. (2007). Association of West Nile virus illness and urban landscapes in Chicago and Detroit. *International Journal of Health Geographics* 6 (1): 10–20.

Ruiz, M.O., Chaves, L.F., Hamer, G.L. et al. (2010). Local impact of temperature and precipitation on West Nile virus infection in Culex species mosquitoes in northeast Illinois, USA. *Parasites & Vectors* 3: 19.

Sanders, L. (1990). L'analyse statistique des données en géographie. Alidade-Reclus, Montpeliers.

Sannier, C.A.D., Taylor, J.C., and Campbell, K. (1998). Compatibility of FAO-ARTEMIS and NASA Pathfinder AVHRR Land NDVI data archives for the African continent. *International Journal of Remote Sensing* 19: 3441–3450.

Savage, H.M., Anderson, M., Gordon, E. et al. (2006). Oviposition activity patterns and West Nile virus infection rates for members of the *Culex pipiens* complex at different habitat types within the hybrid zone, Shelby County, TN, 2002 (Diptera: Culicidae). *Journal of Medical Entomology* 43 (6): 1227–1238.

Savard, J.P.L., Clergeau, P., and Mennechez, G. (2000). Biodiversity concepts and urban ecosystems. *Landscape and Urban Planning* 48: 131–142.

Shannon, C.E. (1948). A mathematical theory of communication. *Bell System Technical Journal* 27: 623–656.

Smith, R.M. (1986). Comparing traditional methods for selecting class intervals on choropleth maps. *Professional Geographer* 38 (1): 62–67.

Soverow, J.E., Wellenius, G.A., Fisman, D.N., and Mittleman, M.A. (2009). Infectious disease in a warming world: how weather influenced West Nile virus in the United States (2001–2005). *Environmental Health Perspectives* 117 (7): 1049–1052.

Turner, M.G. (1990). Spatial and temporal analysis of landscape patterns. *Landscape Ecology* 4 (1): 21–30.

Vogelmann, J.E., Sohl, T., and Howard, S.M. (1998). Regional characterization of land cover using multiple sources of data. *Photogrammetric Engineering & Remote Sensing* 64: 45–57.

Voogt, J.A. and Oke, T.R. (1997). Complete urban surface temperatures. *Journal of Applied Meteorology* 36: 1117–1132.

Voogt, J.A. and Oke, T.R. (2003). Thermal remote sensing of urban climates. *Remote Sensing of Environment* 86: 370–384.

Wan, Z. and Dozier, J. (1996). A generalized split-window algorithm for retrieving land-surface temperature from space. *IEEE Transactions on Geoscience and Remote Sensing* 34 (2): 892–905.

Wan, Z., Zhang, Y., Zhang, Q., and Li, Z.-L. (2004). Quality assessment and validation of the MODIS global land surface temperature. *International Journal of Remote Sensing* 25 (1): 261–274.

Wang, Y., Zhang, X., Liu, H., and Ruthie, H.K. (1999). Landscape characterization of metropolitan Chicago region by Landsat TM. *The Proceeding of ASPRS Annual Conference*, Portland, Oregon (17–21 May 1999), pp. 238–247.

Weng, Q., Lu, D., and Schubring, J. (2004). Estimation of land surface temperature-vegetation abundance relationship for urban heat island studies. *Remote Sensing of Environment* 89: 467–483.

Whitman, L. (1937). The multiplication of the virus of yellow fever in *Aedes aegyoti*. *The Journal of Experimental Medicine* 66: 133–140.

Winters, A.M., Eisen, R.J., Lozano-Fuentes, S. et al. (2008). Predictive spatial models for risk of West Nile virus exposure in eastern and western Colorado. *American Journal of Tropical Medicine and Hygiene* 79 (4): 581–590.

Wu, J., Dennis, E.J., Matt, L., and Paul, T.T. (2000). Multiscale analysis of landscape heterogeneity: scale variance and pattern metrics. *Geographic Information Sciences* 6 (1): 6–19.

Yang, K., Wang, X.H., Yang, G.J. et al. (2008). An integrated approach to identify distribution of *Oncomelania hupensis*, the intermediate host of *Schistosoma japonicum*, in a mountainous region in China. *International Journal for Parasitology* 38: 1007–1016.

Zou, L., Miller, S.N., and Schmidtmann, E.T. (2006). Mosquito larval habitat mapping using remote sensing and GIS: implications of coalbed methane development and West Nile Virus. *Journal of Medical Entomology* 43 (5): 1034–1041.

11

Impacts of Urbanization on Land Surface Temperature and Water Quality

11.1 Introduction

Urban growth induces the replacement of natural land covers with the impervious urban materials, the modifications of the biophysical environment, and the alterations of the land surface energy processes (Lo and Quattrochi 2003). Land surface temperature (LST) derived from satellite remotely sensed thermal infrared (TIR) imagery is a key variable to understand the impacts of urbanization-induced land-use and land-cover (LULC) changes. The LULC changes have been linked to the urban thermal patterns in different ways (Chen et al. 2006; Weng et al. 2006; Rinner and Hussain 2011). Weng et al. (2004) used vegetation fraction derived from the linear spectral mixture analysis and suggested that the unmixed vegetation fraction possessed a direct correlation with the radiance, thermal, and moisture properties of Earth surface that determined LST. Small (2006) performed a comparative analysis of surface reflectance and surface temperature in 24 cities. Their detailed analyses showed that the variability in the urban thermal field depended on biophysical land surface components. Amiri et al. (2009) correlated the temporal LSTs with the LULC by employing the temperature-vegetation index space in Tabriz, Iran. The study suggested that urbanization resulted in the migration of pixels in the feature space from low temperature-dense vegetation condition to the high temperature-sparse vegetation condition. Moreover, quantification of urban-induced LULC changes enabled examination of the influence of the urban landscape patterns on thermal variations. For instance, the impact of the spatial configuration and biophysical composition of LULC on urban warming can be investigated (Deng and Wu 2013; Zheng et al. 2014). These studies, based on landscape ecology, aimed at exploring the causes and consequences of thermal heterogeneity across a range of scales (Turner et al. 2001) and provided abundant information for linking patterns with processes in urban ecological studies (Luck and Wu 2002). Yet it is challenging to automatically characterize urban LULC changes consistently at an acceptable accuracy

Techniques and Methods in Urban Remote Sensing, First Edition. Qihao Weng.

(Loveland and Defries 2004; Sexton et al. 2013; Zhu and Woodcock 2014). Furthermore, thermal characteristics over time may change to respond to land-cover changes and thus become nonstationary, e.g. the mean and yearly amplitude of LSTs may change over time. A temporal analysis of thermal landscapes, therefore, requires the consideration of time-varying thermal characteristics. One way to avoid non-stationarity in modeling the temporal thermal landscape patterns is to divide time-series observations into individual segments that correspond to different land covers. As such, consistent Time Series Land Surface Temperature (TSLST) data sets are called for in revealing the urban thermal dynamics caused by land-cover conversions (Weng 2014). Nonetheless, at present, such data sets at medium spatial resolution and with regular temporal frequency are not available.

A large fraction of water quality studies explores the ramifications of historical changes in LULC on surface runoff and nonpoint source pollution loading and/or concentration (Young et al. 1996; Weng 2001; Ierodiaconou et al. 2004; Goonetilleke et al. 2005; Wilson and Weng 2010). While an adequate amount of research has been conducted on the potential impacts of future climate changes on water resources, most of these studies did not integrate future land-use configurations in their analysis (Stone et al. 2001; Imhoff et al. 2007; Zhang et al. 2007; Abbaspour et al. 2009; Bekele and Knapp 2010). Similarly, many studies that characterized the future land-use composition of an area were standalone and did not incorporate hydrologic or water quality modeling (Turner 1987; Parks 1991; Lee et al. 1992; Landis 1995; Clarke et al. 1997; Lopez et al. 2001; Petit et al. 2001; Zhang and Li 2005; Rounsevell et al. 2006). Notwithstanding the apparent lack of integration, few studies have combined the two in analyzing the potential water quality impacts (Maximov 2003; Chang 2004; Ducharne et al. 2007; Beighley et al. 2008; Tu 2009; Chung et al. 2011; Praskievicz and Chang 2011). Responses of water quality to climate and land-use changes so far remain partly understood, especially at the subbasin level. Furthermore, most of the future land characterization are either oversimplified or are not directly connected to existing land-cover composition when performing the forecasting. As a result, the synergistic impacts of future detailed urban land-use configurations and trends, under various climate emission scenarios, on surface water quality at the subbasin level are currently fuzzy.

In this chapter, two case studies will be introduced to examine the impact of LULC changes on LST and on surface water quality, respectively. First, a TSLST data set from 507 Landsat Thematic Mapper (TM)/Enhanced Thematic Mapper plus (ETM+) images of Atlanta, Georgia, between 1984 and 2011, investigated the impact of urban LULC changes on temporal thermal characteristics by breaking down the time-series observations into temporally homogenous segments. Secondly, future land-use/planning scenarios for the Des Plaines River watershed in the Chicago metropolitan area will be developed in order to evaluate the response of total suspended solids (TSS) to the combined impacts of future land-use and climate scenarios.

11.2 Impact of Urbanization-Induced Land-Use and Land-Cover Change on LST

11.2.1 Study Area

The study area consists of the metropolitan area of Atlanta, Georgia, defined by its 13 urban counties (Figure 11.1). The Atlanta region has a humid subtropical climate with abundant rainfalls evenly distributed throughout a whole year. The area is among the foothills of the Appalachian Mountains and marked by rolling hills and dense tree coverage (Gournay et al. 1993). The identified land covers are water, urban, barren, forest, shrubland, herbaceous area, planted/cultivated, and wetlands based on the National Land-Cover

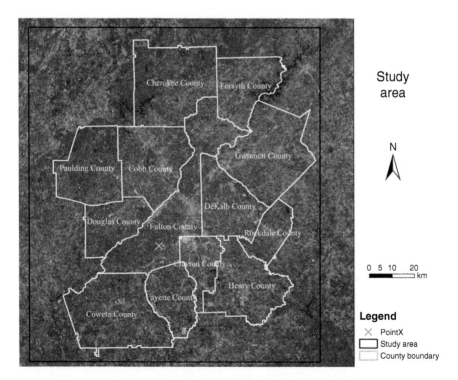

Figure 11.1 The geographical location of the Atlanta metropolitan area defined by its 13 urban counties. The time-series land surface temperatures (LSTs) around the PointX (surrounding 20 pixels) were selected to illustrate the LST decomposition procedure. *Source:* Fu and Weng (2016). Reproduced with permission of Elsevier.

Database (NLCD) 2011 (Jin et al. 2013). According to the US Census Bureau, the population of Atlanta metropolis statistical area is estimated to be 5 522 942 with an annual growth rate of 1.9%. The study area has experienced peri-urban/urban growths at the cost of forested lands in the late twentieth century, while it also witnesses the revitalization of the city's neighborhoods in the new century spurred by the 1996 Olympics and the encouragement of the compact urban growths for a "smart" city. Its emergence as the premier commercial, industrial, and transportation center of Southeastern United States gradually transforms the land covers, hydrological systems, and biodiversity (Rose and Peters 2001; Gillies et al. 2003; Miller 2012). The suburbanization/urbanization contributes to the polycentric structure of the Atlanta region landscape and pushes the rural/urban fringe farther and farther away from the original Atlanta urban core. The adverse consequences of urbanization have aroused the attention of the scientific community to focus on air pollutions, such as nitrogen oxide emissions and PM2.5 concentrations (Lo and Quattrochi 2003; Hu et al. 2013).

11.2.2 Data Sets and Image Preprocessing

All the Landsat images of Level 1T for path 19, row 37 and path 19, row 36 (WRS-2) available from 1984 to 2011 were downloaded through United States Geological Survey (USGS) online portal. The preference of the Level 1T images is because they are more precisely rectified than the Level-1 Systematic (corrected) (L1G) images. The downloaded data sets included surface reflectance and brightness temperature (BT) processed by the Landsat Ecosystem Disturbance Adaptive Processing System (LEDAPS) (Masek et al. 2006). Normalized difference vegetation index (NDVI) was derived from the surface reflectance. The images were then subject to the mosaic operation before tailoring to the study area and resampling to the 120 m. According to the metadata, the images were selected if cloud coverage in the study area was less than 90%. This percentage criterion was set to remove images that were severely contaminated by clouds. Finally, a total of 507 images from Landsat 5 TM (337 images) and Landsat 7 ETM+ (170 images) between 1984 and 2011 for the study area were collected and processed. Since the downloaded images may still contain noisy pixels, the Robust Iteratively Reweighted Least Squares (RIRLS) technique (Zhu and Woodcock 2014) was employed to further screen out the outliers by comparing model estimates with the Landsat observations.

Auxiliary reference data was collected from the NLCD 2006 and 2011 and the high spatial resolution images from Google Earth. Both stable pixels (i.e. pixels did not experience LULC changes from 2006 to 2011, totaling 150 pixels) and unstable pixels (pixels with LULC changes from 2006 to 2011, totaling 150 pixels) were randomly collected from all LULC cover types. The stable pixels were used for training the classifier and validation and then for evaluating the classification accuracy, while the unstable pixels along with the stable pixels are assessing the accuracy of change detection.

11.2.3 LULC Classification and Change Detection

The Continuous Classification and Change Detection (CCDC) algorithm (Zhu and Woodcock 2014) was applied to surface reflectance, BT, and NDVI for LULC classification and change detection. The algorithm identified LULC changes first by virtue of a time-series model consisting of the seasonal and trend components (Eq. 11.1).

$$\hat{P}(i,d) = a_i + b_{1i} \cos\left(\frac{2\pi d}{T}\right) + b_{2i} \sin\left(\frac{2\pi d}{T}\right) + c_i d \tag{11.1}$$

where $\hat{P}(i,d)$ is the predicted value for reflectance, BT, or $NDVI$ at Julian day d (the continuous count of days from the beginning of the Julian Period), a_i and c_i are the coefficients for the trend component, b_{1i} and b_{2i} are the coefficients for the seasonal component, and T is 365. The two coefficients (b_{1i} and b_{2i}) were used to implicitly include the phase (when the sine and cosine functions were combined) and thus the fitting procedure was simplified to the linear optimization. The model began to detect land-cover change when there are 15 clear-sky observations (the model initialization phase). The first 12 observations were used to determine outliers. Since there are four coefficients in Eq. (11.1), 12 clear observations (three times of the number of the model coefficients) made model estimation robust and accurate. The last three observations allowed the model to determine whether changes in land cover occurred. The basis of the change detection algorithm was to compare model predictions with cloud-free satellite observations and to normalize their differences by three times of the root-mean-square error ($RMSE$). The land-cover change was flagged if the normalized difference exceeded the predefined thresholds as showed in Eq. (11.2) in three consecutive days.

$$\frac{1}{k} \sum_{i=1}^{k} \frac{\left| P(i,d) - \hat{P}(i,d) \right|}{3 \times RMSE_i} > 1 \tag{11.2}$$

where k is the dimension of the data sets used in the algorithm. The use of three times $RMSE$ considered that LULC change typically occurred when the spectral signature deviated from model prediction by more than three times of the $RMSE$ (Zhu and Woodcock 2014). The normalization interval was dependent on the number of clear-sky observations that were used in the fitting procedure. Given that LULC change may happen in the first and last observations or during the process of the model initialization for the first 12 observations, another three conditions (Eqs. 11.3a–11.3c) were set to identify abnormal observations for the first and last observations of the model initialization and the observations between them.

$$\frac{1}{k}\sum_{i=1}^{k}\frac{\left|P(i,d_1)-\hat{P}(i,d_1)\right|}{3\times RMSE_i}>1 \tag{11.3a}$$

$$\frac{1}{k}\sum_{i=1}^{k}\frac{\left|P(i,d_n)-\hat{P}(i,d_n)\right|}{3\times RMSE_i}>1 \tag{11.3b}$$

$$\frac{1}{k}\sum_{i=1}^{k}\frac{\left|c_i(d)\right|}{3\times\left(RMSE_i/T_{model}\right)}>1 \tag{11.3c}$$

where d_1 and d_n are the Julian days for the first and last observation during model initialization, and T_{model} is the time range between the first and the last observation of the model initialization. Once the model initialization phase was finished, it was used as the basis for land-cover change detection and classification.

The classification was accomplished through a Random Forest Classifier with inputs of the coefficients derived from the time-series model. The principle of the classification supposed that different land covers showed different modeling characteristics (i.e. different coefficients for seasonal and trend components and modeling errors) for reflectance, *NDVI* and *BTs*. The variables, $\bar{P}$, b_{1i}, b_{2i}, c, and *RMSE*, derived from the time-series model were used as inputs for the classification with $\bar{P}$ representing the mean value:

$$\bar{P}=a+c\times\frac{t_1+t_n}{2} \tag{11.4}$$

where t_1 and t_n are the Julian dates for the starting and ending time of the model, respectively. The Random Forest Classifier technique (Breiman 2001) was utilized to train the reference data, and the trained classifier was applied to the whole time-series data set on a pixel-by-pixel basis to derive LULC maps over the time.

11.2.4 Decomposition Analysis of LST

The occurrence of LULC changes may induce non-stationarity in a TSLST data set since surface thermal responses are different before and after land-cover conversions. To analyze the time-varying surface thermal characteristics, the TSLST data set were first divided into temporally homogeneous segments in which the mean LST value and the amplitude value were constant. Then, a decomposition scheme of a time-series additive model comprising the seasonality component and the trend component (Eq. 11.5) was applied to both stable and unstable pixels. The LOcally wEighted regreSsion Smoother (LOESS) scheme was utilized for decomposition due to its ability to reduce the outlier effects (Cleveland et al. 1990).

$$Y_t = T_t + S_t + \varepsilon$$

$$T_t = a + bt$$

$$S_t = \sum_{n=1}^{N}\left[c_{1n}\sin\left(\frac{2\pi nt}{T}\right) + c_{2n}\cos\left(\frac{2\pi nt}{T}\right)\right] \tag{11.5}$$

where a and b are the coefficients for the fitted linear trend, N is the number of harmonic items used, t is the Julian day, c and T are the coefficient and periodic frequency for the harmonic item, ε is the noise, S_t is the seasonality component, T_t is the trend component, and Y_t is the time-series observations. A linear change was assumed to characterize the gradual change over years (Verbesselt et al. 2010). Eventually, the number of the harmonic terms (variable N) was adopted due to its ability to deal with the irregular spaced remote sensing time series and to characterize the complex periodic patterns. The decomposition procedure was applied to each pixel in the study area. Aiming to globally minimize $RMSEs$ for all pixels, the variable N was set to 2 with the average $RMSE$ of 4.8 K for the whole study area. In addition, amplitude values were calculated for each land cover based on the equation $\sqrt{\sum_{i=1}^{2}\left(c_{1i}^2 + c_{2i}^2\right)}$ (Table 11.3). The annual and semiannual amplitudes (Table 11.4) were calculated using the equation $\sqrt{c_{1i}^2 + c_{2i}^2}$ ($i = 1$ for annual and 2 for semiannual variations). Figure 11.2 shows LST decomposition procedure using mean time-series data at PointX (Figure 11.1) where land cover changed from evergreen forest to urban-low intensity on 12 June 2007. Figure 11.1a shows the satellite observations and modeling results in the two segments (evergreen forest and urban-low intensity). Figure 11.1b–d shows the seasonality component, trend component, and modeling residues (differences between model prediction and satellite observations), respectively.

11.2.5 Results

11.2.5.1 LULC Analysis

The NLCD 2011 classification scheme was adopted in this study. The land covers classified included water, urban-low intensity, urban-medium intensity, urban-high intensity, barren land, shrubland, herbaceous area, pasture, crops, and woody wetlands. The "urban open space" cover, consisting of vegetation in

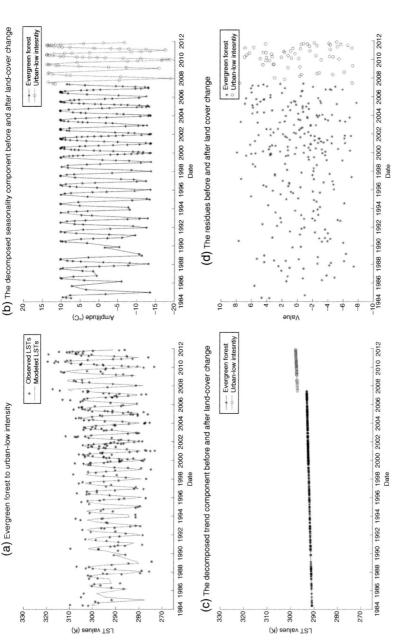

Figure 11.2 LST decomposition procedure using mean time-series data at PointX (Figure 11.1) where land cover changed from evergreen forest to urban-low intensity on 12 June 2007. (a) shows the satellite observations and modeling results in the two segments (evergreen forest and urban-low intensity). (b)–(d) show the seasonality component, trend component, and modeling residues (differences between model prediction and satellite observations), respectively. *Source:* Fu and Weng (2016). Reproduced with permission of Elsevier.

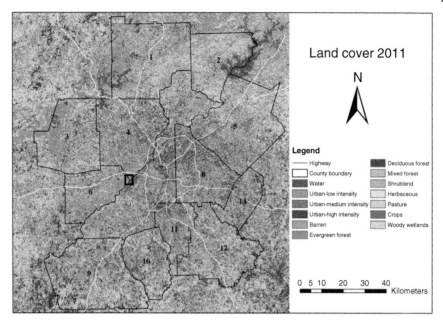

Figure 11.3 Land-use and land-cover (LULC) map in 2011. The rectangle area was selected to examine the typical land-cover change from evergreen forest to urban-medium intensity. 1, Cherokee County; 2, Forysth County; 3, Paulding County; 4, Cobb County; 5, Gwinnett County; 6, Douglas County; 7, Fulton County; 8, Dekalb County; 9, Cowetta County; 10, Fayette County; 11, Clayton County; 12, Henry County; and 13, Rockdale County. *Source:* Fu and Weng (2016). Reproduced with permission of Elsevier.

the form of lawn grasses, was excluded since its selection and identification were contextual (Sexton et al. 2013) and merged into the type of urban-low intensity. The "emergent herbaceous wetlands" was also disregarded because it was rare in the study area. Figure 11.3 shows the spatial patterns of land covers in 2011 derived from the CCDC algorithm. The classification accuracy was assessed by randomly splitting the selected 150 stable pixels into two groups (120 pixels and 30 pixels, respectively) for training and cross validation. The process was repeated for 100 times in the Random Forest. The accuracy of classification is shown in Table 11.1. The land-cover classification yielded the user's accuracy of 89% and the producer's accuracy of 78% and above. The large commission errors occurred among barren land, mixed forest, shrubland, and herbaceous area, while the large omission errors existed among mixed forest and herbaceous area. The 150 unstable pixels, selected between 2006 and 2011, were used with the previous 150 stable pixels to assess the accuracy of change detection. Table 11.2 shows the user's accuracy of 91% and producer's accuracy of 94% for stable pixels, and the user's accuracy of 94% and producer's accuracy of 91% for unstable pixels. The overall accuracy for the change

Table 11.1 Accuracy of land-cover classification.

	W	UL	UM	UH	B	EF	DF	MF	S	H	P	C	WW	UA
W	290	0	0	0	0	0	0	0	0	0	0	0	2	0.98
UL	0	295	8	0	0	0	3	0	0	8	6	0	0	0.92
UM	0	14	186	15	3	0	0	0	0	0	0	0	0	0.85
UH	0	0	14	329	7	0	0	0	0	0	0	0	0	0.94
B	0	9	11	12	140	4	0	0	0	0	0	1	0	0.79
EF	0	0	0	0	0	286	0	7	0	7	0	0	18	0.90
DF	0	2	0	0	0	0	309	4	0	3	0	0	0	0.97
MF	0	0	0	0	0	6	14	113	0	0	0	0	7	0.79
S	0	0	0	0	0	11	0	0	56	4	0	0	0	0.79
H	0	5	0	0	0	4	5	4	3	90	5	0	0	0.78
P	0	7	0	0	0	0	0	0	0	0	127	13	0	0.86
C	0	6	0	0	5	0	0	0	0	0	9	98	0	0.83
WW	2	0	0	0	0	20	0	7	0	0	0	0	173	0.87
PA	0.98	0.87	0.85	0.92	0.90	0.86	0.93	0.85	0.95	0.80	0.86	0.88	0.87	0.89

Source: Fu and Weng (2016). Reproduced with permission of Elsevier.
Note: W, water; UL, urban low intensity; UM, urban medium intensity; UH, urban high intensity; B, barren land; EF, evergreen forest; DF, deciduous forest; MF, mixed forest; S, shrubland; H, herbaceous; P, pasture; C, crops; WW, woody wetlands; PA, producer's accuracy; and UA, user's accuracy.

detection was 94%. The higher user's accuracy than producer's accuracy for the unstable pixels suggested that the algorithm created more commission errors than the omission errors. The omission errors were likely caused by partially changed pixels since changes within these pixels were difficult to detect. Another reason could probably be the model initialization phase, requiring the screening of the cloudy pixels as well as the identification of possible LULC changes. If LULC change occurred within the model initialization, change detection may not work well or even incorrectly. The commission errors may be caused by the consecutive missed cloudy pixels. Consecutive cloudy pixels could result in the inaccurate or even false change detection and thus affect the land-cover classification accuracy.

In addition, both stable (50 pixels) and unstable pixels (50 pixels) were randomly selected from urban covers and used to evaluate the accuracy for the three urban-related land changes. Table 11.2 shows the overall accuracy of 82% for urban change detection, lower than the overall accuracy for all categories of LULC changes. Urban areas were highly heterogeneous; fitting time-series model to urban reflectance, BT, and NDVI can result in large *RMSE* values (*RMSE* value in urban areas is 2.5 K larger on average than that in nonurban areas). Since the change detection was dependent on the normalization of the *RMSE*, large *RMSE* values may lead to the change detection insensitive for urban pixels. Furthermore, urban-low and urban-medium intensity covers were frequently mixed by impervious surfaces and vegetation. The accuracy of change detection depended largely on the proportion of the changes within mixed pixels. In sum, the overall accuracy of 88 and 82% in urban classification and change detection, respectively, suggested that the CCDC can be effectively used to monitor urban expansion and growth over the decadal years.

Table 11.2 Accuracy assessments of change detection.

	Stable pixels	Changed pixels	User's accuracy
All land covers			
Stable pixels	136	14	91%
Unstable pixels	9	141	94%
Producer's accuracy	95%	91%	Overall: 92%
Urban			
Stable pixels	42	8	84%
Unstable pixels	10	40	80%
Producer's accuracy	81%	83%	Overall: 82%

Source: Fu and Weng (2016). Reproduced with permission of Elsevier.
Note: Stable pixels, pixels without land-use and land-cover (LULC) changes over time. Unstable pixels, pixels have LULC changes over time.

Further analysis showed LULC changes occurred almost in the whole study area, except for the central metropolitan and the northwestern areas. Urban-related land-cover conversions had more than 1000 pixels and mainly occurred outside the encircled beltways in Fulton and Gwinnett counties. Here urban-related land conversions mean the land-cover changes to one of the three urban covers. The Fayette and Rockdale Counties had fewer pixels, around 200, converted to urban land covers. The remaining counties, including Cherokee, Clayton, Cobb, DeKalb, Douglas, Henry, Coweta, Forsyth, and Paulding, revealed a relative similar amount of land-cover conversions to urban, ranging from 400 to 700 pixels. The urban-related land-cover conversions accounted for nearly 0.9% of the study area (approximately 16553 ha). The urban growth was mainly conversed from deciduous forest, mixed forest, evergreen forest, and herbaceous areas. Eventually, 28 annual land-cover maps from 1984 to 2011 were compiled and available as video data (currently, it is available at YouTube link https://www.youtube.com/watch?v=NlOVYGfCP54).

11.2.5.2 Thermal Temporal Signatures for LULC Changes

Table 11.3 indicated the existence of the multi-timescale urban heat island intensities as well as the trend difference between urban and nonurban areas, and the impact of urbanization-induced land-cover changes on the LST variations over the time (temporal thermal signatures) has not been evaluated. Different types of urban-related land-cover conversions can be discerned to uncover the temporal thermal signatures corresponding to the major pathways for urban expansion. Table 11.4 shows the mean LST values for different types of urban land conversions. For each conversion type, the first row in the Table 11.4 refers to the mean LST value of nonurban land cover, and the second row refers to the mean LST value for the urban land cover. Each conversion type was characterized by the annual, semiannual, and gradual change coefficients of LST as described by Eq. (11.5). Major urban land-cover change types included: conversions of evergreen forest to urban-low and urban-medium intensity, deciduous forest to urban-low and urban-medium intensity, mixed forest to urban-low and urban-medium intensity, herbaceous area to urban-low intensity, urban-low intensity to urban-high intensity, and urban-medium intensity to urban-high intensity. Table 11.4 suggests that the greatest LST change, in terms of the amplitude of annual coefficients, was observed in the conversion type of evergreen forest to urban-medium intensity with the annual differences of 5.7 K. In contrast, the conversion of urban-medium to urban-high intensity land found the smallest annual variation difference. This was not surprising because both cover types shared similar thermal characteristics as evident in Table 11.3. Table 11.4 further suggests that LST alterations caused by nonurban to urban conversions were always larger than those caused by the conversion of urban-low, or urban-medium intensity, to urban-high intensity. With respect to the semiannual variations, all of the urban land conversions

Table 11.3 Mean seasonality and trend component by land-cover type.

| | Seasonality component | | | | | Trend component | |
| | Annual coefficients | | Semiannual coefficients | | Amplitude | Mean value | Gradual change (e-4) |
	C_{11}	C_{12}	C_{21}	C_{22}	$\sqrt{\sum_{i=1}^{2}(c_{1i}^2 + c_{2i}^2)}$	a	b
Water	10.6 (−6.1)	5.0 (0.7)	0.7 (−1.1)	0.3 (0.2)	11.7	289.9 (−5.0)	3 (6)
Urban-low intensity	13.5 (−3.2)	3.3 (−1.0)	1.8 (0)	0.1 (0)	14.0	292.3 (−2.6)	6 (3)
Urban-medium intensity	15.5 (−1.2)	4.3 (0)	1.8 (0)	0.1 (0)	16.2	293.3 (−1.6)	8 (1)
Urban-high intensity	16.7 (0)	4.3 (0)	1.8 (0)	0.1 (0)	17.3	294.9 (0)	9 (0)
Barren	16.3 (−0.4)	4.7 (0.4)	2.4 (0.6)	0.1 (0)	17.1	294.2 (−0.7)	6 (3)
Evergreen forest	11.3 (−5.4)	2.5 (−1.8)	2.2 (0.4)	0.1 (0)	11.8	290.1 (−4.8)	5 (4)
Deciduous forest	12.3 (−4.4)	3.1 (−1.2)	1.8 (0)	0.3 (0.2)	12.8	291.0 (−3.9)	4 (5)
Mixed forest	11.6 (−5.1)	2.9 (−1.4)	1.9 (0.1)	0.2 (0.1)	12.1	290.3 (−4.6)	4 (5)
Shrubland	12.5 (−4.2)	2.8 (−1.5)	2.3 (0.5)	0.2 (0.1)	13.0	291.6 (−3.3)	7 (2)
Herbaceous	12.5 (−4.2)	3.1 (−1.2)	1.8 (0)	0.2 (0.1)	13.0	290.6 (−4.3)	6 (3)
Pasture	13.2 (−3.5)	3.7 (−0.6)	1.8 (0)	0.1 (0)	13.8	292.2 (−2.7)	6 (3)
Crops	14.3 (−2.4)	3.6 (−0.7)	2.1 (0.3)	0.7 (0.6)	14.9	292.9 (−2.0)	7 (2)
Woody wetlands	11.2 (−5.5)	2.8 (−1.5)	1.9 (0.1)	0.2 (0.1)	11.7	290.4 (−4.5)	4 (5)

Source: Fu and Weng (2016). Reproduced with permission of Elsevier.

Note: The unit is Kelvin (K). The values in the brackets were calculated by subtracting the values of the urban-high intensity, i.e. keep the value of the urban-high intensity as reference (0), and then subtract the reference value for each land cover in each column. The first two coefficients (C_{11}, C_{12}) represent the annual variations, the other two coefficients (C_{21}, C_{22}) characterize the semiannual variations, and the last two coefficients (a, b) are the mean and the gradual variations within the trend component. The number in the "gradual change" column has the exponential notation of e-4.

Table 11.4 Thermal signatures of major urban-related land conversions.

| Change types | Seasonality component | | | | | | Trend component | |
| | Annual coefficients | | | Semiannual coefficients | | | Gradual change | |
	C_{11}	C_{21}	Annual amplitude	C_{21}	C_{22}	Semiannual amplitude	a	b (e-4)
EF to UL	11.8	2.2	12.0	1.8	0.3	1.8	290.8	4
	13.5	3.3	13.9	2.0	0.3	2.0	292.9	5
EF to UM	10.5	3.0	10.9	1.4	0.8	1.6	290.5	4
	16.1	4.2	16.6	2.1	0.4	2.1	293.9	8
DF to UL	12.0	3.3	12.4	1.4	0.5	1.5	289.7	5
	14.8	3.5	15.2	2.6	0.1	2.6	292.0	6
DF to UM	12.3	2.9	12.6	1.9	1.2	1.9	288.5	5
	15.8	4.4	16.4	2.5	0.3	2.5	292.7	7
MF to UL	11.7	2.8	12.0	1.4	0.6	1.5	288.9	4
	14.4	4.0	14.9	1.7	0.2	1.7	292.1	6
MF to UM	12.4	3.3	12.8	1.4	0.3	1.4	289.5	3
	15.7	5.0	16.5	1.6	0.5	1.7	291.0	6
H to UL	11.9	3.1	12.3	1.5	0.1	1.5	291.3	5
	15.3	3.9	15.8	1.6	0.9	1.8	294.2	8
UL to UH	14.8	3.3	15.2	1.5	0.2	1.5	290.8	6
	16.7	5.6	17.6	2.0	0.1	2.0	294.1	8
UM to UH	15.7	3.5	16.1	1.4	0.6	1.5	296.8	5
	16.2	5.6	17.1	2.1	0.1	2.1	299.8	9

Source: Fu and Weng (2016). Reproduced with permission of Elsevier.
Note: EF, evergreen forest; UL, urban low intensity; UM, urban medium intensity; UH, urban high intensity; DF, deciduous forest; MF, mixed forest; and H, herbaceous. To illustrate the impact of urbanization-induced LULC changes, pixels changed to urban land covers, i.e. urban-low intensity, urban-medium intensity, or urban-high intensity are used to show the average thermal alterations corresponding to the individual land-cover change types. C_{11} and C_{21} are the coefficients of the annual variations; C_{21} and C_{22} are the coefficients of the semiannual variations; and a and b are the coefficients for the trend component. The amplitudes of both the annual and semiannual coefficients were calculated according to the descriptions in Section 3.2. The unit is Kelvin (K). For each conversion type, the first row refers to the mean LST value for the nonurban land cover, and the second row refers to the mean LST value for the urban land cover. The coefficient b has the exponential notation e-4.

showed a consistent increase in LST signature, with difference in the magnitude ranging from 0.2 K (evergreen forest to urban-low intensity) to 1.1 K (deciduous forest to urban-medium intensity). The mean value of the trend component suggested that LST increase was dependent on the change type and the largest temperature increase, by 4.2 K, was observed for the conversion of deciduous forest to urban-medium intensity. Since the statistics in Table 11.3 were based on all the pixels, it was observed that the coefficients of the gradual change for different urban land-cover conversions in Table 11.3 were slightly different from those in Table 11.4. Table 11.4 shows that conversion of evergreen forest to urban-medium intensity yielded the largest increase in trend value by 0.0004 K/day.

Figure 11.4 was created to show the thermal trajectories over the time. The representative area (the rectangle area in Figure 11.3) was selected to illustrate the LST change from evergreen forest to urban-medium intensity from 1984 to 2011. Figure 11.4c shows LST variation for that change. The mean LST value over the decadal years increased by from 290.4 to 297.5 K, or by approximately 7 K. Figure 11.4b shows the mean LST value of 297.5 K, which is larger than 293.3 K in Table 11.3 and the overall value of 293.9 K in Table 11.4. This is likely caused by the inclusion of urban-high intensity in the analysis owing to the representative areas selected near the land cover of the urban-high intensity and the similar spectral and thermal characteristics between urban-high and urban-medium intensity. However, the coefficients of the gradual change as well as the seasonality component from Figure 11.4c for the land covers of evergreen forest and urban-medium intensity were in accordance with the statistics in both Tables 11.3 and 11.4. Thus, the impact of urbanization-induced LULC changes on surface thermal properties must be assessed from both seasonality and trend components.

11.2.6 Discussion and Conclusions

With 507 Landsat images, the current study enabled a historical reconstruction of LULC changes from 1984 to 2011 and an examination of the impact of human-induced LULC changes on surface thermal properties from a time-series perspective. Compared with previous studies (Weng et al. 2004; Chen et al. 2006; Small 2006), the present research contributed to the analysis of landscape thermal patterns by using the sequential Landsat images. The LST decomposition analyses within homogenous temporal segments suggested that the multi-timescale UHI effect could be estimated from the four coefficients of the seasonal component derived from an additive model. Urban high-intensity land exhibited the largest intra- and interannual LST variations, by 17.4 and 0.0009 K/day, respectively, indicating that any land conversion to the urban cover could increase LST over the time. In addition to the increase pattern of the trend component for each land cover, this study also revealed

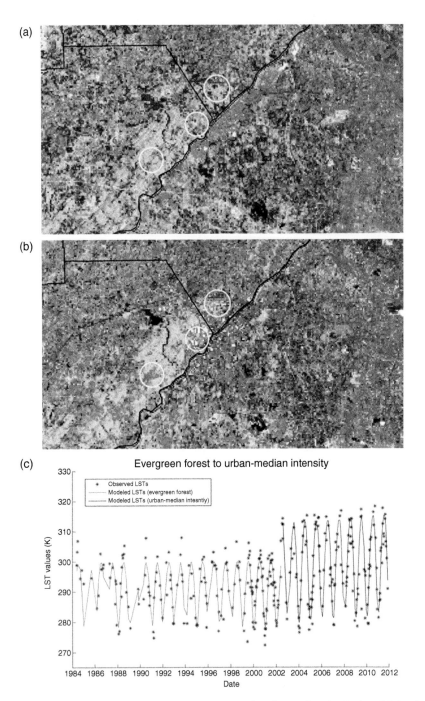

Figure 11.4 Landsat imagery color composite (R = band 5, G = band 4, B = band 3) for date 14 June 1984 (a) and date 30 May 2011 (b), and modeled LST variation for the pixels changes from evergreen-forest to urban medium-intensity land as (c). The area in (a) and (b) refers to the rectangular area shown in Figure 11.3 (black rectangle). The pixels used for (c) come from the circles. These pixels were characterized by the land-cover conversion of evergreen forest to urban-medium intensity. Counties shown in (a) and (b) include 3, Paulding County; 4, Cobb County; 6, Douglas County; and 7, Fulton County. *Source:* Fu and Weng (2016). Reproduced with permission of Elsevier.

that the mean difference of trend component between urban and nonurban areas was 1.8 K/decade. The larger trend change of the urban land covers suggested LSTs were influenced by the urbanization-induced land-cover changes. Furthermore, temporal thermal signatures were created and computed to illustrate how LST pattern corresponded to the major categories of urban LULC changes. The conversion of evergreen forest to urban-medium intensity was found to result in the largest increase in annual temperature amplitude (by 5.7 K) and the largest trend increase by 0.0004 K/day.

The availability of the large amount of Landsat data necessitates reconsidering the traditional algorithms and techniques of land-cover classification and changing detection. This study applied the CCDC algorithm to reflectance, BT, and NDVI for LULC classification and change detection. The algorithm was capable of providing land-cover classification and change detection by using all Landsat images available for the study area. It maximized the number of images that could be used for LULC classification and change detection, and the average accuracy of classification and change detection was 89 and 92%, respectively, for the study area. However, the effectiveness of the algorithm was subject to the number of clear-sky pixels available for the study area. The heterogeneity of urban materials and the large *RMSE* value resulting from the time-series modeling may explain the relative lower accuracy of the urban change detection. Parameters, such as the three times of *RMSE* and the number of harmonic frequency, can also affect the performance of change detection and classification. Therefore, further efforts will be made to evaluate the classification and change detection in different study areas.

11.3 Simulating the Impacts of Future Land-Use and Climate Changes on Surface Water Quality

11.3.1 Study Area

The study was conducted within the Chicago Metropolitan Statistical Area (MSA) located in northeastern Illinois (Figure 11.5). The study examined that portion of Des Plaines River watershed lies within the Chicago MSA. The watershed is 2055.9 km^2 and spans four counties within the Chicago MSA. The watershed lies between 41°11′27.5″ and 42°28′54.2″ north at the border with the state of Wisconsin, while its longitudinal expanse lies between 87°42′24.1″ and 88°15′28.9″ west.

The topography of the watershed is relatively flat and low in elevation. Altitude varies between 164.8 and 244.1 m above sea level. Most of the area is less than 213 m. Chicago MSA is located within the humid continental warm summer climatic region. Summers are warm and humid with an average July temperature of 23.8 °C while January temperature in winter

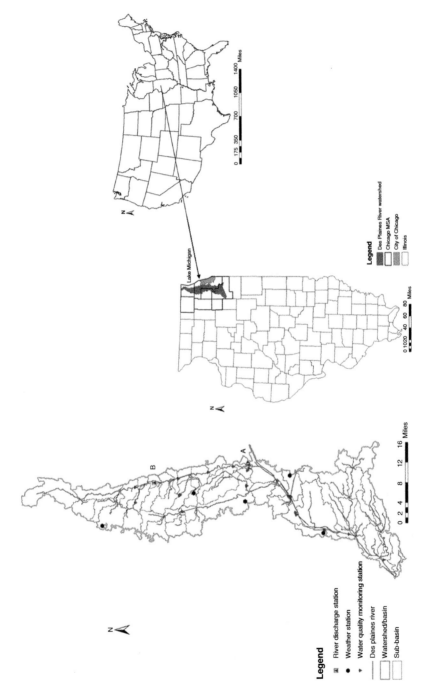

Figure 11.5 Des Plaines River watershed, Chicago, Illinois. *Source:* Wilson and Weng (2011). Reproduced with permission of Elsevier.

averages around −4.2 °C; average precipitation for the region is 863 mm (Cutler 2006). Summer is characterized by rainfall while the region receives more snow fall than rainfall during the winter period.

11.3.2 Land-Use and Climate Data

Three Landsat TM images were used in developing future land-use maps for the Des Plaines River watershed. The images were acquired by Landsat 5 satellite on 1 June 1990, 3 May 2000, and 23 May 2010, respectively. To assess the accuracy of classification of the TM images, high-resolution aerial photographs, thematic topographic maps and Chicago Metropolitan Agency for Planning (CMAP) zoning maps were used (CMAP 2010a). Other data used in the processing of TM images includes digital raster graphic (DRG) images and United States Decennial Census population and housing unit data at the block-group level for 1990 and the block level for 2000.

The Department of Agriculture Natural Resource Conservation Service (NRCS) Soil Survey Geographic (SSURGO) data provided soil data. Elevation data was solicited from USGS 10-m Digital Elevation Model (DEM). Climate data include National Environmental Satellite Data and Information Service (NESDIS) rainfall, snow water equivalent, and temperature data for 2006–2010, while future temperature and precipitation data were obtained from the Engineering Department at the University of Santa Clara at www.engr.scu. edu/~emaurer/global_data. This spatially downscaled climate data is based on global climate model output of the World Climate Research Program's (WCRP's) Coupled Model Intercomparison Project phase 3 (CMIP3) multi-model data set (Meehl et al. 2007). The data is available in a spatially downscaled format as described by Maurer et al. (2009) using the bias-correction/spatial downscaling method (Wood et al. 2004) to a 0.5° grid, based on the 1950–1999 gridded observations of Adam and Lettenmaier (2003). Continuous stream flow and monthly grab samples of water quality data that spans 2006–2010 were obtained from USGS archival station discharge data and Metropolitan Water Reclamation District of Greater Chicago water quality data, respectively.

11.3.3 Landsat Image Processing and Derivation of Future Land-Use Maps

Landsat TM images were corrected for geometric distortion by first-order polynomial method using nearest neighbor algorithm. The images were then corrected for atmospheric interference with the use of dark-object subtraction method (Lu et al. 2002; Jensen 2005). A hybrid approach was implemented in image classification. In stage one, an unsupervised classification using Iterative Self-Organizing Data Analysis (ISODATA) classifier was employed.

ISODATA classifier was used in stage one because the classification accuracy obtained from per-pixel supervised classifiers was slightly lower than that attained from ISODATA clustering. In state two, a decision tree/expert system post-classification sorting and further division of some land classes obtained during stage one of image classification was employed (Lawrence and Wright 2001; Wentz et al. 2008; Kahya et al. 2010). Twelve land-use classes were produced. Image classification accuracy ranged between 89 and 91% and exceeded the minimum threshold established for remotely sensed images (Congalton 1991).

In developing 2020 and 2030 land-use/planning scenario maps for the Des Plaines River watershed, a multi-perceptron neural network built on Markov chain modeling method embedded in Idrisi's Land Change Modeler (LCM) was employed. An urban landscape is characterized by various states or land-use configurations which itself is dynamic through time. In modeling urban LULC over time, a number of transition probabilities have to be developed for each direction of change (Weng 2002). Four drivers of land-use change were included in the prediction of future land-use maps for the study area. The drivers encompass evidence likelihood of change between 2000 and 2010, land value, crime level, and distance from the City of Chicago. Evidence likelihood is an empirical probability of change between an earlier land-use image and a later image. All land-use change drivers were tested for their predictive power to forecast future images. A two-stage approach was adopted in developing future land-use maps. The first stage used the classified land-use map for 1990 and 2000 to simulate a land-use map for 2010. This was done in order to ensure that the multi-perceptron neural network in LCM understood the aforementioned drivers of land-use change in the watershed and also to verify the prediction of the model's land-use maps. The 1990 and 2000 land-use maps served as observed data for calibration of LCM while the 2010 classified land-use map was used to verify the simulated map for 2010. The neural network was retrained until the simulated 2010 land-use map matched with the classified map. At the completion of this exercise, the 2000 and 2010 classified land-use maps were used to simulate three sets of land-use maps for 2020 and 2030, respectively, based on three land-use/planning scenarios.

In Scenario One (SN1), incentive was given to the development of low-density residential (LDR) and open space/vegetation land use. This was done to reflect one major goal of CMAP in the 2040 Regional Plan for Chicago Metropolitan Statistical Area (CMAP 2010b).

In Scenario Two (SN2), the incentive given to the growth of LDR land class in SN1 was removed in order to allow it to respond fully to the land-use change drivers used in the study. Limited incentive was given to the development of commercial/urban land use. Incentive given to the growth of open space/vegetation was reduced compared to that established for SN1.

In Scenario Three (SN3), higher incentive was given to the growth of commercial/urban mix land class. Slight disincentive was given to the transition of agricultural land. All other land-use classes had their model parameters set to normal in order to allow them to respond to the drivers of land-use change.

Owing to the multiplicity of urban land uses and its highly dynamic nature within the Des Plaines River watershed, a total of 132 transitions took place between 2000 and 2010. To simplify the modeling process, only 41 transitions were included and executed in the three separate runs. Each transition of land use was run in separate sub-model under LCM's transition potential routine. A comprehensive land-use predicted map for 2020 and 2030 was executed. At the final stage, the three comprehensive predicted maps that were generated during the forecasting exercise were combined with the use of Expert System in Erdas Imagine to arrive at one comprehensive predicted land-use map for each land-use scenario. This exercise was undertaken in order to capture most of the significant transitions that are likely to take place in the watershed between 2010 and the future.

11.3.4 SWAT Hydrologic and Water Quality Model

11.3.4.1 Model Construction

The Soil and Water Assessment Tool (SWAT) was developed to aid the evaluation of land management practices on water supplies and nonpoint source pollution loading in watersheds and large river basins over long periods of time (Arnold et al. 1993, 1998). SWAT is an operational fully distributed model that operates on a daily time step but can be cumulated to monthly output. The model uses a command structure to route runoff and nonpoint source pollutants through a watershed. SWAT is divided into several subcomponents which encompass hydrology, weather, sedimentation, soil temperature, crop growth, nutrients, pesticides, and agricultural management. The hydrologic subcomponent can be further divided into surface and ground water hydrology. The surface hydrology routine was employed in this study. SWAT model requires numerous data which encompasses land use, soil, elevation, precipitation, temperature, humidity, stream flow, and water quality input to facilitate surface water flow and quality modeling. In this study, we employed SWAT version 2009.93.5. An extensive description of the SWAT model can be found in Arnold et al. (1998).

Contemporary and future climate data were spatially distributed among five weather stations within the Des Plaines River watershed (Figure 11.5). Future climate data that represents the Intergovernmental Panel on Climate Change (IPCC) Special Report on Emission Scenario (SRES) B1 and A1B scenarios were used in model construction and simulation. Each SRES assumes different trajectory of future economic growth and energy consumption with varying effects on the emission of greenhouse gases and aerosol precursor (IPCC 2001). The A1B group assumes that the world will experience rapid economic growth conditioned with the introduction of nonfossil fuel technologies that will create a balance between fossil and nonfossil energy. Whereas the B1 category envisages a world where there is a rapid change in production geared towards reduction in material intensity and the introduction of clean technologies that will replace fossil fuel energy. We employed Panoply program to visualize and extract that portion of future climate data that covers the study area (NASA 2010). Future temperature and precipitation data were transformed before inputting into SWAT to simulating nonpoint source pollution loading for 2020 and 2030. A stochastic weather generator approach was employed to temporally downscale the average monthly WCRP CMIP spatially downscaled climate data to daily precipitation and temperature time step required by SWAT (Wilby et al. 2004; Wilks 2010). In deriving daily future precipitation and temperature, a modified version of the chain-dependent process (a stochastic weather generator method) was employed. Following this step, the daily downscaled climate variables were summed and averaged at a monthly time step to verify its agreement with the spatially downscaled WCRP CMIP data. Areas of disagreement between the two were forced to reflect the WCRP CMIP values by simply adding and subtracting from points of under and over prediction, respectively. This was implemented to avoid the problem of "overdispersion phenomenon," wherein the variability of the two sometimes differ (Qian et al. 2008). For a detailed description of the mathematics behind the weather generator chain-dependent approach to temporal downscaling of climate variables, please refer to the works by Wilby et al. (2004) and Wilks (2010).

Seven USGS stream flow stations along the Des Plaines River were incorporated in the model to facilitate calibration and validation of flow. Twenty-two water quality stations were also included in the model to calibrate, validate, and enable comparison of water quality model results of TSS concentrations at the sub-watershed scale (Figure 11.5). Stream discharge data and water quality data spans 2006–2010 water years. The 2006 and 2007 data were used as "warm up"/initializing years for the SWAT model, while the model was calibrated between 2008 and 2009 and validated for the 2010 water year. SWAT "warm up" is essential to facilitate a fully functional model that reflects real-world basin hydrology. For the 2020 and 2030 water years, model simulation started four years prior to those targeted water years in order to maintain a similar framework adopted for the 2010 water year.

A total of 13 SWAT models were independently constructed in order to effectively evaluate the impacts of land-use and climate change on surface water quality. The temporal distribution of models includes one for 2010, and six for 2020 and 2030. For 2020 and 2030, each model utilized the future land-use/planning scenario maps under the climate variables SRES A1B and B1 scenarios. For example, each land-use/planning scenario (SN1, SN2, and SN3) was modeled under future climate (B1 and A1B) emission scenarios. This approach was adopted to compare the impacts of combined land-use and climate change on surface water quality. The watershed was automatically delineated with the use of a 10-m DEM in SWAT ArcGIS extension resulting to 58 subbasins and 1603 hydrologic response units (HRUs). An HRU is a contiguous land area within each subbasin that has uniform land use, soil, slope, and management combination that drains directly into the subbasin (Neitsch et al. 2004). Following model calibration and validation, final simulation runs were set up to produce mean monthly output in order to remove the biases caused by differences in daily distribution of precipitation and temperature between contemporary and future climate data.

11.3.4.2 Sensitivity Analysis

Sensitivity analysis was performed on SWAT to ascertain model parameters that experienced the largest change with variations in model input. Parameters that control flow, sediment, and water quality were included in the analysis. Forty-one parameters were tested for their level of sensitivity. A hybrid sensitivity analysis method that combines Latin Hypercube (LH) and One-Factor-At-a-Time (OAT) sampling algorithm was employed. LH allows the selection of a set of parameter values within a unique set of parameter space during each loop in the sampling process, while OAT facilitates the random selection of a parameter from each LH loop and the alteration of its value from the previous simulation by a user-defined percentage (Veith and Ghebremichael 2009). During the sensitivity analysis, 320 simulation runs were executed. For a detailed explanation of the mechanics of the algorithm (see Van Griensven 2005). Out of the 41 parameters, 6 were ranked very sensitive and therefore given higher premium during model calibration.

11.3.4.3 Model Calibration and Validation

The model was automatically calibrated for flow, TSS, and phosphorus. Seven model parameters were included in SWAT calibration of flow. The model was calibrated for river flow at the subbasin level for 2008 and 2009 water years based on daily observed river discharges at seven USGS stream discharge stations along the Des Plaines River watershed (Figure 11.5). Model parameter optimization was conducted with the use of parameter solution (ParaSol) using uncertainty analysis algorithm built in SWAT-CUP software. ParaSol has its root on a modified version of the global optimization algorithm SCE-UA (Duan et al. 1993). The method uses the sum of the squares of the residuals (SSQ) as the objective function. For a detailed explanation of the model parameterization philosophy and setup (see Abbaspour 2007; Yang et al. (2008)).

At the end of autocalibration simulation runs, an average daily Nash–Sutcliffe coefficient of 0.73 was achieved for calibration of flow from the seven stream discharge stations. The Nash-Sutcliffe coefficient is a statistics used to ascertain whether a model is properly calibrated and validated (Nash and Sutcliffe 1970). The Nash–Sutcliffe criteria for accepting model calibration and validation was set at 0.6. Following autocalibration, some flow parameters were manually refined to improve model simulation efficiency. SWAT was validated for flow for the 2010 water year in the seven USGS stream discharge stations mentioned above. Table 11.2 gives a detailed description of the efficacy of model calibration and validation of flow within the watershed.

At the completion of flow calibration, a similar but more rigorous approach was adopted in the calibration of TSS and phosphorus within the watershed. The Nash–Sutcliffe relative efficiency coefficient and the relative index of agreement coefficient were also employed in assessing the efficacy of the model prediction of total suspended sediment and phosphorus concentrations (Willmott 1981; Krause et al. 2005). The relative (modified) index of agreement threshold for model calibration and validation completion was set at 0.48.

Observed water quality data were converted from concentration to load to be compatible with the units used by the SWAT model (Wilson and Weng 2011). TSS and phosphorus were independently calibrated for the 2008 and 2009 water years and validated for the 2010 water year at the subbasin scale. Seven parameters that encapsulate water quality and sediment were automatically calibrated followed by manually refining some of them. Three urban land management parameters were also manually refined (Table 11.1). Water quality and sediment parameters were included for calibration because sensitivity analysis pointed out that they were sensitive. Urban land management parameters were manually refined to reflect the configuration of impervious land cover in the watershed. Water quality and sediment model parameters were calibrated at a monthly time step as a result of the lack of continuous daily water quality data. At the end of model calibration and the achievement of model simulation criteria threshold, TSS and total phosphorus were validated at seven water quality stations that are in proximate distance to the USGS stream discharge stations. Simulated water quality results were converted from load to concentration to make comparison with observed water quality data easier.

11.3.5 Results

11.3.5.1 Predicted Land-Use Changes
The Des Plaines River watershed land-use maps for the planning scenarios (SN1, SN2, and SN3) generated for 2020 estimated that the sizes of LDR, open space/vegetation, institutional, transport, communication, and utilities (TCU), commercial/urban mix, and institutional lands increased in relation to the 2010 spatial extent. Other land-use classes that are projected to increase in some of the planning scenarios include high-density residential (HDR) and medium-density residential (MDR) areas. Agriculture, industrial, vacant, and to a lesser extent wetlands are estimated to decrease in spatial extent by 2020.

Table 11.5 Land-use change between 2010 and 2020 (%).

Land use	SN1	SN2	SN3
HDR	−8.9	0.5	6.2
MDR	−3.2	6.5	6.6
LDR	10.1	8.3	9.3
Industrial	−3.5	−5.5	−5.5
TCU	5.6	5.6	5.6
Wetland	−0.82	−0.93	−0.5
Water	0	0	0
Commercial/urban mix	2.1	5.8	5.2
Institutional	0.4	0.4	0.4
Agriculture	−20.3	−24.7	−23.5
Open space/vegetation	6.4	4.8	3.9
Vacant	−1	−5.9	−11

Source: Wilson and Weng (2011). Reproduced with permission of Elsevier.
Note: SN1, SN2, and SN3, land-use/planning scenarios 1, 2, and 3. HDR, high-density residential; MDR, medium-density residential; LDR, low-density residential; TCU, ransport, communication, and utilities.

Land-use change analysis between 2010 and 2030 illustrates a similar trajectory exemplified by 2020 with the exception of slight differences between spatial gains and losses among land-use classes (Tables 11.5 and 11.6). For example, all residential categories demonstrated larger net gains in area by 2030 compared to that exhibited in 2020. Reduction in industrial and vacant land is more pronounced for 2030 compared to the projected value for 2020. The rest of the land covers did not show marked variation for the two future dates (Figure 11.6).

11.3.5.2 Watershed Response of TSS to Land-Use and Climate Changes

TSS responded very differently to changes in land use and climate during the summer and winter periods of 2020 and 2030, respectively. At the general watershed level for both SRES B1 and A1B, TSS concentration was relatively higher during the month of March compared to July. These months were used to understand temporal responses of water quality to future summer and late winter/early spring climates. The difference between the average SRES B1-simulated value of TSS for July 2010 and July 2020 demonstrated a 14% decline under SN1, while the corresponding reductions for SN2 and SN3 are 16 and 19%, respectively. In a similar vein, maximum and average TSS concentrations for the month of July are predicted to wane by 2020 and 2030 for the other land-use and climate scenarios compared to the baseline period (Figure 11.7a).

Table 11.6 Land-use change between 2010 and 2030 (%).

Land use	SN1	SN2	SN3
HDR	−6.8	8.6	14.8
MDR	−2.9	8.1	9.7
LDR	15.5	8.7	10.5
Industrial	−4.2	−5.5	−5.7
TCU	5.8	5.8	5.8
Wetland	−0.82	−0.93	−0.5
Water	0	0	0
Commercial/urban mix	7.3	15	13.9
Institutional	0.4	0.4	0.4
Agriculture	−20.6	−23.7	−25.7
Open space/vegetation	6.4	2.4	3.2
Vacant	−19.3	−26.1	−26.7

Source: Wilson and Weng (2011). Reproduced with permission of Elsevier.
Note: SN1, SN2, and SN3, planning scenarios 1, 2, and 3. HDR, high-density residential; MDR, medium-density residential; LDR, low-density residential; TCU, ransport, communication, and utilities.

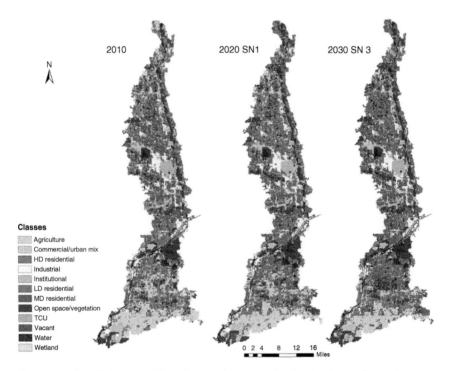

Figure 11.6 Des Plaines River Watershed land-use maps for 2010 and some future dates. *Note:* HD, high density; MD, medium density; LD, low density; TCU, transport, communication, and utilities. *Source:* Wilson and Weng (2011). Reproduced with permission of Elsevier.

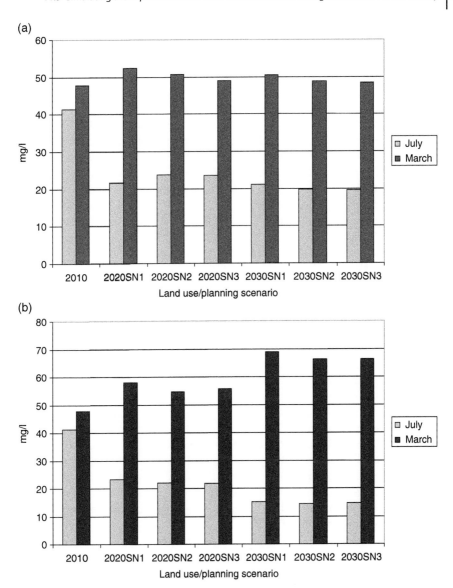

Figure 11.7 Difference between maximum total suspended solids (TSS) in 2010 and future dates under the scenarios of Special Report on Emission Scenario (SRES) B1 (a) and SRES A1B (b). *Source:* Wilson and Weng (2011). Reproduced with permission of Elsevier.

Evaluating the difference between the average SRES A1B-simulated value of TSS for July 2010 and July 2020 demonstrated a 20% decline under SN1, while the values for SN2 and SN3 are 18 and 21%, respectively. By July 2030, the A1B climate scenario under the three land-use/planning scenarios is projected to generate the least TSS concentrations over the evaluation period.

The analysis shows that SN2 is slated to have the lowest TSS concentration over the entire study period (Figure 11.7a). The inter-period decline in TSS concentration can be attributed to the dramatic estimated average reduction in precipitation and increase in temperature for July 2020 and 2030, while the differences exhibited by the various land-use/planning scenarios can be ascribed to the higher reduction in agricultural and vacant lands in SN2 and SN3.

The maximum and average estimated TSS concentrations for March displays higher values for all land-use and climate scenarios compared to the baseline period (Figure 11.7b). Projected increases in average TSS concentrations for March 2020 under the B1 climate scenario are slated at 78, 43, and 45% under land-use/planning scenarios 1–3, respectively, while average TSS concentrations projected for the A1B scenarios by 2020 are slightly higher than that of B1. Dramatic increases (greater than 200%) in TSS concentrations are predicted for the A1B scenario by 2030, while the B1 scenario is expected to exhibit the least accretion in TSS concentration vis-à-vis the baseline period. Increment in average TSS concentration in the watershed can be attributed to increases in precipitation by 2020 and 2030. Average future precipitation for March under the B1 and A1B climate scenarios demonstrate significant increases of 60 and 365%, respectively, while temperature on average displays a decline of about 2 °C compared to the baseline period.

The impact of the above on TSS loading and concentration is additional erosive and transportation mechanism through higher river discharge, while lower temperatures will tend to reduce the rate of potential evapotranspiration (PET) thereby indirectly reducing infiltration rates in pervious areas of the watershed. Potential changes in river discharge between March 2010, and SRES B1 and A1B for the same month in 2020 and 2030, demonstrated an increase of between 46 and 204%, respectively. These inter-seasonal projected surges and abatement in TSS concentration can be attributed more to reduction and increase in precipitation during the months of July and March, respectively, than the changes in the proportion of land-use classes. The model results indicated greater changes in TSS concentration across climate scenarios compared to land-use/planning scenarios.

The general watershed response of TSS to climate and land-use changes has revealed an increase in concentration levels in late winter/early spring period compared to the same season in 2010, while concentration wanes during the summer season. Studies conducted on the impacts of climate change and urban development on hydrology and water quality in temperate regions have revealed higher late winter and early spring runoff compared to summer periods (Tu 2009; Bekele and Knapp 2010). In addition, few studies that have analyzed the response of TSS loads to climate and land-use changes predicted lower loads in the summer, while the winter and early spring periods were observed to demonstrate considerable increases in loads (Asselman et al. 2003; Chang 2004; Praskievicz and Chang 2011). Results of this study with regards to potential TSS concentrations for water quality to a large extent are in consonance with the aforementioned scholarships. However, it is problematic to make a direct comparison to these studies as their results were reported in loads rather than concentration. Our study has revealed that TSS concentration is not always linearly related to river discharge as under certain climate scenarios, TSS concentration can be lower even though loads are projected to be higher than the baseline period. It is important that potential concentrations of constituents be calculated especially for water resource planning and management needs.

11.3.6   TSS Response at the Sub-Watershed Scale SRES B1

At the sub-watershed level, the model predicted that 79% of the watershed area (51 out of 58 subbasins) will have lower TSS concentration by July 2020 under SN1, while the proportion for SN2 and SN3 is 76 and 75%, respectively. Model results for 2030 predicted that 88% of the watershed area will have lower TSS concentration by July 2030, while figures for SN2 and SN3 are estimated at 83 and 84%, respectively. This differential in the spatial distribution of the pollutant between July 2020 and 2030 can be attributed to the further reduction in precipitation and runoff which was observed to affect TSS loading and concentration. Moreover, the 2030 results suggest that larger portion of the watershed area will have lower concentration of TSS concentration compared to the projected value for 2020 (Figure 11.8). Average decline in TSS concentration within these subbasins for all land-use/planning scenarios over the projected period is 14.1 mg/l. General land-use trend in subbasins that are expected to have lower TSS concentrations in July 2020 compared to the baseline period encompasses an average range of decline in agricultural land (4–21%), LDR (1–12%), and slight decrease in industrial land use (1–2%), while spatial gains are demonstrated within MDR (2–8%), open space/vegetation (1–4%), commercial/urban mix (1–2%), and vacant lands. Potential land-use change trajectory for all the 2030 planning scenarios mirrors that demonstrated by the 2020 scenarios.

Analysis of the modeling results for March 2020 under SN1 indicates that TSS concentration will be higher in all but one subbasin (98.3% of watershed area) compared to the 2010 values. The estimated TSS proportions for SN2 and SN3 are 98 and 100%, respectively. Average increase in the pollutant is estimated at 30.5 mg/l. The general subbasin distributional trend of TSS over the evaluation period displays an increase of the pollutant concentration within more subbasins by March 2020 compared to the same period in 2030 (Figure 11.9). The general trend of land use within areas of the watershed that is projected to have acute increase in TSS concentration by March 2020 compared to the baseline period exhibits substantial increase in LDR (6–20%), vacant (3–5%), commercial/urban mix (1–2%), slight increase in TCU, and

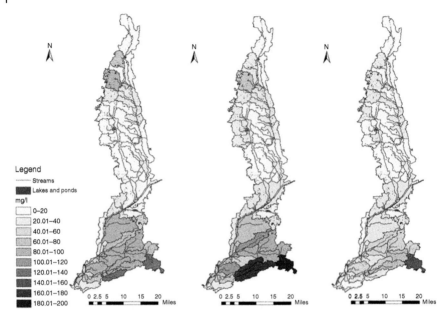

Figure 11.8 Comparison of July TSS concentration between 2010 and future Scenario One (SN1) SRES B1. *Source:* Wilson and Weng (2011). Reproduced with permission of Elsevier.

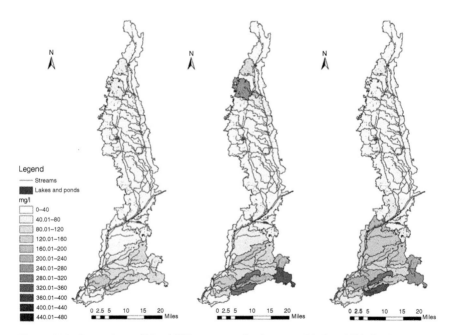

Figure 11.9 Comparison of March TSS concentration between 2010 and SN1 SRES B1. *Source:* Wilson and Weng (2011). Reproduced with permission of Elsevier.

open space/vegetation (1–4%) lands; notable reduction in MDR (greater than 15%) is demonstrated. The other land classes did not show potential significant changes over the 20-year period.

The pattern and trend of land use displayed by the other land-use/planning scenarios for 2030 is similar to that of 2020 with higher increase of additional 2–3% in open space/vegetation, and an inverse relationship between LDR and MDR. In some subbasins, in both 2020 and 2030 land-use/planning scenarios, there is either a large increase in LDR (20–35%) or a relatively smaller decline in MDR, or a reverse wherein MDR increases exponentially and LDR reduces at a lower rate.

11.3.7 TSS Response at the Sub-Watershed Scale SRES A1B

The modeling result for July predicts that between 71 and 73% of the watershed area will have lower concentration of TSS under the land-use/planning scenarios for 2020, while values for 2030 are slated at 92.8% for SN1, and 87% for SN2 and SN3, respectively. Average reduction in TSS levels is projected to be higher for 2030 compared to the simulated values for all 2020 land-use/planning scenarios. Figure 11.10 illustrates reduction in the spatial distribution of TSS concentration between July 2010 and 2020, in addition to the projected

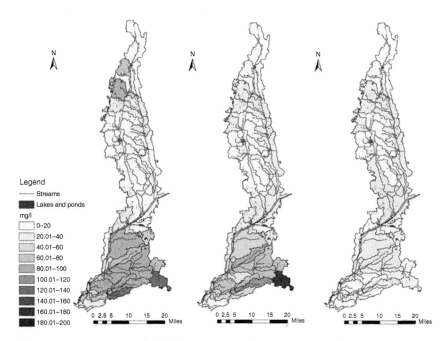

Figure 11.10 Comparison of July TSS concentration between 2010 and future SN1 SRES A1B. *Source:* Wilson and Weng (2011). Reproduced with permission of Elsevier.

concentrations for 2030. The pollutant is projected to continually reduce in the watershed area up to 2030. General land-use trend within the subbasins that are projected to have lower concentration of the pollutant in 2020 and 2030 under various land-use/planning scenarios reveals relatively notable decline in agricultural (4–19%), LDR or MDR, and vacant (3–14%) lands, while smaller increase was observed in open space/vegetation and commercial/urban mix land classes. Land-use trend mirrors that established in SRES B1 with relative imperceptible differences.

Simulation results for March 2020 demonstrate that 93, 92, and 94.7% of the watershed area will have higher TSS concentrations for SN1, SN2, and SN3, respectively, in relation to the same month in 2010. The 2030 analysis predicts that 97% of the watershed area will have higher TSS concentration under SN1 and SN2, while the equivalent for SN3 is 95%. Greater numbers of subbasins are expected to have higher concentrations of TSS by March 2030 compared to 2020 and 2010 (Figure 11.11). The general land-use trend within subbasins, that are expected to have higher levels of TSS, demonstrates increase in agriculture (1–4%), and either high increase in LDR (7–34%) and relatively lower increase in MDR (3–8%), or vice versa. Furthermore, marginal increase and decrease in commercial/urban mix and industrial lands, respectively, are also displayed. The land-use pattern and trend mirrors land-use scenarios under SRES B1 climate emission scenario.

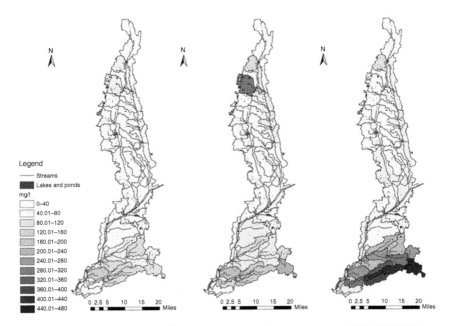

Figure 11.11 Comparison of March TSS concentration between 2010 and SN1 SRES A1B. *Source:* Wilson and Weng (2011). Reproduced with permission of Elsevier.

The response of TSS to climate and land-use changes demonstrates that the higher the precipitation level coupled with relatively low temperature, the larger the concentration of TSS in a watershed. This arises because TSS is a pollutant that responds to flushing more than dilution within a watershed and therefore, higher river discharge will mostly facilitate greater erosion and flushing/transportation of the pollutant compared to dilution (Walling and Webb 1992; Yusop et al. 2005). The modeling results further suggest that TSS will be a major problem in most areas of the watershed during the month of March. Furthermore, the literature on historical and contemporary water quality modeling has pointed out that certain pollutants sometimes attach to suspended sediments (Bradley and Lewin 1982; Freeman and Fox 1995; Qu and Kelderman 2001; Zonta et al. 2005). TSS might not only be the pollutant that will increase during the month of March but can also contribute to the transportation of other nonpoint source pollutants. The study has also revealed that not all subbasins in a watershed will increased their TSS levels during periods of high river discharge in the future. The significance of this is that land-use composition and slope are also important determinants of TSS concentrations within a watershed characterized by a relatively flat topography as displayed by the Des Plaines River watershed. The relatively flat topography of the watershed especially in the southern portion causes runoff to be sluggish in that zone and increase in TSS load compared to the background of the water, thereby facilitating an increase in its concentration level. Analysis of TSS has also demonstrated that areas of a watershed that has MDR and HDR land are better for mitigating high levels of TSS loading during the late winter and early spring periods. Furthermore, the development of a mixed land configuration wherein MDR, HDR, open space/vegetation, and other land-use composition are combined will augur better for TSS levels even under heavy precipitation conditions. We suggests the construction of additional reservoirs along Des Plaines River and major tributaries that can regulate excess down stream flow of water and hence reduce TSS loading at the subbasin level during winter and early spring periods in the future. The reservoir can be structured in a way that allows normal flow of the river during the summer periods characterized by very low river discharge.

11.4 Summary

This study has revealed that future land-use and climate changes have the potential of dramatically changing the concentration levels of total suspended sediments at both the general watershed and subbasin scales in the Des Plaines River watershed. Future climate change exerts a larger impact on the concentration of pollutants than the potential impact of land-use change. The various land-use/planning scenarios demonstrated slight differences between nonpoint source pollutant concentrations under constant climate emission scenario, but this trend significantly veered when inter-seasonal and inter-climate emission types were examined. The response of future water quality to climate and land-use heavily depends on the type of climate emission scenario being evaluated. At the subbasin scale, the growth of middle and HDR land use at the expense of LDR would help reduce late winter and early spring loadings and the concentration of TSS. This modeling approach developed in this study can be applied to watersheds in other metropolitan settings provided the data and expertise are available. It would also be interesting to assess TSS response to other climate emission scenarios and watersheds that have more abrupt changes in elevation compared to the Des Plaines River watershed. Finally, a longer-term analysis of future land and climate changes is recommended to compare possible changes in water quality beyond 2030.

References

Abbaspour, K.C. (2007). User manual for SWAT-CUP. SWAT calibration and uncertainty analysis programs. Swiss Federal Institute of Aquatic Science and Technology, Eawag, Dübendorf, Switzerland. http://www.eawag.ch/organisation/abteilungen/siam/software/swat/index_EN (accessed 5 July 2019). Abbaspour, K.C., Faramarzi, M., Ghasemi, S.S., and Yang, H. (2009). Assessing the impact of climate change on water resources in Iran. *Water Resources Research* 45: 1–16.

Adam, J.C. and Lettenmaier, D.P. (2003). Adjustment of global gridded precipitation for systematic bias. *Journal of Geophysical Research* 108: 1–14.

Amiri, R., Weng, Q.H., Alimohammadi, A., and Alavipanah, S.K. (2009). Spatial-temporal dynamics of land surface temperature in relation to fractional vegetation cover and land use/cover in the Tabriz urban area, Iran. *Remote Sensing of Environment* 113 (12): 2606–2617.

Arnold, J.G., Allen, P.M., and Bernhardt, G. (1993). A comprehensive surface-ground flow model. *Journal of Hydrology* 142: 47–69.

Arnold, J.G., Srinivasan, R., Muttiah, R.S., and Williams, J.R. (1998). Large area hydrologic modeling and assessment part 1: model development. *Journal of the American Water Resources Association* 34 (1): 73–89.

Asselman, N.E., Middelkoop, H., and Dijk, P.M. (2003). The impacts of changes in climate and land use on soil erosion, transportation and deposition of suspended sediment in the River Rhine. *Hydrological Processes* 17: 3225–3244.

Beighley, R.E., Dunne, T., and Melack, J.M. (2008). Impacts of climate variability and land use alterations on frequency distributions of terrestrial runoff loading to coastal waters in Southern California. *Journal of the American Water Resource Association* 44 (1): 62–74.

Bekele, E.G. and Knapp, V. (2010). Watershed modeling to assessing impacts of potential climate change on water supply availability. *Water Resources Management* 24: 3299–3320.

Bradley, S.B. and Lewin, J. (1982). Transport of heavy metals on suspended sediments under high flow conditions in a mineralized region of Wales. *Environmental Pollution Series B Chemical and Physical* 4 (4): 257–267.

Breiman, L. (2001). Random forests. *Machine Learning* 45 (1): 5–32.

Chang, H. (2004). Water quality impacts of climate and land use changes in Southern Pennsylvania. *The Professional Geographer* 56 (2): 240–257.

Chen, X.L., Zhao, H.M., Li, P.X., and Yin, Z.Y. (2006). Remote sensing image-based analysis of the relationship between urban heat island and land use/cover changes. *Remote Sensing of Environment* 104 (2): 133–146.

Chicago Metropolitan Agency for Planning (2010a). *Land Use Inventory (versions 1.0 and 2.1), 2010*. Chicago, IL.

Chicago Metropolitan Agency for Planning (2010b). *Go To 2040 Comprehensive Regional Plan*. Full version (October 2010). Chicago, IL.

Chung, E., Park, K., and Lee, K.S. (2011). The relative impacts of climate change and urbanization on the hydrologic response of a Korean urban watershed. *Hydrological Processes* 25: 544–560.

Clarke, K.C., Hoppen, S., and Gaydos, L. (1997). A self-modifying cellular automaton model of historical urbanization in the San Francisco Bay Area. *Environment and Planning B; Planning and Design* 24: 247–261.

Cleveland, R.B., Cleveland, W.S., McRae, J.E., and Terpenning, I. (1990). STL: a seasonal-trend decomposition procedure based on loess. *Journal of Official Statistics* 6 (1): 3–73.

Congalton, R.G. (1991). A review of assessing the accuracy of classification of remotely sensed data. *Remote Sensing of the Environment* 37: 35–46.

Cutler, I. (2006). *Chicago: Metropolis of the Mid-Continent*, 4e. Chicago, IL: The Geographic Society of Chicago.

Deng, C.B. and Wu, C.S. (2013). Examining the impacts of urban biophysical compositions on surface urban heat island: a spectral unmixing and thermal mixing approach. *Remote Sensing of Environment* 131: 262–274.

Duan, Q., Gupta, V.K., and Sorooshian, S. (1993). Shuffled complex evolution approach for effective and efficient global minimization. *Journal of Optimization Theory Application* 76: 501–521.

Ducharne, A., Baubion, C., Beudoin, N. et al. (2007). Long-term perspective of the Seine River system: confronting climate and direct anthropogenic changes. *Science of the Total Environment* 375: 292–311.

Freeman, W. and Fox, J. (1995). ALAWAT: a spatially allocated watershed model for approximating stream, sediment, and pollutant flows in Hawaii, USA. *Environmental Management* 19 (4): 567–577.

Fu, P. and Weng, Q. (2016). A time series analysis of urbanization induced land use and land cover change and its impact on land surface temperature with Landsat imagery. *Remote Sensing of Environment* 175 (4): 205–214.

Gillies, R.R., Box, J.B., Symanzik, J., and Rodemaker, E.J. (2003). Effects of urbanization on the aquatic fauna of the Line Creek watershed, Atlanta – a satellite perspective. *Remote Sensing of Environment* 86 (3): 411–422.

Goonetilleke, A., Thomas, E., Ginn, S., and Gilbert, D. (2005). Understanding the role of land use in urban storm water quality management. *Journal of Environmental Management* 74 (1): 31–42.

Gournay, I., Beswick, P.G., Sams, G.W., and Architects, A.I. (1993). *AIA Guide to the Architecture of Atlanta*. University of Georgia Press.

Hu, X., Waller, L.A., Al-Hamdan, M.Z. et al. (2013). Estimating ground-level PM2.5 concentrations in the southeastern U.S. using geographically weighted regression. *Environmental Research* 121: 1–10.

Imhoff, J.C., Kittle, J.L.J., Gray, M.R., and Johnson, T.E. (2007). Using the Climate Assessment Tool (CAT) in USA EPA BASINS integrated modeling system to assess watershed vulnerability to climate change. *Water Science and Technology* 56: 49–56.

IPCC (2001). *Climate Change 2001: The Scientific Basis*. Contribution of Working Group 1 to the Third Assessment Report of the Intergovernmental Panel on Climate Change. J.T. Houghton, Y. Ding, D.J. Griggs, et al., eds. Cambridge, UK and New York, NY: Cambridge University Press.

Jensen, J.R. (2005). *Introductory Digital Image Processing: A Remote Sensing Perspective*, 3e. Upper Saddle River, NJ: Prentice Hall.

Jin, S., Yang, L., Danielson, P. et al. (2013). A comprehensive change detection method for updating the National Land Cover Database to circa 2011. *Remote Sensing of Environment* 132: 159–175.

Kahya, O., Bayram, B., and Reis, S. (2010). Land cover classification with an expert system approach using Landsat ETM imagery: a case study of Trabzon. *Environmental Monitoring and Assessment* 160: 431–438.

Krause, P., Boyle, D.P., and Base, F. (2005). Comparison of different efficiency criteria for hydrological model assessment. *Advances in Geosciences* 5: 89–97.

Landis, J.D. (1995). Imagining land use futures: applying the California futures model. *Journal of the American Planning Association* 61 (4): 438–457.

Lawrence, R.L. and Wright, A. (2001). Rule-based classification systems using classification and regression tree (CART) analysis. *Photogrammetric Engineering & Remote Sensing* 67 (10): 1137–1142.

Lee, R.G., Flamm, R., Turner, M.G. et al. (1992). Integrating sustainable development and environmental vitality, a landscape ecology approach. In: *Watershed Management: Balancing Sustainability and Environmental Change* (ed. R.J. Naiman), 499–521. New York, NY: Springer-Verlag.

Lo, C.P. and Quattrochi, D.A. (2003). Land-use and land-cover change, urban heat island phenomenon, and health implications: a remote sensing approach. *Photogrammetric Engineering & Remote Sensing* 69 (9): 1053–1063.

Lopez, E., Bocco, G., Mendoza, M., and Duhau, E. (2001). Predicting land cover and land use change in the urban fringe: a case in Morelia city, Mexico. *Landscape and Urban Planning* 55 (4): 271–285.

Loveland, T.R. and Defries, R.S. (2004). *Observing and Monitoring Land Use and Land Cover Change*. *Ecosystems and Land Use Change*, 231–246. American Geophysical Union.

Lu, D., Mausel, P., Brondizio, E., and Moran, E. (2002). Assessment of atmospheric correction methods for Landsat TM data applicable to Amazon basin LBA research. *International Journal of Remote Sensing* 23 (13): 2651–2671.

Luck, M. and Wu, J.G. (2002). A gradient analysis of urban landscape pattern: a case study from the Phoenix metropolitan region, Arizona, USA. *Landscape Ecology* 17 (4): 327–339.

Masek, J.G., Vermote, E.F., Saleous, N.E. et al. (2006). A Landsat surface reflectance dataset for North America, 1990–2000. *IEEE Geoscience and Remote Sensing Letters* 3 (1): 68–72.

Maurer, E.P., Adam, J.C., and Wood, A.W. (2009). Climate model based consensus on the hydrological impacts of climate change to the Rio Lempa basin of Central America. *Hydrology and Earth System Sciences* 13: 183–194.

Maximov, I.A. (2003). Integrated assessment of climate and land use change effects on hydrology and water quality of the Upper and Lower Great Miami River. PhD dissertation. Department of Geography, University of Cincinnati.

Meehl, G.A., Covey, C., Delworth, T. et al. (2007). The WCRP CMIP3 multi-model dataset: a new era in climate change research. *Bulletin of the American Meteorological Society* 88: 1383–1394.

Miller, M.D. (2012). The impacts of Atlanta's urban sprawl on forest cover and fragmentation. *Applied Geography* 34: 171–179.

NASA (2010). *Panoply Version 2.9.2 User Guide*. New York, NY: Goddard Institute for Space Studies.

Nash, J.E. and Sutcliffe, J.V. (1970). River flow forecasting through conceptual models: part 1 – a discussion of principles. *Journal of Hydrology* 10 (3): 282–290.

Neitsch, S.L., Arnold, J.G., Kiniry, J.R. et al. (2004). *Soil and Water Assessment Tool Input/Output File Documentation*. https://swat.tamu.edu/media/1291/SWAT2005io.pdf (accessed 3 July 2019).

Parks, R.J. (1991). Models of forested and agricultural landscapes: integrating economics. In: *Quantitative Methods in Landscape Ecology: The Analysis and Interpretation of Landscape Heterogeneity* (eds. M.G. Turner and R.H. Gardner), 309–322. New York, NY: Springer-Verlag.

Petit, C., Scudder, T., and Lambin, E. (2001). Quantifying processes of land-cover change by remote sensing: resettlement and rapid land-cover change in southeastern Zambia. *International Journal of Remote Sensing* 22 (17): 3435–3456.

Praskievicz, S. and Chang, H. (2011). Impacts of climate change and urban development on water resources in the Tualatin River Basin, Oregon. *Annals of the Association of American Geographers* 101 (2): 249–271.

Qian, B., Gameda, S., and Hayhoe, H. (2008). Performance of stochastic weather generators LARS-WG and AAFC-WG for reproducing daily extremes of diverse Canadian climates. *Climate Research* 37: 17–33.

Qu, W. and Kelderman, P. (2001). Heavy metal contents in the Delft canal sediments and suspended solids of the River Rhine: multivariate analysis for source tracing. *Chemosphere* 45 (6–7): 919–925.

Rinner, C. and Hussain, M. (2011). Toronto's urban heat island-exploring the relationship between land use and surface temperature. *Remote Sensing* 3 (6): 1251–1265.

Rose, S. and Peters, N.E. (2001). Effects of urbanization on streamflow in the Atlanta area (Georgia, USA): a comparative hydrological approach. *Hydrological Processes* 15 (8): 1441–1457.

Rounsevell, M.D., Reginster, I., Araujo, M.B. et al. (2006). A coherent set of future land use change scenarios for Europe. *Agriculture, Ecosystems and Environment* 114: 57–68.

Sexton, J.O., Urban, D.L., Donohue, M.J., and Song, C. (2013). Long-term land cover dynamics by multi-temporal classification across the Landsat-5 record. *Remote Sensing of Environment* 128: 246–258.

Small, C. (2006). Comparative analysis of urban reflectance and surface temperature. *Remote Sensing of Environment* 104 (2): 168–189.

Stone, M.C., Hotchkiss, R.H., Hubbard, C.M. et al. (2001). Impacts of climate change on Missouri River basin water yield. *Journal of the American Water Resources Association* 37 (5): 1119–1129.

Tu, J. (2009). Combined impact of climate and land use changes on streamflow and water quality in eastern Massachusetts, USA. *Journal of Hydrology* 379: 268–283.

Turner, M.G. (1987). Spatial simulation of landscape change in Georgia: a comparison of 3 transition models. *Landscape Ecology* 1: 29–36.

Turner, M.G., Gardner, R.H., and O'Neill, R.V. (2001). *Landscape Ecology in Theory and Practice: Pattern and Process.* Springer.

Van Griensven, A. (2005). Sensitivity, auto-calibration, and model evaluation in SWAT 2009. User guide distributed with ArcSWAT program.

Veith, T.L. and Ghebremichael, L.T. (2009). How to: applying and interpreting the SWAT auto-calibration tools. *2009 International SWAT Conference proceedings*, University of Colorado at Boulder, Boulder, Colorado (5–7 August 2009). Texas Water Resources Institute Technical Report No. 356, Texas A&M University System, pp. 26–33.

Verbesselt, J., Hyndman, R., Newnham, G., and Culvenor, D. (2010). Detecting trend and seasonal changes in satellite image time series. *Remote sensing of Environment* 114 (1): 106–115.

Walling, D.E. and Webb, B.W. (1992). Chapter 3: water quality. 1: physical characteristics. In: *The Rivers Handbook*, vol. 1 (eds. P. Calow and G.E. Pettse), 48–72. Oxford: Blackwell.

Weng, Q. (2001). Modeling urban growth effects on surface runoff with the integration of remote sensing and GIS. *Environmental Management* 28 (6): 737–748.

Weng, Q. (2002). Land use change analysis in the Zhujiang Delta of China using satellite remote sensing, GIS, and stochastic modeling. *Journal of Environmental Management* 64 (3): 273–284.

Weng, Q. (2014). *Global Urban Monitoring and Assessment Through Earth Observation*, 424. Boca Raton, FL: CRC Press/Taylor and Francis.

Weng, Q.H., Lu, D.S., and Liang, B.Q. (2006). Urban surface biophysical descriptors and land surface temperature variations. *Photogrammetric Engineering & Remote Sensing* 72 (11): 1275–1286.

Weng, Q.H., Lu, D.S., and Schubring, J. (2004). Estimation of land surface temperature-vegetation abundance relationship for urban heat island studies. *Remote Sensing of Environment* 89 (4): 467–483.

Wentz, E.A., Nelson, D., Rahman, A. et al. (2008). Expert system classification of urban land use/land cover for Delhi, India. *International Journal of Remote Sensing* 29 (15): 4405–4427.

Wilby, R.L., Charles, S.P., Zorita, E. et al. (2004). Guidelines for use of climate scenarios developed from statistical downscaling methods [Online]. http://www. ipcc-data.org/guidelines/dgm_no2_v1_09_2004.pdf (accessed 3 July 2019).

Wilks, D.S. (2010). Use of stochastic weather generators for precipitation downscaling. *Wiley Interdisciplinary Reviews: Climate Change* 1 (6): 898–907.

Willmott, C.J. (1981). On the validation of models. *Physical Geography* 2: 184–194.

Wilson, C.O. and Weng, Q. (2010). Assessing surface water quality and its relations with urban land cover changes in the Lake Calumet Area, Greater Chicago. *Environmental Management* 45: 1096–1111.

Wilson, C.O. and Weng, Q. (2011). Simulating the impacts of future land use and climate changes on surface water quality in the Des Plaines River Watershed, Chicago Metropolitan Statistical Area, Illinois. *Science of the Total Environment* 409 (20): 4387–4405.

Wood, A.W., Leung, L.R., Sridhar, V., and Lettenmeier, D.P. (2004). Hydrological implications of dynamical and statistical approaches to downscaling climate model outputs. *Climate Change* 62: 189–216.

Yang, J., Reichert, P., Abbaspour, K.C., and Yang, H. (2008). Comparing uncertainty analysis techniques for SWAT application to Chaohe Basin in China. *Journal of Hydrology* 358: 1–23.

Young, W.J., Marston, F.M., and Davis, J.R. (1996). Nutrient export and land-use in Australian Catchments. *Journal of Environmental Management* 47: 165–183.

Yusop, Z., Tan, L.W., Ujang, Z. et al. (2005). Runoff quality and pollution loadings from a tropical urban catchment. *Water Science and Technology* 52 (9): 125–132.

Zhang, C. and Li, W. (2005). Markov chain modeling of multinomial land-cover classes. *GIScience and Remote Sensing* 42: 1–18.

Zhang, X., Srinivasan, R., and Hao, F. (2007). Predicting hydrologic response to climate change in the Louhe River basin using the SWAT model. *American Society of Agricultural and Biological Engineers* 50 (3): 901–910.

Zheng, B., Myint, S.W., and Fan, C. (2014). Spatial configuration of anthropogenic land cover impacts on urban warming. *Landscape and Urban Planning* 130: 104–111.

Zhu, Z. and Woodcock, C.E. (2014). Continuous change detection and classification of land cover using all available Landsat data. *Remote Sensing of Environment* 144: 152–171.

Zonta, R., Collavini, F., Zaggia, L., and Zuliani, A. (2005). The effect of floods on the transport of suspended sediments and contaminants: a case study from the estuary of the Dese River (Venice Lagoon, Italy). *Environment International* 31 (7): 948–958.

12

Remote Sensing of Socioeconomic Attributes

12.1 Introduction

Population growth and urbanization in the past decades have experienced an unprecedented magnitude in human history (UNFPA 2008), producing considerable impacts on socioeconomic development, availability of resources, and environmental protection at the local, regional, and global scales. Obtaining timely and accurate information of urban characteristics and socioeconomic attributes is valuable for both urban management and decision-making (Fabos and Petrasovits 1984) and for urban studies. Traditional census and survey methods cannot meet the needs of acquiring this type of information. It is important to develop suitable approaches for estimating population and mapping its distribution in timely and cost-effective manner. Remote sensing data with their advantages in spectral, spatial, and temporal resolutions have demonstrated to be effective in providing information of physical characteristics of urban areas, including its size, shape, and rates of change, and have been widely used for mapping and monitoring of urban biophysical features (Haack et al. 1997; Jensen and Cowen 1999; Weng et al. 2006, 2008). However, the use of remote sensing method to estimate and map socioeconomic attributes has not been explored fully.

Population is the socioeconomic attribute that has been mostly studied by remote sensing methods. Population estimation with remote sensing data has long been explored and many approaches based on aerial photographs and satellite imageries, including Landsat, SPOT, night-light of Defense Meteorological Satellite Program (DMSP)-operational Line Scanner, Radar SIR-A, Radar QuickSCAT, have been developed (Lindgren 1971; Lo and Welch 1977; Iisaka and Hegedus 1982; Lo 1986a, 1986b, 1995, 2001, 2008; Meliá and Sobrino 1987; Langford et al. 1991; Elvidge et al. 1995, 1997; Sutton 1997; Sutton et al. 1997, 2001; Harvey 2000, 2002a, 2002b; Li and Weng 2005; Wu and Murray 2005, 2007; Lu et al. 2006; Briggs et al. 2007; Nghiem et al. 2009; Viel and Tran 2009). Wu et al. (2005) grouped population estimation methods into two categories: statistical modeling and areal interpolation. The statistical modeling method

Techniques and Methods in Urban Remote Sensing, First Edition. Qihao Weng.
© 2020 by The Institute of Electrical and Electronics Engineers, Inc.
Published 2020 by John Wiley & Sons, Inc.

established an estimation model based on the relationship between population and remote sensing-derived variables, while the areal interpolation method generated a refined population surface by redistributing census population data to specific pixel sizes. The statistical modeling approach can be further categorized into four types depending on the information extracted from remote sensing data: (i) measurement of built-up areas, (ii) counting of dwelling units, (iii) measurements of land-use (LU) areas, and (iv) image pixel characteristics. Each method has its own advantages and disadvantages, depending upon the geographical scale of study and remote sensing data used (Lo 1986b; Li and Weng 2005). Statistical models are usually used to estimate population for a specific areal unit such as census unit, which have arbitrary boundaries. However, some researchers have attempted to integrate census data with data from different spatial divisions, which are rarely consistent with each other. This process is called areal interpolation or cross-area estimation (Goodchild and Lam 1980; Langford et al. 1991; Goodchild et al. 1993).

Different methods of areal interpolation have been developed (Goodchild and Lam 1980; Goodchild et al. 1993), including dasymetric method. The dasymetric method is defined as to utilize ancillary data to redistribute population data from arbitrarily delineated enumeration districts like census unit into zones of increased homogeneity with the purpose of better representation of the underlying statistical surface (Eicher and Brewer 2001). Previous studies show that with ancillary data such as remote sensing-derived products the dasymetric mapping technique can provide a visual and statistically more accurate representation of population distribution than that represented by aggregated units, such as Census block group (BG) or tract (Eicher and Brewer 2001; Mennis 2003; Briggs et al. 2007; Maantay et al. 2008).

On the other hand, urbanization has triggered a number of environmental problems at multiple spatial scales, including the loss of natural vegetation, open spaces, and wetlands; the adverse effect on local and regional climates; and the increase of pressure on water use, energy consumption, and infrastructure. Timely information on the temporal and spatial patterns of urban environmental quality (UEQ) is the prerequisite for the formation of new policies that support sustainable development and smart cities. Therefore, it is essential to assess UEQ at multiple temporal and spatial scales for effective urban planning and management. The UEQ has long been studied in geographical literature. It has been described using quantitative measures (Bederman and Hartshorn 1984; Fabos and Petrasovits 1984; Akbari et al. 1990, 2003; Fotheringham and Wong 1991; Comrey and Lee 1992; Adams et al. 1995; Elvidge et al. 1995; Arnold and Gibbons 1996; Elvidge et al. 1997; Eicher and Brewer 2001; Fung and Siu 2001; Abolina and Zilans 2002; Bonaiuto et al. 2003; Chander and Markham 2003; Chen et al. 2006; Briggs et al. 2007), qualitative descriptions (Goodchild and Lam 1980; Goodchild et al. 1993; Fung and Siu 2000), attitudinal explanations (Green 1957; Adams et al. 1995), and landscape features (Fabos and Petrasovits 1984; Elvidge et al. 1995; Elvidge et al. 1997; Bonaiuto et al. 2003; Chander Markham 2003; Chen et al. 2006; Briggs et al. 2007). However, UEQ is complex in practice and a complete understanding of it is still lacking. It is essentially a multidimensional concept comprising physical, spatial, economic, and social aspects of the urban environment. UEQ can be assessed from a variety of perspectives such as physical urban layout, infrastructure, economic effect, government policy, public opinion, and social consideration. The challenge is that no simple way is available to model and to predict the interaction of all aspects of UEQ. A key issue is how to derive effective UEQ indicators that not only enable people better map the phenomenon but also serve as an effective management tool in urban planning and sustainable development. Like most other geographical phenomena and environmental processes, UEQ is dependent on both spatial and temporal scales. Nevertheless, there are few attempts to have focused on these two properties of UEQ.

The technology of remote sensing presents advantages over conventional data collection methods and shows great potential in UEQ studies (Fabos and Petrasovits 1984; Elvidge et al. 1995, 1997; Akbari et al. 1990, 2003; Bonaiuto et al. 2003; Chander Markham 2003; Chen et al. 2006; Briggs et al. 2007). The main attraction of this technology primarily results from the provision of time-synchronized data coverage over a large area with both high spatial detail and high temporal frequency at a low cost and its integration with Geographic Information System (GIS). Research on UEQ on the basis of remotely sensed data dates back to the 1950s (Arnold and Gibbons 1996). Among all kinds of digital images collected by different sensors, medium resolution satellite images such as SPOT and Landsat are the primary data sources (Akbari et al. 1990, 2003; Haack et al. 1997; Bonaiuto et al. 2003; Chander Markham 2003; Chen et al. 2006; Briggs et al. 2007). Recently, images having very high spatial and spectral resolutions become available (like IKONOS) and have been applied to conduct UEQ studies at a microscale level (Fabos and Petrasovits 1984; Elvidge et al. 1995, 1997). Information derived from remote sensing data used for UEQ research is mainly concerned with four types of environmental variables, i.e. vegetation index, land surface temperature (LST), impervious surfaces, and land-use and land-cover (LULC) types. Among all the variables used to evaluate UEQ, vegetation is recognized as the key one for many reasons: filtering air, water, and sunlight; cooling urban heat; recycling pollutants; moderating local urban climate; and providing shelters to animals and recreational areas for people (Bederman and Hartshorn 1984; Elvidge et al. 1997;Harvey 2000, 2002a, 2002b; Heynen 2006). This greenness information derived from remotely sensed images to assess UEQ is typically measured by a vegetation index rather than the amount of vegetation cover (Elvidge et al. 1997). The vegetation indices have been used as the primary indicators of UEQ in different ways. Fung and Siu studied temporal change of UEQ in Hong Kong by normalized difference vegetation index (NDVI) derived from SPOT (Akbari et al. 1990, 2003). Based on the vegetation density map produced from the IKONOS image, Nichol and Wong developed a 3D Virtual Reality model to depict UEQ conditions of Hong Kong at a much detailed level (Elvidge et al. 1997). Temperature is a major concern in urban climate since it relates to the urban heat island phenomenon (Hsu 1971; Iisaka and Hegedus 1982; Jensen 2000; Irvine et al. 2009) and directly affects the thermal comfort and health of urban dwellers (Jensen and Cowen 1999). It hence provides a unique perspective of UEQ for a given area. High temperatures are often regarded as undesirable by most people. Several UEQ studies have employed LST extracted from satellite images as an indicator of urban environment (Bonaiuto et al. 2003; Chander Markham 2003; Chen et al. 2006; Briggs et al. 2007). These studies frequently used medium resolution data like Landsat imagery, but high-resolution images like IKONOS have also applied to provide detailed temperature information at a microscale level through data fusion (Fabos Petrasovits 1984; Elvidge et al. 1997). Impervious surfaces have been emerging as a key environmental indicator for sustainable urban development in recent years (Ji and Jensen 1999; Keiner 2004). Impervious surface maps are also helpful for estimating socioeconomic factors like population density and social conditions (Langford et al. 1991; Landorf et al. 2008). Li and Weng incorporated the impervious surface data derived from a Landsat Enhanced Thematic Mapper plus (ETM+) image to successfully model the quality of life in Indianapolis, Indiana (Bonaiuto et al. 2003; Briggs et al. 2007).

For a complete assessment of UEQ, socioeconomic variables, especially those derived from census data, must be combined with physical variables extracted from remote sensing images through the technique of GIS. However, those biophysical variables are measured on a per-pixel scale in raster format, while those socioeconomic variables are measured as an areal unit in vector format. Because of the difference in data format and scale of measurement, methods for integrating the two groups of variables are sought after. Two general approaches have been employed in previous UEQ studies: principal component analysis (PCA) (Akbari et al. 1990; Bonaiuto et al. 2003; Chander Markham 2003; Chen et al. 2006; Briggs et al. 2007) and GIS overlay (Fabos Petrasovits 1984; Elvidge et al. 1997; Chen et al. 2006). These two methods have also been employed as prerequisite to generate a synthetic index to quantify UEQ. Recently, Nichol and Wong have proposed using multiple-criteria queries for data integration. However, the result is strongly subject to the choice of specific environmental thresholds (Fabos and Petrasovits 1984; Elvidge et al. 1995, 1997).

In this chapter, methods for estimating population and assessing UEQ will be demonstrated through two case studies. The population estimation study intended to combine the statistical-based and dasymetric-based methods (Langford et al. 1991; Yuan et al. 1997; Harvey 2000) and to redistribute census population. The case study of population estimation aimed to compare the effectiveness of the spectral response-based and the LU-based methods for population estimation of US census BGs and to produce a more accurate presentation of population distribution by combining the dasymetric mapping with LU-based methods. On the other hand, the UEQ study intended to evaluate the UEQ changes from 1990 to 2000 in Indianapolis, Indiana, using the integrated techniques of remote sensing and GIS. Specifically, this study intends to derive an UEQ index based on the synthetic indicators of physical variables extracted from Landsat images and socioeconomic variables derived from US census data and to develop a new method for assessing UEQ change over the time.

The study area is Indianapolis/Marion County, Indiana, USA (Figure 12.1). According to the US Census Bureau, Marion County has a total area of $1044\,km^2$, 860 454 people with density of $838/km^2$, and 387 183 housing units. With its large population, Indianapolis ranks as the twelfth largest city in the United States in 2000 (US Census Bureau, Census 200 Summary file). In recent decades, the city has been experiencing areal expansion by encroaching agricultural land and other nonurban lands due to population increase and economic growth. From 1990 to 2000, the population of Marion County increased by approximately 7.9%.

The data sources for this research primarily came from two census data and two Landsat images, with the former used to extract socioeconomic variables and the latter for the environmental variables. The two census data were Census 1990 and

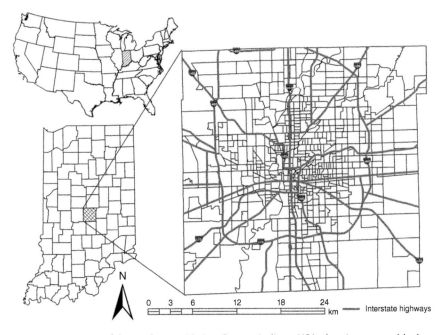

Figure 12.1 A map of the study area: Marion County, Indiana, USA, showing census block groups and major highways.

2000. The two Landsat images included one Thematic Mapper (TM) collected on 6 June 1991 (TM91) and one ETM+ acquired on 22 June 2000 (ETM+00). Both images were first georectified to a common Universal Transverse Mercator (UTM) coordinate system using 1 : 24000-scale topographic maps as reference. For each image, 25 ground control points were selected to generate coefficients for a first-order polynomial, and a nearest-neighbor method was applied to resample the image according to their original theoretical spatial resolution. The resultant values of root mean square were all found to be less than 0.4 pixel.

12.2 Population Estimation Using Landsat ETM+ Imagery

12.2.1 Methodology

Two methods for population estimation were explored (Li and Weng 2010), including (i) the combination of spectral bands, LST, vegetation abundance, and impervious surface; (ii) residential LU and dasymetric mapping based on land use. Figure 12.2 illustrates the analytical procedures for population

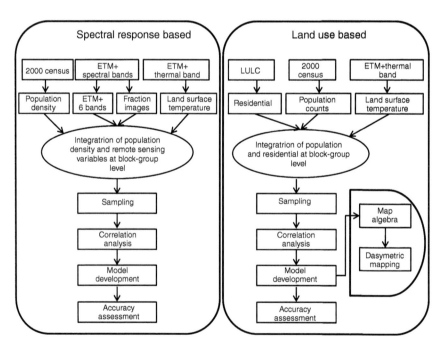

Figure 12.2 Procedures for population estimation. *Source:* Li and Weng (2010). Reproduced with permission of Taylor & Francis.

estimation used in this study. Four steps were involved in population estimation: (i) extraction of population from the census data; (ii) development of remote sensing variables; (iii) development of population estimation models by the integration of population and remote sensing-derived variables; and finally (iv) accuracy assessment.

12.2.1.1 Extraction of Population Parameters

The population data of Marion County at the BG level were extracted from the 2000 census data. Two population parameters were used in this research: total population and population density. Population density was calculated by dividing the total population in a BG by the corresponding area. Marion County had 658 BGs, with an average area of $1.59 \, \text{km}^2$ and an average population density of $1587/\text{km}^2$.

12.2.1.2 Extraction of Remote Sensing Variables

Linear Spectral Mixture Analysis (LSMA) approach has been widely used for deriving biophysical parameters from medium-resolution satellite images (Weng et al. 2004; Weng and Hu 2008; Weng and Lu 2009). It assumes that the spectral signature of the mixed pixel measured by a sensor is a linear combination of the spectral signatures of all pure components, called end-members, within the pixel (Adams et al. 1995; Roberts et al. 1998). In this study, LSMA was used to derive green vegetation and impervious surface fraction images. The detailed procedures were described by Lu and Weng (2006). Then, LST image was derived from ETM+ thermal band (thermal infrared [TIR]) (10.44–12.42 µm) with a spatial resolution of 60 m. A detailed description of the LST calculation can be found in Weng et al. (2004). Finally, the Maximum likelihood classifier was used to classify the ETM+ image. Housing density at block level, zoning data, and high spatial resolution aerial photographs was used to assist the selection of training samples. Eleven LULC classes (i.e. commercial areas, transportation, industrial areas, water, low-density residential, medium-density residential, high-density residential, grass, crop land, fallow, and forest) were initially classified. In the final classification result, six LULC classes were produced by combining commercial, transportation, and industrial as urban, and combining crop, fallow, forest, and grass as vegetation (Figure 12.3). Since no universal standards exist for separating residential density into high-, medium- and low classes, this study established its own criteria based on housing density and zoning data. The residential lands having housing units of more than $1300/\text{km}^2$ were assigned as high-density residential areas, those having housing units of less than $400/\text{km}^2$ as low-density residential areas, and those in between as medium-density residential areas. The resulting image was refined by developing a decision tree based on housing density at the block level. Fifty samples for each LULC type were randomly selected and compared to references which were collected from high spatial resolution aerial

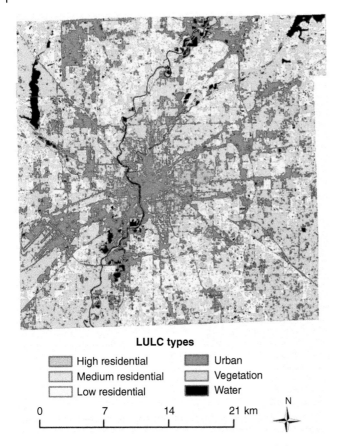

Figure 12.3 LULC map extracted from Landsat Enhanced Thematic Mapper plus (ETM+) imagery with the maximum likelihood classifier. *Source:* Li and Weng (2010). Reproduced with permission of Taylor & Francis.

photographs. Overall accuracy, producer's accuracy, user's accuracy, and kappa statistic were calculated based on the error matrices (Table 12.1). The overall classification accuracy reached 86%.

12.2.1.3 Data Integration and Model Development

Since population data at BG had different data format and spatial resolution with remote sensing data, it is necessary to conduct data conversion before implementing further analyses. Remote sensing-derived variables (e.g. spectral bands, fraction images, and LST) were aggregated to the BG level and the mean values of these variables for BG were then computed. LULC types were summarized as the count of pixels falling within a BG. In addition, the sum of temperature values for residential LU types within a BG was also calculated. All

Table 12.1 Accuracy assessment of land-use and land-cover classification result.

Land-use and land-cover (LULC) type	Producer's accuracy (%)	User's accuracy (%)
Water	97.9	95.9
Urban	86.6	80.6
Residential-H	51.6	72.7
Residential-M	81.4	74.5
Residential-L	76.3	63.0
Vegetation	91.9	97.0
Overall accuracy	86.0%	
Overall kappa	0.81	

Source: Li and Weng (2010). Reproduced with permission of Taylor & Francis.

these remote sensing-derived variables, as well as population data, were stored as the attributes of BGs and used for statistical analyses.

A random sampling technique was used to extract samples based on 25% of the total number of BGs. Initially, a total of 174 from 658 BGs were selected. A threshold of 2.5 standard deviations was used to identify the outliers for the samples during regression processing. Pearson's correlation coefficient was used to explore the correlation between population parameters and remote sensing variables. Furthermore, stepwise regression analysis was conducted to identify suitable variables for development of population estimation models.

There are two classes of statistical regression models depending on population parameters. The first category is designed to indicate the relationship between average population density in each census unit and scale-invariant indicators derived from remote sensing imagery (Harvey 2002a; Wu and Murray 2005). The spectral response-based method belongs to this category. Spectral response-based population estimation was carried out with independent variable of six Landsat ETM+ spectral bands, fraction images, and LST. Population density was used as the dependent variable. Six ETM+ spectral bands were tested first, vegetation fraction, impervious surface, and LST were then incorporated, and a series of regression models were developed. Another category of regression models aims at showing the relationship between population counts in each census unit and a number of scale-dependent indicators (e.g. pixel counts) from remote sensing data for the corresponding census unit (Langford et al. 1991; Sutton et al. 1997). LU-based population estimation in this study belongs to this category. It was conducted based on the urban LULC classification image. Population count was used as the dependent variable, and the areas of each type of residential land were used as independent variables. Three types of residential use were examined by using following equation:

$$P_i = D_l * A_{li} + D_m * A_{mi} + D_h * A_{hi} \qquad (12.1)$$

where P_i is total population in census BG i; A_{li}, A_{mi}, and A_{hi} are pixel counts of low, medium, and high residential land falling within the BG i, respectively; D_l, D_m, and D_h are coefficients of A_{li}, A_{mi}, and A_{hi}, respectively. Because the Eq. (12.1) did not take account of the variation within a particular residential LU type, LST was then incorporated into the regression model with an assumption that population had strong correlation with LST.

$$P_i = D_l * A_{li} + D_m * A_{mi} + D_h * A_{hi} + \alpha * T_i \qquad (12.2)$$

where T_i is the sum of temperature of residential pixels in BG i, and α is coefficient of T.

The population estimation for each BG was conducted by the models described above. The dasymetric mapping method was used to redistribute population onto the LULC map. Different types of residential land were first extracted from the LULC map as binary images. For each type of residential land, population for each pixel was calculated by multiplying a binary image with the corresponding coefficients obtained from the regression models. When LST was incorporated, the temperature of residential pixels within a BG was also used as a coefficient. Dasymetric population map was obtained by adding these images together. On the dasymetric population map, each pixel value was population count. Because population was redistributed only onto residential lands, other LULC types had a zero value. In addition, due to constant pixel size, the population count can be considered as "population density" too, except that the total area was $900\,m^2$, not $1\,km^2$.

12.2.1.4 Accuracy Assessment

When a developed model is applied for prediction, there are some discrepancies between estimates and true values, and the differences are called residuals. It is necessary to examine whether the model fits with training data sets (the process of internal validation) or fits with other data sets that are not used as training sets (the process of external validation) (Harvey 2002a). Mean relative error (RE) is often calculated to indicate estimation accuracy.

$$\text{Mean relative error}\,(RE) = \frac{\sum_{k=1}^{n} \left| (Pg - Pe)/Pg \right|}{n} \times 100 \qquad (12.3)$$

where Pg and Pe are the referenced and the estimated values of either the population or population density, respectively, and n is the number of BGs used for accuracy assessment. The smaller the RE is, the better a model is. In this study, all BGs were used to assess population estimation models. In addition, a residual distribution map was generated for the selected model and is used to show the geographical distribution of residuals. Higher positive residuals indicate larger underestimation, while higher negative residuals imply larger overestimation.

12.2.2 Results

12.2.2.1 Population Estimation with Spectral Response-Based Methods

It is found that Landsat ETM+ bands 4 (near infrared) and 5 (middle infrared) were significantly correlated with population density. Impervious surface had better correlation with population density than vegetation abundance. Of the nine selected variables, LST was the most strongly correlated with population density. The stepwise regression analysis based on ETM+ reflective bands shows that the coefficient of determination (R^2) was only 0.18 with bands 1 and 5 as explanatory variables (Table 12.2). Obviously, ETM+ bands alone were not sufficient to accurately estimate population density. Table 12.2 further indicates that the incorporation of LST into the regression model can significantly increase R^2 value (from 0.18 to 0.48). The following was the best estimation model based on these variables:

$$PD = -112582.438 + 388.179 * TEMP - 61.511 * B7 \tag{12.4}$$

where *TEMP* is the mean value of temperature of BG, and *B7* is the mean value of band 7 (middle infrared) of BG. This model was then used to calculate population density for each BG. The RE was computed.

The distribution of residuals from this model was illustrated in Figure 12.4. The BGs with extreme high population density (especially greater than 3000/km^2) were highly underestimated, while those BGs with extreme low density (less than 400/km^2) were largely overestimated, leading to high RE.

Table 12.2 The best population estimation models based on spectral data.

Model	Potential variables	Explanatory variables	R^2	Relative error (RE) (%)
1	Enhanced Thematic Mapper plus (ETM+) bands	B1-Mean, B5-Mean	0.18	
2	ETM+ bands, fractions, and land surface temperature (LST)	TEMP-Mean, B7-Mean	0.48	237
Model 1	PD = 4083.082 – 61.875 * B5-Mean + 36.677 * B1-Mean			
Model 2	PD = –112 582.438 + 388.179 * TEMP-Mean – 61.511*B7-Mean			

Source: Li and Weng (2010). Reproduced with permission of Taylor & Francis.

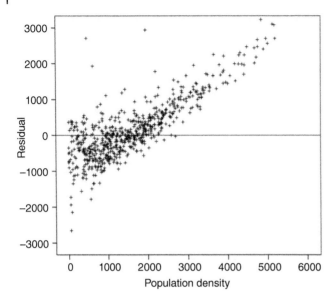

Figure 12.4 Residual distribution with population density based on population estimation model 2. *Source:* Li and Weng (2010)). Reproduced with permission of Taylor & Francis.

12.2.2.2 Population Estimation with the LU-Based Method

It is found that correlation between population counts and residential LU became stronger from low- to medium- and high-density residential LU (from 0.392 to 0.758 and 0.824). The sum of LST had the strongest correlation with population (0.905). Two models based on these variables were then developed and summarized in Table 12.3. These models were forced through the origin point based on the assumption that no people resided in BGs that had no residential LU. Model 3 was generated by using only residential LUs as explanatory variables, and Model 4 was developed by adding LST as an additional independent variable. The RE shows that the two models yielded a similar accuracy.

A comparative analysis indicates that the LU-based method provided better estimation than the spectral response-based method in terms of RE. The Model 3, which was based on residential LUs, provided the best result for this study. Therefore, it was chosen to estimate population for the study area.

A residual distribution map based on Model 3 was created (Figure 12.5 left panel). Blue and cyan indicated the overestimated population, and red and orange indicated the underestimated population. Comparing the residual map to the population density distribution from census (Figure 12.5 right panel), it is found that high-density BGs located in the center of the city tended to be underestimated. However, for the low-density BGs, especially in rural areas such as the northwestern, southwestern, and southeastern corners of the city, it appeared that overestimate was a tendency. In such areas, residential

Table 12.3 Population estimation models based on land-use data.

Model	Explanatory variables	R^{2a}	*RE* (%)
3	Residential-L	0.98	21.4
	Residential-M		
	Residential-H		
4	Residential-H	0.98	21.4
	Land surface temperature		
Model 3	$P = 0.654445*\text{Residential-L} + 0.711886*\text{Residential-M} + 2.319887*\text{Residential-H}$		
Model 4	$P = 0.002334*\text{TEMP-sum}^{b} + 1.524954*\text{Residential-H}$		

Source: Li and Weng (2010). Reproduced with permission of Taylor & Francis.
[a] For regression through the origin (the no-intercept model), R^2 measures the proportion of the variability in the dependent variable about the origin explained by regression. This cannot be compared to R^2 for models which include an intercept.
[b] TEMP-sum is the sum of temperature of residential pixels within a block group.

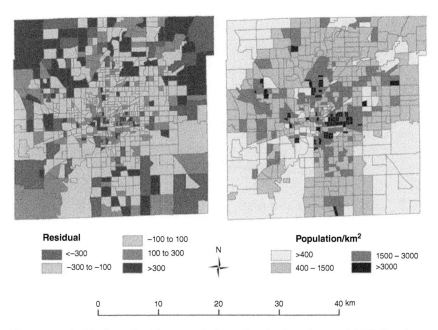

Figure 12.5 Residuals resulted from population estimation based on model 3 (left) and population density from census (right). *Source:* Li and Weng (2010). Reproduced with permission of Taylor & Francis.

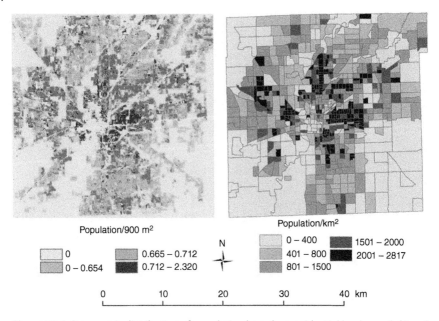

Figure 12.6 Dasymetric distribution of population based on residential land uses (left) and estimated population density based on model 3 (right). *Source:* Li and Weng (2010)). Reproduced with permission of Taylor & Francis.

dwellings were scattered and highly mixed with vegetation, thus were difficult to separate from surrounding vegetation.

The estimated population based on Model 3 for each BG was redistributed within residential classes using the dasymetric method. The dasymetric map shows the real population distribution pattern, on which LULC types such as agriculture, forest, and water, had no people (Figure 12.6 left panel), while the presentation of population distribution by a choropleth map (Figure 12.6 right panel) did not conform to the population distribution patterns in the real world.

12.2.3 Discussion and Conclusions

This research compared two types of population estimation methods, spectral response based and LU based. In the spectral response method, impervious surface and vegetation image fractions and LST were used as explanatory variables in addition to traditional variables using spectral radiance. The results indicate that when applied to the census BG, Landsat ETM+ spectral bands alone cannot provide satisfied estimation results, but the combination of LST and spectral bands can improve the estimation performance. The LU-based method provided better estimation performance than the spectral-based method. The LU-based prediction either by stratifying the density of residential

LUs or by the combination of residential LU and LST can produce reasonable estimation. Dasymetric mapping provided a vivid representation of spatial distribution of population.

The use of remotely sensed data to estimate population is still a challenging task both on theory and methodology due to remote sensing data per se, the complexity of urban landscape, and the complexity of population distribution. An important concern is the selection of population indicators from remote sensing data. The most commonly applied indicators in previous studies are spectral radiance and their transforms. Since Iisaka and Hegedus (1982) first used spectral radiance based on SPOT image to estimate Tokyo population, there were several attempts to explore the spectral radiance and its transforms as population predictors in zonal statistical models (Lo 1995; Harvey 2002a, 2002b; Li and Weng 2005; Wu and Murray 2005). However, spectral radiances are not directly related to population. For example, the areas with low population density could be located in industrial or commercial areas or in forest-/agriculture-dominated areas, but the spectral characteristics of these landscapes are significantly different. In addition, the aggregation of spectral radiances to the zonal unit level on which the population information are collected altered the inherent relationships between population and spectral radiance due to the effect of modifiable areal unit problem (MAUP). Studies by Lo (1995), Harvey (2002a) this study showed that direct use of mean values of spectral radiance could produce significant errors in population/population density. Using the transforms of spectral radiance in population estimation models (Harvey 2002a; Wu and Murray 2005) significantly increased the correlations between population density and remote sensing variables. However, the selection of transforms seemed to be arbitrary, and the variables used in regression model may highly correlate to each other, thus multicollinearity may exist in multiple regression. Incorporation of impervious surface fraction, vegetation surface fraction, and LST into population estimation models improved model performance, which may also strengthen the rationale for indicator selection and eliminated collinear variables from the model. It appears that residential impervious surface image fraction is an excellent population indicator due to both its stability and underlying relationship with population (Wu and Murray 2005; Lu et al. 2006). However, the development of high quality of impervious surface data and the differentiation of nonresidential (e.g. commercial, industrial, and transportation) from residential impervious surfaces warrant further studies. Moreover, under certain circumstances, the same amount of impervious surface may have significantly different population density because of different patterns of residential uses. Thus, LULC data, especially those with high categorical resolution, possesses irreplaceable advantages in population estimation. In this research, different classes of residential use (high, medium, and low) were distinguished by the aid of ancillary data (i.e. housing density and zoning data) and regression models were then developed using these residential classes as explanatory variables. This procedure has demonstrated to be able to provide a good estimation result. This finding is agreement with the studies of Lo (1995) and Wu and Murray (2005). Both studies applied high and low residential LUs in development of population estimation models. Another advantage for using LULC data is that they are more stable predictors compared with spectral responses and can be obtained from different remote sensing data sources. Therefore, the LU-based method exemplified in this study may be easily transferred to other data sets or other study areas for population estimation. However, the definitions of different LULC types, especially different residential LU, are somewhat subjective. Moreover, extraction of accurate urban LULC information from moderate spatial resolution satellite images presents another challenge. The areas with high population density are mainly located in urban regions with multistory buildings, with which optical remote sensing data may not be capable for identifying vertical information such as height and structure of multistory buildings. In addition, low residential and residential areas scattered in the forest and agricultural areas can have highly similar spectral responses making them difficult to separate.

A major uncertainty of population estimation comes from low-density residential areas with overestimation, and from high-density residential areas with underestimation. This study has shown that the use of residential impervious surface, different densities of residential LULC classes, and LST can partially solve this problem. The incorporation of building height information seems to be another useful approach. Since LiDAR data can provide building height information, incorporation of LiDAR data with impervious surface or with residential classes may provide new insights for population estimation, especially for high-density residential areas.

12.3 Assessing Urban Environmental Quality Change

12.3.1 Methodology

Figure 12.7 illustrates the procedure of developing two UEQ synthetic indexes and evaluating the UEQ changes. This section explains the details of the analysis (Liang and Weng 2011).

12.3.1.1 Extraction of Environmental Variables

In order to provide a complete representation of the UEQ condition in the physical environmental aspect for the study area, two LULC maps were first derived from each Landsat image with different classifiers. The land-cover (LC) map was given by performing the unsupervised classification using the ISODATA (Interactive Self-Organizing Data Analysis Technique) clustering

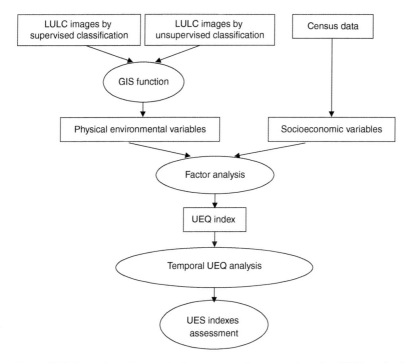

Figure 12.7 Procedures for developing urban environmental quality (UEQ) synthetic index and evaluating the UEQ changes. *Source:* Liang and Weng (2011). Reproduced with permission of IEEE.

method and the maximum likelihood decision rule. The resultant LC categories included cropland and pasture (UCroPas), water (UWater), urban built-up lands (UBuiltup), forest (UForest), grass (UGrass), and barren lands (UBarren). The overall accuracy obtained from these two maps was 88.80% for TM91 and 89.00% for ETM+00. The LU map was produced using a supervised maximum likelihood classifier. Eight categories were identified: cropland and pasture (SCroPas), water (SWater), commercial and industrial (SComInd), high-density residential (SHDR), medium-density residential (SMDR), low-density residential (SLDR), forest (SForest), and grass (SGrass). The selection of training areas used aerial photographs that were taken around the time of image acquisitions and the results of the unsupervised classification. Zoning data, which is primarily related to land use, was also used in choosing training and test sites for high-, medium-, and low-density residential (Fung and Siu 2001). The overall accuracy for the two maps was found to be 76.25% for TM91 and 75.50% for ETM+00.

Because temperature is also recognized as an important index of human thermal comfort, it was also derived from the TIR bands of TM91 and ETM+00 using the Weng et al. method (Hsu 1971). Basically, three steps were involved in the procedure: first, converting the digital number (DN) values of the thermal band into spectral radiance; second, converting the spectral radiance to at-satellite brightness temperature, namely, blackbody temperature; and finally adjusting the blackbody temperature to LST by incorporating emissivity biases due to LC differences. The last step was accomplished using the LULC maps produced by the unsupervised spectral classification.

Several transformed bands on the basis of reflective channels were additionally created from the two satellite images. They included NDVI and two more indexes that were essentially derived from the Landsat reflective bands (e.g. Normalized Difference Water Index [NDWI] and Normalized Difference Built-up Index [NDBI]) (Li and Weng 2005). NDWI is an indicator for water content within vegetation and it is ideal to be combined with NDVI for representing the state of vegetation (Li and Weng 2005). By setting suitable threshold values, it is suggested all these indexes can effectively differentiate different LULC types (e.g. vegetation and builtup) (Li and Weng 2005). Besides, the second principal components (PC2) extracted from PCA were also produced to serve as an additional vegetation indicator for extracting environmental variables. Although the derived first principal component contains the majority information of all of the original bands, it does not specifically refer to any indicative environmental information and thus was not used in the study.

12.3.1.2 Extraction of Socioeconomic Variables from Census Data

The social-economic indicators were derived from the census data for 1990 and 2000 at the BG level. These data together with the related Topologically Integrated Geographic Encoding and Referencing system (TIGER) shape files were downloaded from US Census Bureau (Liang et al. 2007). Since census BG data contained up to hundreds of variables regarding all aspects of socioeconomic condition, a total of 30 variables relating to the variables commonly used in previous UEQ studies (i.e. population density, housing density, and median family income) were initially extracted from the two census data sets, respectively (Haack et al. 1997; Bonaiuto et al. 2003; Chander Markham 2003; Chen et al. 2006; Briggs et al. 2007). After a series of calculation procedure, up to 13 socioeconomic variables were determined for this research: population density (PD), median age population (MedAge), households (HH), house units (HU), vacant house units (VHU), owner-occupied house units(OHU), median house value (HV), medium house income (MHI), median family income (MFI), per capita income (PCI), percentage of families under poverty (POV), unemployment rate (UNEMP), and percentage of college graduates (PCG). All these variables were computed for each BG of census 1990 and 2000. It was noted that the two census data sets do not contain the same number of BGs – census 1990 has 703 BGs and census 2000 has 658. Nor do they share a consistent coverage for all areal units. Therefore, the two data sets were preprocessed to remove those "different" BGs to maintain a common 614 analytical units. The modified census data were utilized as the basis for later analysis.

12.3.1.3 Data Integration of Environmental and Socioeconomic Variables

Factor analysis is one of the major approaches to integrate environmental parameters derived from remotely sensed images and socioeconomic variables extracted from census data (Fabos and Petrasovits 1984; Bonaiuto et al. 2003; Chander Markham 2003; Chen et al. 2006; Briggs et al. 2007; Liang and Weng 2011). Among all, PCA is used the most for UEQ analysis and was also employed in this study. Before running PCA, all the derived environmental and socioeconomic variables were aggregated to all 614 BGs. Pearson's correlation coefficients (r) were then computed to provide a preliminary analysis of the relationships among the variables. Major principal components representing the majority variance of the original data sets were thus extracted.

12.3.1.4 UEQ Index Development and Validation

The derived principal components were used to develop a synthetic UEQ index after being weighted by the percentage of variance that each component explained. Accuracy assessment is needed for validating the effectiveness of the developed UEQ synthetic indexes. A few approaches have been proposed by using regression modeling (Bonaiuto et al. 2003; Briggs et al. 2007) and sociological methods (Fabos and Petrasovits 1984). In this study, the linear regression analysis was performed to assess the robust of the UEQ indexes by using the original variables that showed the strongest correlation with the considered components as predictors. A coefficient of determination (R^2) was calculated to examine the effectiveness of the developed regression models.

12.3.2 Results

12.3.2.1 Factor Analysis Results

Factor analysis was used to reduce the initial 32 UEQ variables. The principal component method of extraction aims at identifying uncorrelated components that account for variation in the original variables and ordering them based on their variation. Since only a few components will account for most of the variation, they are identified as the key factors to represent the original variables. In this way, the number of variables can be reduced.

For all input variables depicting the condition of the UEQ in 1990 and 2000 for the study area, the initial results of factor analysis concerned that of Kaiser–Meyer–Olkin (KMO) and Bartlett's test values. These two indexes are often used to examine the suitability of data for factor analysis (Bonaiuto et al. 2003). The data are acceptable for factor analysis only when their KMO is larger than 0.50 and the significant levels of Bartlett's test is less than 0.10 (Bonaiuto et al. 2003). With the significant levels of Bartlett's test all equaled to 0.000, the KMOs of the two data sets were 0.78 and 0.50, respectively, implying their suitability for factor analysis. All the 32 variables were then checked by their communalities in order to further validate their capabilities in factor analysis (Table 12.4).

Table 12.4 Summary of communality for 33 variables and 31 variables.

	1990 Communality		2000 Communality	
	32 Variables	28 Variables	32 Variables	29 Variables
SCroPas	0.95	0.95	0.95	0.95
SWater	0.92	0.92	0.94	0.94
SComInd	0.78	0.78	0.92	0.92
SHDR	0.47		0.47	
SMDR	0.48		0.40	
SLDR	0.78	0.78	0.75	0.75
SForest	0.86	0.86	0.81	0.81
SGrass	0.98	0.98	0.97	0.97
UCroPas	0.93	0.93	0.91	0.91
UWater	0.94	0.94	0.93	0.93
UBuiltup	0.85	0.85	0.91	0.91
UForest	0.88	0.88	0.90	0.90
UGrass	0.99	0.99	0.98	0.98
UBarren	0.67	0.67	0.85	0.85
Temperature	0.41		0.86	0.86
NDBI	0.84	0.84	0.84	0.84
NDWI	0.90	0.90	0.94	0.94
NDVI	0.95	0.95	0.75	0.75
PC2	0.88	0.88	0.85	0.85
Population density	0.70	0.70	0.72	0.72
Median age population	0.93	0.93	0.95	0.95
Households	0.97	0.97	0.96	0.96
House unit	0.98	0.98	0.96	0.96
Occupied house unit	0.97	0.97	0.89	0.89
Vacant house unit	0.72	0.72	0.62	0.62
Median house income	0.88	0.88	0.83	0.83
Median family income	0.92	0.92	0.88	0.88
Per capital income	0.93	0.93	0.89	0.89
Median house value	0.77	0.77	0.74	0.74
Percentage of college graduates	0.78	0.78	0.77	0.77
Percentage of families under poverty	0.55	0.55	0.57	0.57
Unemployment rate	0.47		0.41	

Source: Liang and Weng (2011). Reproduced with permission of IEEE.

Communality serves as an indicator of the amount of variance that is accounted by each variable with the consideration of the factors in the factor solution. Generally speaking, small communalities correspond to variables which do not fit well with the factor solution and hence should be removed from the analysis. Table 12.4 indicates that three variables, SHDR, SMDR, and UNEMP detected small communalities (less than 0.50) for both data sets. Temperature also observed a small communality for the 1990 data set. All these variables were thus dropped in the subsequent analysis. At the end, only 28 variables of the 1990 data set but 29 of the 2000 were used for the factor analysis.

Using the rule of minimum eigenvalue should be larger than 1.00 (Elvidge et al. 1995), four factors were extracted from each data set (Table 12.5). To assist the interpretation of resultant factors, factor solution was rotated using the varimax approach. Overall, for the 1990 data set, the first factor accounted about 34.38% of the total variance; the second factor 18.40%, the third factor 13.08%, and the fourth factor 8.11%. Together, the first four factors explained approximately 74% of the variance. For the 2000 data set, the first factor accounted about 35.44% of the total variance; the second factor 18.21%, the third factor 9.68%, and the fourth factor 8.53%. Together 72.86% of the total variance was accounted for by the first four factors.

Table 12.5 reports the factor loadings for each variable on the four selected factors after rotation. In this study, the analysis of factor loadings only considers variables with loadings larger than 0.71, which is suggested as excellent (Lindgren 1971). For the 1990 data set, Factor 1 had large positive loadings with four LULC variables – SCroPas (0.96), SGrass (0.88), UCroPas (0.96), and UGrass (0.87). Clearly, this factor was strongly linked to the greenness of environment. The higher the score in Factor 1 is, the better the UEQ in the greenness aspect is. Factor 2 had high positive loadings with two population characteristics (HH – 0.95 and MedAge – 0.92) and three house density properties (HU – 0.97, OHU – 0.95, and VHU – 0.79). Consequently, this factor indicated the degree of crowdedness of the study area. The higher the score of Factor 2 is, the less the space for people to reside is available, thus leading to a poor living environment. Apparently, Factor 3 exhibited a strong positive correlation with all three income variables (MHI – 0.90, MFI – 0.94, and PCI – 0.95), the education variable (PCG – 0.87), and the house value variable (HV – 0.82). This was the factor related to material welfare. The higher the score in Factor 3 is, the better the city in its economic status is. Only three variables were found to have loadings larger than 0.71 in Factor 4 and they were all derived from the images' transformed index bands: NDWI (0.83), NDVI (0.93), and PC2 (0.83). Factor 4 was obviously associated to the scenic amenity of the region. The larger the score of Factor 4 is, the better the physical environmental quality is. Figure 12.8 shows the geographic distribution of the four factors. Overall, their distribution is subject to the spatial pattern possessed by the original variable that had the highest loadings with them. In 1990, the four

Table 12.5 Rotated factor loading matrix for the 1990 and 2000 data.

	1990				2000			
	Factor 1	Factor 2	Factor 3	Factor 4	Factor 1	Factor 2	Factor 3	Factor 4
SCroPas	**0.96**	0.15	0.02	-0.05	0.18	**0.95**	0.07	0.00
SWater	0.14	0.11	0.12	-0.04	0.24	0.12	0.11	-0.06
SComInd	0.31	0.28	-0.01	-0.25	0.34	0.39	0.00	-0.27
SLDR	0.14	0.51	0.29	0.38	0.67	0.10	0.25	0.22
SForest	0.63	0.07	0.14	0.15	0.10	0.67	0.16	0.05
SGrass	**0.88**	0.22	0.11	0.10	0.38	**0.83**	0.08	0.03
UCroPas	**0.96**	0.08	0.02	-0.05	0.11	**0.95**	0.01	-0.02
UWater	0.18	0.15	0.13	-0.03	0.25	0.13	0.13	-0.05
UBuildup	0.36	0.39	0.02	-0.16	0.57	0.40	0.02	-0.22
UForest	0.61	0.08	0.20	0.19	0.19	0.70	0.24	0.11
UGras	**0.87**	0.20	0.09	0.04	0.39	**0.81**	0.11	0.06
UBarren	0.61	0.11	0.04	-0.04	0.01	0.16	0.01	0.04
Temp					**-0.09**	-0.24	-0.31	**-0.72**
NDBI	0.38	0.02	0.00	-0.58	**0.01**	**0.14**	-0.17	**-0.83**
NDWI	-0.14	-0.01	0.28	**0.83**	0.04	0.01	0.33	**0.89**
NDVI	0.04	0.07	0.28	**0.93**	-0.02	-0.08	0.13	**0.86**
PC2	0.25	0.11	0.20	**0.83**	0.14	0.24	0.01	**0.76**

PD	-0.14	0.15	-0.32	-0.13	0.07	-0.16	-0.19	0.00
MedAge	0.24	**0.92**	0.08	0.10	**0.94**	0.22	0.08	0.01
HH	0.16	**0.95**	0.12	0.10	**0.95**	0.20	0.12	0.04
HU	0.14	**0.97**	0.09	0.06	**0.96**	0.19	0.10	0.03
OHU	0.16	**0.95**	0.11	0.09	**0.78**	0.32	0.17	**0.14**
VHU	-0.06	**0.79**	-0.11	-0.19	0.70	0.03	-0.12	-0.07
MHI	0.13	-0.04	**0.90**	0.20	0.06	0.15	**0.84**	0.25
MFI	0.08	0.00	**0.94**	0.16	0.07	0.10	**0.90**	0.20
PCI	0.03	0.04	**0.95**	0.07	0.05	0.04	**0.93**	0.10
HV	0.10	0.15	**0.82**	0.19	0.07	0.09	**0.85**	0.12
PCG	-0.03	0.15	**0.87**	0.04	0.10	0.00	**0.86**	0.04
POV	-0.04	-0.04	-0.55	-0.35	-0.11	-0.06	-0.49	-0.20
Initial eigenvalue	9.63	5.15	3.66	2.27	10.28	5.28	2.81	2.48
% of variance	34.38	18.40	13.08	8.11	32.44	18.21	9.68	8.53
Cumulative %	34.38	52.78	65.85	73.96	35.44	53.64	63.32	71.86

Source: Liang and Weng (2011). Reproduced with permission of IEEE.
Note: Loading larger than 0.750 are shown in bold.

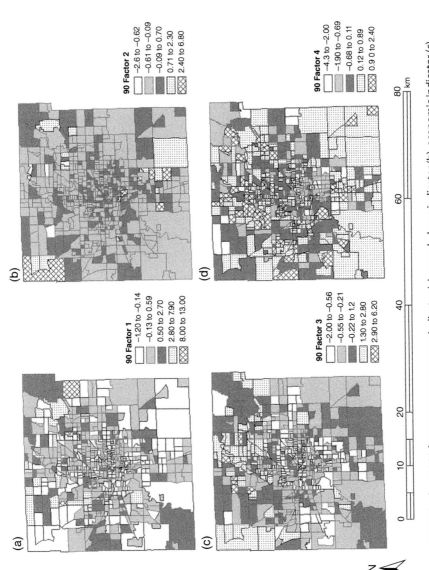

Figure 12.8 The four 1990 factor scores: greenness indicator (a), crowdedness indicator (b), economic indicator (c), and scenic amenity indicator (d). *Source:* Liang and Weng (2011). Reproduced with permission of IEEE.

factors had the highest loadings with cropland and pasture (SCroPas and UCroPas), HU, PCI, and NDVI, respectively, and they hence displayed a similar pattern to the ones relative to these variables.

The examination of the factor loadings reported by the 2000 data set demonstrated similar results. The only exception was the corresponding information represented by different factors was rearranged in the 2000 data. In this case, the first factor was directed to crowdedness aspect of the city since it had strong positive loadings with HH (0.95), MedAge (0.94), HU (0.96), and OHU (0.78). Factor 2 could be viewed as the greenness indicator since its higher loadings were all provided by LULC types of SCroPas (0.96), UCroPas (0.65), SGrass (0.88), and UGrass (0.87). The interpretation of Factors 3 and 4 revealed comparable results that have been observed from the 1990 data. With high loadings of all income variables (MHI – 0.84, MFI – 0.90, and PCI – 0.93), the education variable (PCG – 0.86), and HV (0.85), Factor 3 was also an indicator of depicting the economic image of the region. Finally, Factor 4 had significant positive loadings with NDWI (0.89), NDVI (0.86), and PC2 (0.76), but strong negative loadings with temperature (–0.72) and NDBI (–0.83). Consequently, this factor was the best in illustrating the overall scenic amenity status of the study area. The higher the score in Factor 4 is, the better the UEQ in its physical environment aspect is. Figure 12.9 shows the geographic distribution of the four factors. Just like what have revealed from the 1990 data, the distribution of the four factors was comparable to those of the HU, cropland, and pasture (SCroPas and UCroPas), PCI, and NDWI, since they had the largest loadings with these variables.

Generally speaking, the extracted factors were all suited to be used as the indicators of the environmental quality of the study area in different dimensions in the two years: greenness, crowdedness, economic status, and scenic amenity. The difference of the weight of individual indicators in describing the general UEQ of the study area of that period demonstrated a significant change over the 10 years. In the 1990 when the city was less developed, the overall UEQ was primarily depicted by green vegetation. However, after 10 years of construction and development, the environmental quality of the study area was mainly accounted by house units, which indicated the environment had been significantly affected by socioeconomic factors.

12.3.2.2 UEQ Synthetic Index and UEQ Change Analysis

Since different factors represented UEQ in various aspects, it is necessary to combine all them into a synthetic index. Two UEQ models were thus developed and utilized in this work. Because the greenness, economic, and scenic amenity indicators were contributing positively to UEQ, they were assigned a positive sign. The crowdedness indicator, however, was treated as a negative contributor to UEQ and a negative sign was hence attached to it.

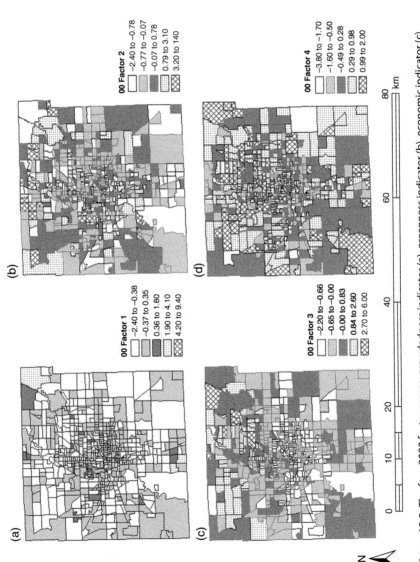

Figure 12.9 The four 2000 factor scores: crowdedness indicator (a), greenness indicator (b), economic indicator (c), and scenic amenity indicator (d). *Source:* Liang and Weng (2011). Reproduced with permission of IEEE.

The percentage of variance that each factor explained was used as weight to create the following two equations:

$$UEQ_{1990} = \frac{\left(34.38 \times \text{Factor}\,1 - 18.40 \times \text{Factor}\,2 + 13.08 \times \text{Factor}\,3 + 8.11 \times \text{Factor}\,4\right)}{100}$$

$$(12.5)$$

$$UEQ_{2000} = \frac{\left(-35.44 \times \text{Factor}\,1 + 18.21 \times \text{Factor}\,2 + 9.68 \times \text{Factor}\,3 + 8.54 \times \text{Factor}\,4\right)}{100}$$

$$(12.6)$$

The UEQ score for each BG was then calculated and the results are illustrated in Figure 12.5. In general, the UEQ score of 1990 ranged from −1.64 to 4.02 with a standard deviation of 0.42. This data were then normalized between 0 and 1 (Figure 12.10a). Nearly 3% of BGs had UEQ scores higher than 0.50 at the new scale. These BGs scattered around the study area from the inner city to the city edge with some located in the northeast side and some centered in the middle south. They correspond to the rich UEQ zones characterized with stronger greenness, larger living space, and higher income status. As approaching the city's downtown, the UEQ conditions became poorer. BGs having the lowest UEQ scores were randomly found throughout the county. These poor UEQ zones were primarily linked to low capital income together with either predominant little greenness or high population density or the mixture of both. With a range of −3.06 to 2.89, the original UEQ score of 2000 also resulted in a moderate standard deviation (0.42) across the region. After being standardized to the new scale of 0–1 range, approximately 3% of BGs had UEQ scores greater than 0.6 (but over 75% were higher than 0.5) and most of them were found in the city's suburban area with some centered the mid-north and northwest corner and some located in the mid-south and southeast corner (Figure 12.10b). Just like those good UEQ zones identified in 1990, these BGs were characterized with lower population density, more abundant greenness, and better economic status. Most BGs with poorer UEQ scores (less than 0.3) were presented in the northern part of the county outside the downtown fringe mainly due to the cluster of houses.

Significant environmental quality changes were uncovered when the two maps were compared. It is found that the best UEQ zones in 1990 were all transformed to less favorable zones in 2000. The environmental quality of most BGs in the south and the southeast was generally improved during the 10 years of development. In 1990, the city center was recognized as the mixture of all kinds of UEQ conditions with tendency to poor environment. In 2000, most BGs in this area were classified as medium-level UEQ zones. Spatially, the whole city suffered from more poor UEQ zones in 1990 than in

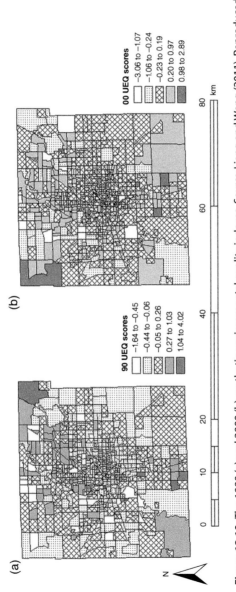

Figure 12.10 The 1990 (a) and 2000 (b) synthetic environmental quality indexes. *Source:* Liang and Weng (2011). Reproduced with permission of IEEE.

the year of 2000. Most of the BGs identified as poor in 1990 have been classified as medium-level UEQ in the 2000 map. Most of the medium-level UEQ zones in the 1990 map, however, were identified as poor zones in the 2000 map. It should be noted that the determination of UEQ zones in the two maps was somewhat different, subject to the UEQ models that were applied. Although UEQ synthetic indexes were developed using the same indicators (greenness, crowdedness, economic, and scenic amenity) and the same modeling procedure, each indicator was weighted differently. This difference in weight further suggests the temporal change of UEQ's connotation. As the urban environment changed over the time, the components previously recognized as important in constituting the synthetic UEQ index may now become less important. In this study, greenness was recognized as the most crucial factor in 1990. Nevertheless, it became less crucial in 2000, and the most critical factor for 2000 was crowdedness.

For the sake of evaluating the effectiveness of the developed UEQ models, regression analyses were performed with the original variables that had the highest loadings with the corresponding factors as the independent variable and the synthetic indexes as the dependent variables and the results are reported in Table 12.6. The R^2 values showed that the two constructed UEQ models could account for more than 94% of the variance in the UEQ at the BG level in the two years. Overall, all four predictors did well in explaining UEQ variances in the two models, since the significance values of F were all as small as 0.00 (smaller than 0.05 – the confidence level chosen in this study). All these results demonstrated the strength of the linear regression relationships between the predictor and outcome variables and thus the validity of the two UEQ models.

Table 12.6 Summary of all regression models.

Models	R^2	Constant	Predictors	Coefficients
90 synthetic UEQ	0.94	−0.50	UCroPas	0.85
			House unit	−0.53
			Per capital income	0.28
			NDVI	0.25
00 synthetic UEQ	0.95	−2.24	House unit	−0.98
			SCroPas	0.61
			Per capital income	0.24
			NDWI	0.21

Source: Liang and Weng (2011). Reproduced with permission of IEEE.

12.3.3   Discussion and Conclusions

This research explored the potential of the integration of satellite images and census data in assessing UEQ within a GIS framework base on a case study in Indianapolis, Indiana, USA. For each year, four major factors were identified: greenness, crowdedness, economic status, and scenic amenity. By assigning different weights to each factor, the synthetic UEQ indexes were created in 1990 and 2000. The investigation of the two synthetic indexes revealed great changes of UEQ in the study area between 1990 and 2000.

In this study, since the UEQ input variables were mainly in two formats – raster images and vector GIS data – one critical issue centered on the integration of remote sensing and GIS. Ideally, a total integration is expected. Yet the accomplishment of this task requires the understanding of two technical impediments: the raster-vector data model dichotomy and the problem of data uniformity. Remote sensing is raster-oriented data collection while GIS is vector based. Both raster and vector data models represent two distinct approaches to geospatial data processing and analysis with their own pros and cons. To blur the raster-vector dichotomy, possible solutions often involve the conversion of both types of data into a common format either in raster's or in vector's. However, the conversion between the two formats can introduce significant errors (Lo 1997). In practice, the conversion of vector data into individual pixels seems to be less realistic, since technically disaggregating vector data to pixels is more difficult to obtain. Besides, GIS is increasingly applied in decision-making, planning, and environmental management. Hence, a vector format is more preferred in data integration for urban planning and environmental applications. The second technical obstacle for the data integration regards the problem of data uniformity. Remote sensing images are primarily used to depict gradients of continuous spectral information over space by changes in DNs recorded at each pixel. The census GIS data, however, are collected at an administrative unit level (e.g. census blocks, BGs, and tracts) where the spatial continuity property of recorded data is no longer kept throughout space. In other words, two kinds of data are acquired at different scales. Yet there is no single answer as to what levels (scales) the two data types can be best integrated. The current research took the census BGs as the basic aggregation unit with a raster to vector conversion performed. The application of aggregation data then introduces the notable MAUP that has been addressed in many previous literatures (Lo 1986a; Lo and Faber 1997; Jensen and Cowen 1999).

Since UEQ is the result of all factors (physical, ecological, and socioeconomic, etc.) that are spatially interacted with each other and this relationship changes with scales, the phenomenon is scale dependent in nature. Based on the scale of observation and measurement, its report may suggest promising at one scale but undesirable at another. Obviously, there remains difficult to identify the "optimal" scale for UEQ analysis. Further research efforts are needed to establish a commonly accepted UEQ definition, which would help to develop methods for repeatedly modeling the spatial context of UEQ and scale dependency of the phenomenon.

In general, the results of this study could impact policy assessment instruments and assist local governments and environmental agencies in monitoring UEQ. The two UEQ models established in the research can be applied to assist urban planners not only to evaluate the city's current UEQ condition, but also to devise efficient development polices to construct a more desirable future UEQ environment at the census BG level. LU planners and policy makers can thus make policy adjustments to encourage effective urban landscape and economic development for building a "green" and "prosperous" city (and hence a better UEQ environment) rather than a "crowded" city (and thus a poor UEQ environment). This then helps making essential progress towards achieving greater urban sustainability that emphasizes the long-term integration of environmental protection, economic development, and social health. Many studies have suggested a more sustainable city is featured by green space (Lo 1986b), economic growth (Lo 1995), and continuity of open spaces (Lo 2001), which correspond well with the key factors in determining a city's UEQ condition according to the current study. Evidently, cities with favorable UEQ environment hold the possibility of delivering a more sustainable future. Besides, the methodology developed in the research can be easily adopted to conduct cross-case and cross-time comparisons at a meaningful level. This leads to a better understanding of the relationship between UEQ and urban physical and socioeconomic condition at a larger scale. It hence lays the groundwork for expediting the realization of the long-term and holistic decision-making process of urban sustainability.

References

Abolina, K. and Zilans, A. (2002). Evaluation of urban sustainability in specific sectors in Latvia. *Environment, Development and Sustainability* 4: 299–314.

Adams, J.B., Sabol, D.E., Kapos, V. et al. (1995). Classification of multispectral images based on fractions of endmembers: application to land cover change in the Brazilian Amazon. *Remote Sensing of Environment* 52: 137–154.

Akbari, H., Rosenfeld, A.H., and Taha, H. (1990). Summer heat islands, urban trees and white surfaces. *ASHRAE Transactions* 96: 1381–1388.

Akbari, H., Shea Rose, L., and Haider, T. (2003). Analyzing the land cover of an urban environment using high resolution orthophotos. *Landscape and Urban Planning* 63: 1–14.

Arnold, C.L. and Gibbons, C.J. (1996). Impervious surface coverage: the emergence of a key environmental indicator. *Journal of American Planning Association* 62: 243–258.

Bederman, S.H. and Hartshorn, T.A. (1984). Quality of life in Georgia: the 1980 experience. *Southeastern Geographer* 24: 78–98.

Bonaiuto, M., Fornara, F., and Bonnes, M. (2003). Indexes of perceived residential environment quality and neighborhood attachment in urban environments: a confirmation study on the city of Rome. *Landscape and Urban Planning* 65: 41–52.

Briggs, D., Gulliver, J., Fecht, D., and Vienneau, D. (2007). Dasymetric modeling of small-area population distribution using land cover and light emission data. *Remote Sensing of Environment* 108: 451–466.

Chander, G. and Markham, B. (2003). Revised Landsat-5 TM radiometric calibration procedures and postcalibration dynamic ranges. *IEEE Transactions on Geoscience and Remote Sensing* 41: 2674–2677.

Chen, X.L., Zhao, H.M., Li, P.X., and Yin, Z.Y. (2006). Remote sensing image-based analysis of the relationship between urban heat island and land use/cover changes. *Remote Sensing of Environment* 104: 133–146.

Comrey, A.L. and Lee, H.B. (1992). *A First Course in Factor Analysis*, 2e. Hillsdale, NJ: Erlbaum.

Eicher, C.L. and Brewer, C.A. (2001). Dasymetric mapping and areal interpolation: implementation and evaluation. *Cartography and Geographic Information Science* 28: 125–138.

Elvidge, C.D., Baugh, K.E., Hobson, V.R. et al. (1997). Satellite inventory of human settlements using nocturnal radiation emissions: a contribution for the global toolchest. *Global Change Biology* 3: 387–395.

Elvidge, C.D., Baugh, K.E., Kihn, E.A. et al. (1995). Mapping city lights with nighttime data from the DMSP operational linescan system. *Photogrammetric Engineering & Remote Sensing* 63: 727–734.

Fabos, J.G. and Petrasovits, I. (1984). Computer-aided land use planning and management. *Research Bulletin* No. 693. Massachusetts Agricultural Experiment Station, pp. 25–45.

Fotheringham, A.S. and Wong, D.W.S. (1991). The modifiable areal unit problem in multivariate statistical analysis. *Environment and Planning A* 23: 1025–1044.

Fung, T. and Siu, W. (2000). Environmental quality and its changes, an analysis using NDVI. *International Journal of Remote Sensing* 21: 1011–1024.

Fung, T. and Siu, W. (2001). A study of green space and its changes in Hong Kong using NDVI. *Geographical and Environmental Modelling* 5: 111–122.

Goodchild, M.F., Anselin, L., and Deichmann, U. (1993). A framework for the areal interpolation of socioeconomic data. *Environment and Planning A* 25: 383–397.

Goodchild, M.F. and Lam, N.S. (1980). Areal interpolation: a variant of the traditional spatial problem. *Geoprocessing* 1: 297–312.

Green, N.E. (1957). Aerial photographic interpretation and the social structure of the city. *Photogrammetric Engineering* 23: 89–96.

Haack, B.N., Guptill, S.C., Holz, R.K. et al. (1997). Urban analysis and planning. In: *Manual of Photographic Interpretation*, 2e (ed. W. Philipson), 517–554. American Society for Photogrammetry and Remote Sensing.

Harvey, J. (2000). Small area population estimation using satellite imagery. *Statistics in Transition* 4: 611–633.

Harvey, J. (2002a). Estimation census district population from satellite imagery: some approaches and limitations. *International Journal of Remote Sensing* 23: 2071–2095.

Harvey, J. (2002b). Population estimation models based on individual TM pixels. *Photogrammetric Engineering & Remote Sensing* 68: 1181–1192.

Heynen, N. (2006). Green urban political ecologies: toward a better understanding of inner-city environmental change. *Environment and Planning A* 38: 499–516.

Hsu, S.Y. (1971). Population estimation. *Photogrammetric Engineering* 37: 449–454.

Iisaka, J. and Hegedus, E. (1982). Population estimation from Landsat imagery. *Remote Sensing of Environment* 12: 259–272.

Irvine, K.N., Devine-Wright, P., Payne, S.R. et al. (2009). Green space, soundscape and urban sustainability: an interdisciplinary, empirical study. *Local Environment* 14: 155–172.

Jensen, J.R. (2000). *Remote Sensing of the Environment: An Earth Resource Perspective*. Pearson Education, Inc.

Jensen, J.R. and Cowen, D.C. (1999). Remote sensing of urban/suburban infrastructure and socio-economic attributes. *Photogrammetric Engineering & Remote Sensing* 65: 611–622.

Ji, M. and Jensen, J.R. (1999). Effectiveness of subpixel analysis in detecting and quantifying urban imperviousness from Landsat Thematic Mapper imagery. *Geocarto International* 14: 33–41.

Keiner, M. (2004). Re-emphasizing sustainable development – the concept of "evolutionability". *Environment, Development and Sustainability* 6: 379–392.

Landorf, C., Brewer, G., and Sheppard, L.A. (2008). The urban environment and sustainable ageing: critical issues and assessment indicators. *Local Environment* 13: 497–514.

Langford, M., Maguire, D.J., and Unwin, D.J. (1991). The areal interpolation problem: estimating population using remote sensing in a GIS framework. In: *Handling Geographical Information: Methodology and Potential Applications* (eds. L. Masser and M. Blakemore), 55–77. New York: Longman Scientific and Technical co-published in the United States with Wiley.

Li, G. and Weng, Q. (2005). Using Landsat ETM+ imagery to measure population density in Indianapolis, Indiana, USA. *Photogrammetric Engineering and Remote Sensing* 71: 947–958.

Li, G. and Weng, Q. (2010). Fine-scale population estimation: how Landsat ETM+ imagery can improve population distribution mapping. *Canadian Journal of Remote Sensing* 36 (3): 155–165.

Liang, B. and Weng, Q. (2011). Assessing urban environmental quality change of Indianapolis, United States, by the remote sensing and GIS integration. *IEEE Journal of Selected Topics in Applied Earth Observations & Remote Sensing* 4 (1): 43–55.

Liang, B., Weng, Q., and Lu, D. (2007). Census-based multiple scale residential population modeling with impervious surface. In: *Remote Sensing of Impervious Surfaces* (ed. Q. Weng), 409–430. CRC Press: Taylor & Francis.

Lindgren, D.T. (1971). Dwelling unit estimation with color-IR photos. *Photogrammetric Engineering* 37: 373–378.

Lo, C.P. (1986a). *Applied Remote Sensing*. New York: Longman.

Lo, C.P. (1986b). Accuracy of population estimation from medium-scale aerial photography. *Photogrammetric Engineering & Remote Sensing* 52: 1859–1869.

Lo, C.P. (1995). Automated population and dwelling unit estimation from high resolution satellite images: a GIS approach. *International Journal of Remote Sensing* 16: 17–34.

Lo, C.P. (1997). Application of Landsat TM data for quality of life assessment in an urban environment. *Computer, Environment, and Urban Systems* 21: 259–276.

Lo, C.P. (2001). Modeling the population of China using DMSP Operational Linescan System nighttime data. *Photogrammetric Engineering & Remote Sensing* 67: 1037–1047.

Lo, C.P. (2008). Population estimation using geographically weighted regression. *GIScience and Remote sensing* 45: 131–148.

Lo, C.P. and Faber, B.J. (1997). Integration of Landsat Thematic Mapper and census data for quality of life assessment. *Remote Sensing of Environment* 62: 143–157.

Lo, C.P. and Welch, R. (1977). Chinese urban population estimation. *Annals of the Association of American Geographers* 67: 246–253.

Lu, D. and Weng, Q. (2006). Use of impervious surface in urban land use classification. *Remote Sensing of Environment* 102: 146–160.

Lu, D., Weng, Q., and Li, G. (2006). Residential population estimation using a remote sensing derived impervious surface approach. *International Journal of Remote Sensing* 27: 3553–3570.

Maantay, J.A., Maroko, A., and Porter-Morgan, H. (2008). A New method for population mapping and understanding the spatial dynamics of disease in urban areas. *Urban Geography* 29: 724–738.

Meliá, J. and Sobrino, J.A. (1987). Study on the utilization of SIR-A data for population estimation in the eastern part of Spain. *Geocarto International* 2: 33–37.

Mennis, J. (2003). Generating surface models of population using dasymetric mapping. *Professional Geographer* 55: 31–42.

Nghiem, S.V., Balk, D., Rodriguez, E. et al. (2009). Observations of urban and suburban environments with global satellite scatterometer data. *ISPRS Journal of Photogrammetry and Remote Sensing* 64: 367–380.

Roberts, D.A., Batista, G.T., Pereira, J.L.G. et al. (1998). Change identification using multitemporal spectral mixture analysis: applications in eastern Amazônia. In: *Remote Sensing Change Detection: Environmental Monitoring*

Methods and Applications (eds. R.S. Lunetta and C.D. Elvidge), 137–161. Ann Arbor Press.

Sutton, P. (1997). Modeling population density with nighttime satellite imagery and GIS. *Computers, Environment, and Urban System* 21: 227–244.

Sutton, P., Roberts, D., Elvidge, C.D., and Baugh, K. (2001). Census from heaven: an estimate of the global human population using night-time satellite imagery. *International Journal of Remote Sensing* 22: 3061–3076.

Sutton, P., Roberts, D., Elvidge, C.D., and Meij, H. (1997). A comparison of nighttime satellite imagery and population density for the continental United States. *Photogrammetric Engineering & Remote Sensing* 63: 1303–1313.

United Nations Population Fund (UNFPA) (2008). The State of World Population 2008. *United States Census Bureau.* https://www.unfpa.org/publications/ state-world-population-2008 (accessed 3 July 2019).

Viel, J.F. and Tran, A. (2009). Estimating denominators: satellite-based population estimates at a fine spatial resolution in a European Urban area. *Epidemiology* 20: 214–222.

Weng, Q. and Hu, X. (2008). Medium spatial resolution satellite imagery for estimating and mapping urban impervious surfaces using LSMA and ANN. *IEEE Transaction on Geosciences and Remote Sensing* 46 (8): 2397–2406.

Weng, Q., Hu, X., and Lu, D. (2008). Extracting impervious surface from medium spatial resolution multispectral and hyperspectral imagery: a comparison. *International Journal of Remote Sensing* 29: 3209–3232.

Weng, Q. and Lu, D. (2009). Landscape as a continuum: an examination of the urban landscape structures and dynamics of Indianapolis city, 1991–2000. *International Journal of Remote Sensing* 30 (10): 2547–2577.

Weng, Q., Lu, D., and Liang, B. (2006). Urban surface biophysical descriptors and land surface temperature variations. *Photogrammetric Engineering & Remote Sensing* 72: 1275–1286.

Weng, Q., Lu, D., and Schubring, J. (2004). Estimation of land surface temperature–vegetation abundance relationship for urban heat island studies. *Remote Sensing of Environment* 89: 467–483.

Wu, C. and Murray, A. (2005). A cokriging method for estimating population density in urban areas. *Computers, Environment and Urban Systems* 29: 558–579.

Wu, C. and Murray, A. (2007). Population estimation using Landsat Enhanced Thematic Mapper imagery. *Geographical Analysis* 39: 26–43.

Wu, S., Qiu, X., and Wang, L. (2005). Population estimation methods in GIS and remote sensing: a review. *GIScience and Remote Sensing* 42: 58–74.

Yuan, Y., Smith, R.M., and Limp, W.F. (1997). Remodeling census population with spatial information from Landsat imagery. *Computers, Environment, and Urban Systems* 21: 245–258.

Index

Techniques and Methods in Urban Remote Sensing, First Edition. Qihao Weng.
© 2020 by The Institute of Electrical and Electronics Engineers, Inc.
Published 2020 by John Wiley & Sons, Inc.